THE

IMS

THE

IMS

IP Multimedia Concepts and Services

SECOND EDITION

Miikka Poikselkä

Nokia, Finland

Georg Mayer

Nokia, Finland

Hisham Khartabil

Telio, Norway

Aki Niemi

Nokia, Finland

JOHN WILEY & SONS, LTD

Other Wiley Editorial Offices

John Wiley & Sons, Inc., 111 River Street, Hoboken, NJ 07030, USA

Jossey-Bass, 989 Market Street, San Francisco, CA 94103-1741, USA

Wiley-VCH Verlag GmbH, Boschstr. 12, D-69469 Weinheim, Germany

John Wiley & Sons Australia Ltd, 42 McDougall Street, Milton, Queensland 4064, Australia

John Wiley & Sons (Asia) Pte Ltd, 2 Clementi Loop #02-01, Jin Xing Distripark, Singapore 129809

John Wiley & Sons Canada Ltd, 22 Worcester Road, Etobicoke, Ontario, Canada M9W 1L1

Wiley also publishes its books in a variety of electronic formats. Some content that appears
in print may not be available in electronic books.

Library of Congress Cataloging-in-Publication Data

The IMS: IP multimedia concepts and services / Miikka Poikselkä ... [et al.]. – 2nd ed.
    p. cm.
Includes bibliographical references and index.
ISBN-13: 978-0-470-01906-1 (cloth : alk. paper)
ISBN-10: 0-470-01906-9 (cloth : alk paper)
1. Wireless communication systems. 2. Mobile communication systems. 3. Multimedia communications.
I. Title: IP Multimedia Subsystem. II. Poikselkä, Miikka.
TK5103.2.I48 2006
621.382'12 – dc22   2005026755

British Library Cataloguing in Publication Data

A catalogue record for this book is available from the British Library

ISBN-13 978-0-470-01906-1 (HB)
ISBN-10 0-470-01906-9 (HB)

Project management by Originator, Gt Yarmouth, Norfolk (typeset in 10/12pt Times).
Printed and bound in Great Britain by Antony Rowe Ltd, Chippenham, Wiltshire.
This book is printed on acid-free paper responsibly manufactured from sustainable forestry
in which at least two trees are planted for each one used for paper production.

Contents

Foreword by Petri Pöyhönen

We have telephony so that we can talk to each other. We have messaging so that we can dispatch mail or instant notes. We have browsing so that we can read published content on known sites. We even have search engines to locate content sites, which may have content relevant to us. This may look as if we have a lot on our plate; so, do we need Internet Protocol (IP) Multimedia Subsystem (IMS)?

The problem is that we have no practical mechanism to engage another application-rich terminal in a peer-to-peer session. Enormously successful mobile telephony shows that there is immense value in sharing with peers. With increasingly attractive terminals, the sharing experience will become much more than just exchanging voice.

We will be sharing real-time video (see what I see), an MP3-coded music stream,* a whiteboard to present objects and we will be exchanging real-time game data. Many of these will take place simultaneously. There is undoubtedly the need to break into this completely new ground of communication.

Telephony is sufficient for telephones. Multimedia terminals need IMS networks.

Session Initiation Protocol (SIP) enables clients to invite others to a session and negotiate control information about the media channels needed for the session. IMS builds on top of this and provides a full suite of network operator capabilities enabling authentication of clients, network-to-network interfaces and administration capabilities like charging. All this is essential in order to build interoperating networks that, when combined, can provide a truly global service, in the spirit of good old telephony. This will enable a global market of multimedia terminals.

As IMS is now emerging as the key driver for renewal of maturing mass-market communication services, several technical audiences have an urgent need to understand how it works. Georg Mayer, Aki Niemi, Hisham Khartabil and Miikka Poikselkä are major contributors to IMS industry development through their work in the standard-ization arena. This book provides an essential insight into the architecture and structure of these new networks.

Petri Pöyhönen
Vice President
Nokia Networks

---

*MP3 is the voice compression method developed by the Moving Picture Experts Group (MPEG), by means of which the size of a voice-containing file can be reduced to one-tenth of the original without significantly affecting the quality of voice.

Foreword by Alan Duric

For centuries, communications were known as the ways in which people, goods and ideas move from one place to another. It usually referred to roads and railways and their way of operating, and, more recently, letter writing and telephone calls.

With advances in naval, terrestrial and aerial infrastructure as well as in their *modus operandi* we have come a long way. The advent of electronic personal and business communications, and the way that electronic goods are being exchanged, has revolutionized the communications industry, but, as seen in the last few years with the rise of real-time IP communications, we have barely scratched the surface of what is about to come. Not only do people want to be mobile and achieve seamless integration of all applications (voice, video, TV, IM, chat, conferencing, presence, e-mail, file sharing, contact information, identity, etc.) on a single device, but they also want freedom to change this device for different use scenarios.

The emergence of a new set of intelligent, interactive and location-based broadband services (video, push to talk, mobile access to speech-enabled services) is currently occurring and, along with them, the perceived quality of services provided are moving to new heights. These services, along with their personalization by matching the behavioral characteristics of their users, will bring to service providers new business models and opportunities, creating and shaping the future for competition.

During the last few years, IMS has attracted a lot of attention from the real-time communications industry, being the focal point of the standardization and development efforts of the numerous vendors, and has been followed with great expectations by a number of operators and service providers, regardless of the nature of access networks they operate currently. The fact that the IMS service control architecture is standardized over SIP-based services makes SIP the ideal tool for the new world of electronic communications, making such an emergence of rich, intelligent services possible as well as providing the necessary freedom of choice of device, access network and, ultimately, place and time of electronic communications.

By using an holistic approach, the authors have made this book 'The SwAK' (Swiss Army Knife) of IMS. The knowledge held within its chapters will benefit anyone dealing or working with IMS, providing information that has up to now been missing and giving a solid foundation to those that are just about to start exploring the world of IMS, which is fostering innovative, rich communications in the way it intersects with entertainment services and, thereby, optimizes services for the customer and strengthens the economy.

I am privileged and fortunate to work and cooperate, through standardization work, with the authors and have witnessed first-hand their passion in creating the *modus operandi* of these rich electronic communications services – which have now reached a pinnacle by bringing such efforts to fruition.

Alan Duric
Chief Technology Officer and Co-founder of Telio

Preface

Internet Protocol (IP) Multimedia Subsystem, better known as "IMS", is based on the specification of Session Initiation Protocol (SIP) as standardized by Internet Engineering Task Force (IETF). But SIP as a protocol is only one part of it. IMS is more than just a protocol; it is an architecture for the convergence of data, speech, fixed and mobile networks and is based on a wide range of protocols, most of which have been developed by IETF. IMS combines and enhances them to allow real-time services on top of Universal Mobile Telecommunications System (UMTS) packet-switched domain.

This book was written to provide a detailed insight into what IMS is – i.e., its concepts, architecture, service and protocols. Its intended audience ranges from marketing managers, research engineers, development and test engineers to university students. The book is written in a manner that allows readers to choose the level of knowledge they need and the depth of understanding of IMS they desire to achieve. The book is also very well suited as a reference.

The first few chapters in Part I provide a detailed overview of the system architecture and the entities that, when combined, are necessary to provide IMS. These chapters also present the reference points (interfaces) between these entities and introduces the protocols assigned to these interfaces.

As with every communication system, IMS is built on concepts that offer basic and advanced services to its users. Security is a concept that is required by any communication architecture. In this book we describe security threats and the models used to secure communications in IMS. IMS security, along with concepts such as registration, session establishment, charging and service provisioning, are explained in Chapter 3.

Since both wireless and wireline networks are involved in IMS architecture, IMS becomes an inexpensive medium for Fixed to Mobile Convergence (FMC). Chapter 3 also describes FMC, its benefits and importance.

In IMS, services are not limited to audio, but also include presence, messaging, Push to talk over Cellular, conferencing and group management. In Part II of this book, we introduce some of these advanced services in IMS, including call flows. This part proves that the convergence of services and networks is not a myth, but will have real added value for the user.

SIP and SDP are two of the main building blocks within IMS and their usage gets complemented by a large number of important extensions. Part III goes step by step through an example IMS registration and session establishment at the protocol level, detailing the procedures taken at every entity.

The final part of the book, Part IV, describes the protocols used within IMS in more detail, paying special attention to signalling as well as security protocols. This part of the book shows how different protocols are built up, how they work and why they are applied within IMS.

Third Generation Partnership Project (3GPP) and IETF have worked together during recent years in an amazing way to bring about IMS and the protocols used by it. We, the authors, have had the chance to participate in many technical discussions regarding the architecture and protocols and are still very active in further discussions on the ever-improving protocols and communication systems. Some of these discussions, which often can be described as debates or negotiations, frequently take a long time to conclude and even more frequently do not result in an agreement or consensus on the technical solutions. We want to thank all the people in these standardization bodies as well as those in our own company who have come up with ideas, have shown great patience and have worked hard to standardize this communication system of the future called IMS.

Acknowledgements

The authors of this book would like to extend their thanks to colleagues working in 3GPP and IETF for their great efforts in creating the IMS specifications and related protocols. The authors would also like to give special thanks to the following who helped in the writing of this book by providing excellent review comments and suggestions:

Erkki Koivusalo
Hannu Hietalahti
Tao Haukka
Risto Mononen
Kalle Tammi
Risto Kauppinen
Marco Stura
Ralitsa Gateva
Juha-Pekka Koskinen
Markku Tuohino
Juha Räsänen
Peter Vestergaard
John Loughney
Tapio Paavonen
Kalle Luukkainen
Alan Duric

The authors welcome any comments and suggestions for improvements or changes that could be used to improve future editions of this book. Our email addresses are:

miikka.poikselka@nokia.com
georg.mayer@nokia.com
hisham.khartabil@telio.no
aki.niemi@nokia.com

List of figures

List of tables

Part I

IMS Architecture and Concepts

1

Introduction

1.1 What is the Internet Protocol Multimedia Subsystem (IMS)?

Fixed and mobile networks have gone through a major transition in the past 20 years. In the mobile world, first-generation (1G) systems were introduced in the mid-1980s. These networks offered basic services for users. The main emphasis was on speech and speech-related services. Second-generation (2G) systems in the 1990s brought some data services and more sophisticated supplementary services to the users. The third generation (3G) is now enabling faster data rates and various multimedia services. In the fixed side, traditional Public Switched Telephone Network (PSTN) and Integrated Services Digital Network (ISDN) networks have dominated traditional voice and video communication. In recent years the usage of the Internet has exploded and more and more users are taking advantage of faster and cheaper Internet connection such as Asymmetric Digital Subscriber Line (ADSL). These types of Internet connections enable always-on connectivity, which is a necessity for people to start using real-time communication means – e.g., chatting applications, online gaming, Voice over IP (VoIP).

At the moment we are experiencing the fast convergence of fixed and mobile worlds as the penetration of mobile devices is increasing on a yearly basis and soon we will have more than 2 billion mobile device users. These mobile devices have large, high-precision displays, they have built-in cameras and a lot of resources for applications. They are always-on always-connected application devices. This redefines applications. Applications are no longer isolated entities exchanging information only with the user interface. The next generation of more exciting applications are peer-to-peer entities, which facilitate sharing: shared browsing, shared whiteboard, shared game experience, shared two-way radio session (i.e., Push to talk Over Cellular). The concept of being connected will be redefined. Dialling a number and talking will soon be seen as a narrow subset of networking. The ability to establish a peer-to-peer connection between the new Internet Protocol (IP) enabled devices is the key required ingredient. This new paradigm of communications reaches far beyond the capabilities of the Plain Old Telephone Service (POTS).

In order to communicate, IP-based applications must have a mechanism to reach the correspondent. The telephone network currently provides this critical task of establishing a connection. By dialing the peer, the network can establish an *ad hoc* connection between any two terminals over the IP network. This critical IP connectivity capability is offered only in isolated and single-service provider environments in the Internet;

The IMS: IP Multimedia Concepts and Services Second Edition
Miikka Poikselkä, Georg Mayer, Hisham Khartabil and Aki Niemi © 2006 John Wiley & Sons, Ltd

Figure 1.1   IMS in converged networks.

closed systems compete on user base, where user lock-in is key and interworking between service providers is an unwelcome feature. Therefore, we need a global system – the IP Multimedia Subsystem (IMS). It allows applications in IP-enabled devices to establish peer-to-peer and peer-to-content connections easily and securely. Our definition for the IMS is:

> *IMS is a global, access-independent and standard-based IP connectivity and service control architecture that enables various types of multimedia services to end-users using common Internet-based protocols.*

True integration of voice and data services increases productivity and overall effectiveness, while the development of innovative applications integrating voice, data and multimedia will create demands for new services, such as presence, multimedia chat, push to talk and conferencing. The skill to combine mobility and the IP network will be crucial to service success in the future.

Figure 1.1 shows a converged communication network for the fixed mobile environment. It is the IMS which introduces multimedia session control in the packet-switched domain and at the same time brings circuit-switched functionality in the packet-switched domain. The IMS is a key technology for such network consolidation.

1.2   Example of IMS services

Switching on my Internet Protocol Multimedia Subsystem (IMS) enabled device, it will automatically register to the IMS network using information in the identity module (such as USIM). During registration both device and network are authenticated and my device will get my user identities from the network. After this single registration, all my services will be available, including push to talk, presence, voice and video sessions, messaging and multiplayer games. Moreover, my availability information is updated at the presence server as being "online" and listing my current applications.

When I need to contact my friend Bob, I select Bob from my device's phone book and, based on his presence information, I see immediately that he is available. After pressing the "green button" on my device it will place an "ordinary" call to him. The IMS network will take care of finding and setting up a Session Initiation Protocol (SIP) session between our devices, even though Bob is currently abroad. When my call reaches Bob's terminal he will see that the call is coming from me and, additionally, he sees a text string inserted by me ("Free tickets to movie next Wednesday"). Bob answers, but tells me that he's not sure whether he is able to come. We decide to check the issue again on Sunday. Before hanging up, Bob says to me, "You won't believe what I saw today but just wait a second, I'll show you." Bob starts streaming a video clip to me, and while I'm watching the video, Bob keeps explaining what happened in the zoo earlier that day.

Mike realizes that today is the birthday of his good friend Jill. Although he's travelling and can't meet her today, he wants to send Jill a personal birthday message. While Mike is sitting in a local coffee shop enjoying coffee and reading the latest news from the Internet using his brand-new Wireless Local Area Network (WLAN) device, he decides to send her a video clip as a birthday greeting. Jill is having a bath when she hears her phone ringing. She sees that she has received a message and checks it. She saves the video clip and decides to send something in return. Knowing that Mike knows her weird sense of humour, she sends a picture of herself taking a bath (Figure 1.2).

Peter Simpson is a Londoner and a die-hard Arsenal fan. With sheer luck he has managed to get tickets to an Arsenal–Tottenham derby and sets off to see the game. There he is, sitting at the stadium during the match, when suddenly he gets an

Figure 1.2   Multimedia messaging.

irresistible urge to make his friend envious. He gets his mobile phone and makes a call to his friend John Clark, a Tottenham supporter. John is sitting at his desk and receives an incoming call pop-up on his PC screen, informing him that Peter is calling. He answers and they start to talk. Peter can't contain himself and starts the video-sharing application while zooming onto the field. John receives an incoming video request and accepts the stream. The PC client starts to show the game, and with a pang of jealousy and disappointment John watches Arsenal score. "Nice goal, huh?" asks Peter. "It ain't over yet," says John, gritting his teeth, and ends the video stream. They continue to argue good-naturedly about the game and their teams over the phone.

All the required communication takes place using the IP connectivity provided by the IMS. The IMS offers the capability to select the best and most suitable communication media, to change the media during the session spontaneously, and use the preferred (SIP-capable) communication device over any IP access.

1.3   Where did it come from?

1.3.1   From GSM to 3GPP Release 7

The European Telecommunications Standards Institute (ETSI) was the standardization organization that defined the Global System for Mobile Communications (GSM) during the late 1980s and 1990s. ETSI also defined the General Packet Radio Service (GPRS) network architecture. The last GSM-only standard was produced in 1998, and in the same year the 3GPP was founded by standardization bodies from Europe, Japan, South Korea, the USA and China to specify a 3G mobile system comprising Wideband Code Division Multiple Access (WCDMA) and Time Division/Code Division Multiple Access (TD-CDMA) radio access and an evolved GSM core network (*www.3gpp.org/ About/3gppagre.pdf*). Most of the work and cornerstone specifications were inherited from the ETSI Special Mobile Group (SMG). The 3GPP originally decided to prepare specifications on a yearly basis, the first specification release being Release 99.

1.3.2   3GPP Release 99 (3GPP R99)

It took barely a year to produce the first release – Release 1999. The functionality of the release was frozen in December 1999 although some base specifications were frozen afterward – in March 2001. Fast completion was possible because the actual work was divided between two organizations: 3GPP and ETSI SMG. 3GPP developed the services, system architecture, WCDMA and TD-CDMA radio accesses, and the common core network. ETSI SMG developed the GSM/Enhanced Data Rates for Global Evolution (EDGE) radio access.

WCDMA radio access was the most significant enhancement to the GSM-based 3G system in Release 1999. In addition to WCDMA, UMTS Terrestrial Radio Access Network (UTRAN) introduced the Iu interface as well. Compared with the A and Gb interfaces, there are two significant differences. First, speech transcoding for Iu is performed in the core network. In the GSM it was logically a Base Transceiver Station

Figure 1.3   The role of IMS in packet-switched networks.

(BTS) functionality. Second, encryption and cell-level mobility management for Iu are done in the Radio Network Controller (RNC). In GSM they were done in the Serving GPRS Support Node (SGSN) for GPRS services.

The Open Service Architecture (OSA) was introduced for service creation. On the service side the target was to stop standardizing new services and to concentrate on service capabilities, such as toolkits (CAMEL, SIM Application Toolkit and OSA). This principle was followed quite well, even though the Virtual Home Environment (VHE), an umbrella concept that covers all service creation, still lacks a good definition.

1.3.3   3GPP Release 4

After Release 1999, 3GPP started to specify Release 2000, including the so-called All-IP that was later renamed as the IMS. During 2000 it was realized that the development of IMS could not be completed during the year. Therefore, Release 2000 was split into Release 4 and Release 5.

It was decided that Release 4 would be completed without the IMS. The most significant new functionalities in 3GPP Release 4 were: the Mobile Switching Centre (MSC) Server–Media Gateway (MGW) concept, IP transport of core network protocols, Location Services (LCS) enhancements for UTRAN and multimedia messaging and IP transport for the Gb user plane.

3GPP Release 4 was functionally frozen and officially completed in March 2001. The backward compatibility requirement for changes, essential for the radio interface, was enforced as late as September 2002.

1.3.4   3GPP Releases 5, 6 and 7

Release 5 finally introduced the IMS as part of 3GPP specifications. The IMS is supposed to be a standardized access-independent IP-based architecture that inter-works with existing voice and data networks for both fixed (e.g., PSTN, ISDN, Internet) and mobile users (e.g., GSM, CDMA). The IMS architecture makes it possible to establish peer-to-peer IP communications with all types of clients with the requisite quality of services. In addition to session management, the IMS architecture also addresses functionalities that are necessary for complete service delivery (e.g., registration, security, charging, bearer control, roaming). All in all, the IMS will form the heart of the IP core network.

The content of Release 5 was heavily discussed and, finally, the functional content of 3GPP Release 5 was frozen in March 2002. The consequence of this decision was that many features were postponed to the next release – Release 6. After freezing the content, the work continued and reached stability at the beginning of 2004. Release 6 IMS fixes the shortcomings in Release 5 IMS and also contains novel features. Release 6 was completed in September 2005. Table 1.1 shows the most important features of Release 5 and Release 6. The table also contains Release 7 feature candidates. Work on Release 7 is ongoing and its features are expected to be ready during 2006.

From Table 1.1 you can see that 3GPP has defined a finite architecture for SIP-based IP multimedia service machinery. It contains a functionality of logical elements, a description of how elements are connected, selected protocols and procedures. It is important to realize that optimization for the mobile communication environment has been designed in the form of user authentication and authorization based on mobile identities, definite rules at the user network interface for compressing SIP messages and security and policy control mechanisms that allow radio loss and recovery detection. Moreover, important aspects from the operator point of view are addressed while developing the architecture, such as the charging framework and policy, and service control. This book explains how these aspects have been defined.

Table 1.1 IMS features.

Release 5	Release 6	Release 7
Architecture: network entities and reference points including charging functions.	*Architecture:* interworking (CS, other IP networks, WLAN) and a few, new entities and reference points.	*Architecture:* voice call continuity between CS and PS domains, fixed broadband connection to the IMS.
Signalling: general routing principles, registration, session initiation, session modification, session teardown, network-initiated session release/deregistration flows: • SIP compression between UE and IMS network; • data transfer between user information storage (HSS) and session control entities (CSCF); • data transfer between user information storage (HSS) and application server (AS).	*Signalling:* routing of group identities, multiple registration.	*Signalling:* emergency sessions, SMS support using SIP, combining CS calls and IMS sessions.
Security: IMS AKA for authenticating users and network, integrity protection of SIP messages between UE and IMS network, network domain security.	*Security:* confidentiality protection of SIP messages, IP address-based authentication, Generic Authentication Architecture?	*Security:* adaptation to broadband access, TLS support.
Quality of Service: policy control between IMS and GPRS access network, preconditions and authorization token.	*Quality of Service:* multiplexing media flows of separate sessions on the same PDP context.	*Quality of Service:* policy and charging control harmonization, tokenless QoS authorization.
Services: usage of applications servers and IMS Service Control reference point.	*Services:* presence, messaging, conferencing, Push to talk Over Cellular, group management, local services.	*Services:* supplementary services in SIP.
General: ISIM		*Miscellaneous:* WLAN–UMTS mobility

2

IP Multimedia Subsystem architecture

This chapter introduces the reader to the Internet Protocol (IP) Multimedia Subsystem (IMS). Section 2.1 explains basic architectural concepts: for instance, we explain why bearers are separated and why the home control model was selected. Section 2.2 gives a wide overview of IMS architecture, including an introduction to different network entities and main functionalities. Section 2.3 goes deeper and shows how the entities are connected and what protocols are used between them; it also describes their relationships to other domains: IP networks, Universal Mobile Telecommunications System (UMTS) and Circuit Switched Core Network (CS CN).

2.1 Architectural requirements

There is a set of basic requirements which guides the way in which the IMS architecture has been created and how it should evolve in the future. This section covers the most significant requirements. Third Generation Partnership Project (3GPP) IMS requirements are documented in [3GPP TS 22.228].

2.1.1 IP multimedia sessions

Existing communication networks are able to offer voice, video and messaging type of services using circuit-switched bearers. Naturally, end-users' service offerings should not decline when users move to the packet-switched domain and start using the IMS. The IMS will take communication to the next level by offering enriched communication means. IMS users are able to mix and match a variety of IP-based services in any way they choose during a single communication session. Users can integrate voice, video and text, content sharing and presence as part of their communication and can add or drop services as and when they choose. For example, two people can start a session as a voice session and later on add a game or video component to the same session.

The IMS: IP Multimedia Concepts and Services Second Edition
Miikka Poikselkä, Georg Mayer, Hisham Khartabil and Aki Niemi © 2006 John Wiley & Sons, Ltd

2.1.2   IP connectivity

As the name IP Multimedia Subsystem implies, a fundamental requirement is that a device has to have IP connectivity to access it. Peer-to-peer applications require end-to-end reachability and this connectivity is easiest attainable with IP version 6 (IPv6) because IPv6 does not have address shortage. Therefore, 3GPP has arranged matters so that the IMS exclusively supports IPv6 [3GPP TS 23.221]. However, early IMS implementations and deployments may use IP version 4 (IPv4). 3GPP has created recommendations about how IP version interworking is handled in the IMS [3GPP TR 23.981]. This is further described in Section 3.17.

IP connectivity can be obtained either from the home network or the visited network. The leftmost part of Figure 2.1 presents an option in which User Equipment (UE) has obtained an IP address from a visited network. In the UMTS network, this means that the Radio Access Network (RAN), Serving GPRS Support Node (SGSN) and Gateway GPRS Support Node (GGSN) are located in the visited network when a user is roaming in the visited network. The rightmost part of Figure 2.1 presents an option in which a UE has obtained an IP address from the home network. In the UMTS network this means that the RAN and SGSN are located in the visited network when a user is roaming in the visited network. Obviously, when a user is located in the home network all necessary elements are in the home network and IP connectivity is obtained in that network.

It is important to note that a user can roam and obtain IP connectivity from the home network as shown in the figure. This would allow users to use new, fancy IMS services even when they are roaming in an area that does not have an IMS network but provides IP connectivity. In theory, it is possible to deploy an IMS network in a single area/country and use, say, General Packet Radio Service (GPRS) roaming to connect

------- = Signalling traffic
▬▬▬ = Signalling and User Plane traffic
------ = User Plane traffic

Figure 2.1   IMS connectivity options when a user is roaming.

customers to the home network. In practice, this would not happen because routing efficiency would not be high enough. Consider routing Real-time Transport Protocol (RTP) voice packets from the USA to Europe and then back to the USA. However, this deployment model is important when operators are ramping up IMS networks or, in an initial phase, when they are offering non- or near real-time multimedia services.

2.1.3 Ensuring quality of service for IP multimedia services

On the public Internet, delays tend to be high and variable, packets arrive out of order and some packets are lost or discarded. This will no longer be the case with the IMS. The underlying access and transport networks together with the IMS provide end-to-end Quality of Service (QoS). Via the IMS, the UE negotiates its capabilities and expresses its QoS requirements during a Session Initiation Protocol (SIP) session setup or session modification procedure. The UE is able to negotiate such parameters as:

- Media type, direction of traffic.
- Media type bit rate, packet size, packet transport frequency.
- Usage of RTP payload for media types.
- Bandwidth adaptation.

After negotiating the parameters at the application level, UEs reserve suitable resources from the access network. When end-to-end QoS is created, the UEs encode and packetize individual media types with an appropriate protocol (e.g., RTP) and send these media packets to the access and transport network by using a transport layer protocol (e.g., TCP or UDP) over IP. It is assumed that operators negotiate service-level agreements for guaranteeing the required QoS in the interconnection backbone. In the case of UMTS, operators could utilize the GPRS Roaming Exchange backbone.

2.1.4 IP policy control for ensuring correct usage of media resources

IP policy control means the capability to authorize and control the usage of bearer traffic intended for IMS media, based on the signalling parameters at the IMS session. This requires interaction between the IP connectivity access network and the IMS. The means of setting up interaction can be divided into three different categories [3GPP TS 22.228, 23.207, 23.228]:

- The policy control element is able to verify that values negotiated in SIP signalling are used when activating bearers for media traffic. This allows an operator to verify that its bearer resources are not misused (e.g., the source and destination IP address and bandwidth in the bearer level are exactly the same as used in SIP session establishment).
- The policy control element is able to enforce when media traffic between the end points of a SIP session start or stop. This makes it possible to prevent the use of the bearer until session establishment is completed and allows traffic to start/stop in synchronization with the start/stop of charging for a session in IMS.

- The policy control element is able to receive notifications when the IP connectivity access network service has either modified, suspended or released the bearer(s) of a user associated with a session. This allows IMS to release an ongoing session because, for instance, the user is no longer in the coverage area.

Service-Based Local Policy (SBLP) is used as a synonym for IP policy control in the IMS and it is described further in Section 3.9.

2.1.5 Secure communication

Security is a fundamental requirement in every telecommunication system and the IMS is not an exception. The IMS has its own authentication and authorization mechanisms between the UE and the IMS network in addition to access network procedures (e.g., GPRS network). Moreover, the integrity and optional confidentiality of the SIP messages is provided between the UE and the IMS network and between IMS network entities regardless of the underlaying core network (e.g., RAN and GPRS). Therefore, the IMS provides at least a similar level of security as the corresponding GPRS and circuit-switched networks: for example, the IMS ensures that users are authenticated before they can start using services, and users are able to request privacy when engaged in a session. Section 3.6 will discuss security features in more detail. An overview of applied security solutions is depicted in Figure 2.2.

Figure 2.2   Overview of IMS security.

2.1.6 Charging arrangements

From an operator or service provider perspective the ability to charge users is a must in any network. The IMS architecture allows different charging models to be used. This includes, say, the capability to charge just the calling party or to charge both the calling party and the called party based on used resources in the transport level. In the latter case the calling party could be charged entirely on an IMS-level session: that is, it is possible to use different charging schemes at the transport and IMS level. However, an operator might be interested to correlate charging information generated at transport and IMS (service and content) charging levels. This capability is provided if an operator utilizes a policy control reference point. The charging correlation mechanism is further described in Section 3.11.6 and policy control is explained in Section 3.10.

As IMS sessions may include multiple media components (e.g., audio and video), it is required that the IMS provides a means for charging per media component. This would

Figure 2.3   IMS charging overview.

allow a possibility to charge the called party if she adds a new media component in a session. It is also required that different IMS networks are able to exchange information on the charging to be applied to a current session [3GPP TS 22.101, TR 23.815].

The IMS architecture supports both online and offline charging capabilities. Online charging is a charging process in which the charging information can affect in real time the service rendered and, therefore, directly interacts with session/service control. In practice, an operator could check the user's account before allowing the user to engage a session and to stop a session when all credits are consumed. Prepaid services are applications that need online charging capabilities. Offline charging is a charging process in which the charging information does not affect in real time the service rendered. This is the traditional model in which the charging information is collected over a particular period and, at the end of the period, the operator posts a bill to the customer.

Figure 2.3 shows a simplified view of general charging arrangements in an IMS environment. The key observation is: the IMS adds the possibility to charge for user IP traffic in a more granular manner than before.

2.1.7   Support of roaming

From a user point of view it is important to get access to her services regardless of her geographical location. The roaming feature makes it possible to use services even though the user is not geographically located in the service area of the home

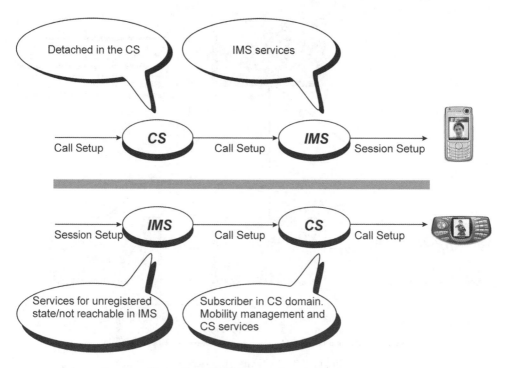

Figure 2.4 IMS/CS roaming alternatives.

network. Section 2.1.2 has already described two instances of roaming: namely, GPRS roaming and IMS roaming. In addition to these two there exists an IMS circuit-switched roaming case. GPRS roaming means the capability to access the IMS when the visited network provides the RAN and SGSN and the home network provides the GGSN and IMS. The IMS roaming model refers to a network configuration in which the visited network provides IP connectivity (e.g., RAN, SGSN, GGSN) and the IMS entry point (i.e., P-CSCF) and the home network provides the remaining IMS func-tionalities. The main benefit of this roaming model compared with the GPRS roaming model is optimum usage of user-plane resources. Roaming between the IMS and the CS CN domain refers to inter-domain roaming between IMS and CS. When a user is not registered or reachable in one domain a session can be routed to the other domain. It is important to note that both the CS CN domain and the IMS domain have their own services and cannot be used from another domain. Some services are similar and available in both domains (e.g., Voice over IP in IMS and speech telephony in CS CN). Figure 2.4 shows different IMS/CS roaming cases.

2.1.8 Interworking with other networks

It is evident that the IMS is not deployed over the world at the same time. Moreover, people may not be able to switch terminals or subscriptions very rapidly. This will raise the issue of being able to reach people regardless of what kind of terminals they have or where they live. To be a new, successful communication network technology and architecture the IMS has to be able to connect to as many users as possible. Therefore, the IMS supports communication with PSTN, ISDN, mobile and Internet users. Additionally, it will be possible to support sessions with Internet applications that have been developed outside the 3GPP community [3GPP TS 22.228].

2.1.9 Service control model

In 2G mobile networks the visited service control is in use. This means that, when a user is roaming, an entity in the visited network provides services and controls the traffic for the user. This entity in the second generation (2G) is called a visited mobile service switching centre. In the early days of Release 5 both visited and home service control models were supported. Supporting two models would have required that every problem have more than one solution; moreover, it would reduce the number of optimal architecture solutions, as simple solutions may not fit both models. Supporting both models would have meant additional extensions for Internet Engineering Task Force (IETF) protocols and increased the work involved in registration and session flows. The visited service control was dropped because it was a complex solution and did not provide any noticeable added value compared with the home service control. On the contrary, the visited service control imposes some limitations. It requires a multiple relationship and roaming models between operators. Service development is slower as both the visited and home network would need to support similar services, otherwise roaming users would experience service degradations. In addition, the number of interoperator reference points increase, which requires complicated solutions (e.g., in terms of security and charging). Therefore, home service control was selected; this means that the entity that has access to the subscriber database and interacts directly with service platforms is always located at the user's home network.

2.1.10 Service development

The importance of having a scalable service platform and the possibility to launch new services rapidly has meant that the old way of standardizing complete sets of teleservices, applications and supplementary services is no longer acceptable. Therefore, 3GPP is standardizing service capabilities and not the services themselves [3GPP TS 22.101]. The IMS architecture should actually include a service framework that provides the necessary capabilities to support speech, video, multimedia, messaging, file sharing, data transfer, gaming and basic supplementary services within the IMS. Section 3.13 further describes how the IMS service control works and Chapters 4–8 explain in more detail how presence, messaging, Push to talk Over Cellular, conferencing and group management services are offered.

2.1.11 Layered design

3GPP has decided to use a layered approach to architectural design. This means that transport and bearer services are separated from the IMS signalling network and

Figure 2.5 IMS and layered architecture.

session management services. Further services are run on top of the IMS signalling network. Figure 2.5 shows the design.

In some cases it may be impossible to distinguish between functionality at the upper and lower layers. The layered approach aims at a minimum dependence between layers. A benefit is that it facilitates the addition of new access networks to the system later on. Wireless Local Area Network (WLAN) access to the IMS was added in 3GPP Release 6 and fixed broadband access to the IMS is being standardized in Release 7.

The layered approach increases the importance of the application layer as services are designed to work independent of the access network and the IMS is equipped to bridge the gap between them. Whether the subscriber is using a mobile phone or a PC client to communicate, the same presence and group list functions in IMS will be used. Different services have different requirements. These include:

- bandwidth;
- latency;
- processing power in the device.

This means that in order for different services to be executed properly, the network has to be equipped with access-aware control and service logic for multimedia services. Multi-access functionality is built into the IMS architecture, which offers a way for fixed and mobile operators to deliver a fixed to mobile convergence solution. This will enable service providers to use the characteristics and capabilities of the currently selected device and its network access method and adapt to it dynamically.

2.1.12   Access independence

The IMS was originally designed to be access-independent so that IMS services can be provided over any IP connectivity networks (e.g., GPRS, WLAN, broadband access x-Digital Subscriber Line). Unfortunately, Release 5 IMS specifications contain some GPRS-specific features. In Release 6 (e.g., GPRS) access-specific issues are separated from the core IMS description and the IMS architecture returned to its born state (i.e., access independent). Figure 2.6 demonstrates the different types of access-independent networks that the IMS can run on. These include Fixed Broadband, WLAN, GPRS and UMTS. In this book we use GPRS as an example.

2.2   Description of IMS-related entities and functionalities

This section discusses IMS entities and key functionalities. These entities can be roughly classified in six main categories:

- Session management and routing family (CSCFs).
- Databases (HSS, SLF).
- Services (application server, MRFC, MRFP).
- Interworking functions (BGCF, MGCF, IMS-MGW, SGW).
- Support functions (PDF, SEG, THIG).
- Charging.

Figure 2.6   Access-independent IMS.

It is important to understand that IMS specifications are set up so that the internal functionality of the network entities is not specified in detail. Instead, specifications describe reference points between entities and functionalities supported at the reference points. For instance, how does CSCF obtain user data from databases? Different reference points will be described in Section 2.3. Additionally, GPRS functions are described at the end of this section.

2.2.1 Call Session Control Functions (CSCF)

There are three different kind of Call Session Control Functions (CSCF): Proxy-CSCF (P-CSCF), Serving-CSCF (S-CSCF) and Interrogating-CSCF (I-CSCF). Each CSCF has its own special tasks and these tasks are described in the following subsections. Common to all CSCFs is that they all play a role during registration and session establishment and form the SIP routing machinery. Moreover, all functions are able to send charging data to an offline charging function. There are some common functions that P-CSCF and S-CSCF are able to perform. Both entities are able to release sessions on behalf of the user (e.g., when S-CSCF detects a hanging session or P-CSCF receives a notification that a media bearer is lost) and are able to check the content of the Session Description Protocol (SDP) payload and to check whether it contains media types or codecs, which are not allowed for a user. When the proposed SDP does not fit the operator's policy, the CSCF rejects the request and sends a SIP error message to the UE.

2.2.1.1 Proxy Call Session Control Function (P-CSCF)

Proxy Call Session Control Function (P-CSCF) is the first contact point for users within the IMS. It means that all SIP signalling traffic from the UE will be sent to the P-CSCF. Similarly, all terminating SIP signalling from the network is sent from the P-CSCF to the UE. There are four unique tasks assigned for the P-CSCF: SIP compression, IPSec security association, interaction with Policy Decision Function (PDF) and emergency session detection.

As the SIP protocol is a text-based signalling protocol, it contains a large number of headers and header parameters, including extensions and security-related information which means that their message sizes are larger than with binary-encoded protocols. For speeding up the session establishment 3GPP has mandated the support of SIP compression between the UE and P-CSCF. The P-CSCF needs to compress messages if the UE has indicated that it wants to receive signalling messages compressed. SIP compression is described in Sections 3.16 and 12.13.15 and in Chapter 23.

P-CSCF is responsible for maintaining Security Associations (SAs) and applying integrity and confidential protection for SIP signalling. This is achieved during SIP registration as the UE and P-CSCF negotiate IPSec SAs. After the initial registration the P-CSCF is able to apply integrity and confidential protection of SIP signalling. See Sections 10.6 and 10.7 for more detailed descriptions.

The P-CSCF is tasked to relay session and media-related information to the PDF when an operator wants to apply SBLP. Based on the received information the PDF is able to derive authorized IP QoS information that will be passed to the GGSN when the GGSN needs to perform Service-Based Local Policy prior to accepting a secondary PDP context activation. This concept is covered in Section 3.10. Moreover, via the PDF the IMS is able to deliver IMS charging correlation information to the GPRS network and, similarly, via the PDF the IMS is able to receive GPRS charging correlation information from the GPRS network. This makes it possible to merge charging data records coming from the IMS and GPRS networks in the billing system. How this is done is shown in Section 3.11.6.

IMS emergency sessions are not yet fully specified (work ongoing in Release 7), therefore it is important that the IMS network detects emergency session attempts and guides a UMTS UE to use the CS network for emergency sessions. Such detection is a task of the P-CSCF. This functionality will not vanish when IMS emergency sessions are supported, as in certain roaming cases it is possible that the UE does not itself recognize that the user has dialled an emergency number. The planned function in Release 7 is that the P-CSCF is able to select an emergency CSCF to handle an emergency session. Such a selection is needed because in IMS roaming cases the assigned S-CSCF is in the home network and the home S-CSCF is unable to route the request to the correct emergency centre.

2.2.1.2   Interrogating Call Session Control Function (I-CSCF)

Interrogating Call Session Control Function (I-CSCF) is a contact point within an operator's network for all connections destined to a subscriber of that network operator. There are four unique tasks assigned for the I-CSCF:

- Obtaining the name of the next hop (either S-CSCF or application server) from the Home Subscriber Server (HSS).
- Assigning an S-CSCF based on received capabilities from the HSS. The assignment of the S-CSCF will take place when a user is registering with the network or a user receives a SIP request while she is unregistered from the network but has services related to an unregistered state (e.g., voice mail). This procedure is described in more detail in Section 3.9.
- Routing incoming requests further to an assigned S-CSCF or the application server (in case of public service identity see Section 11.11).
- Providing Topology Hiding Inter-network Gateway (THIG) functionality. THIG is further explained in Section 2.2.5.

2.2.1.3   Serving Call Session Control Function (S-CSCF)

Serving Call Session Control Function (S-CSCF) is the focal point of the IMS as it is responsible for handling registration processes, making routing decisions and maintaining session states, and storing the service profile(s). When a user sends a registration request it will be routed to the S-CSCF, which downloads authentication data from the HSS. Based on the authentication data it generates a challenge to the UE. After receiving the response and verifying it the S-CSCF accepts the registration and starts supervising the registration status. After this procedure the user is able to initiate and receive IMS services. Moreover, the S-CSCF downloads a service profile from the HSS as part of the registration process.

A service profile is a collection of user-specific information that is permanently stored in the HSS. The S-CSCF downloads the service profile associated with a particular public user identity (e.g., *joe.doe@ims.example.com*) when this particular public user identity is registered in the IMS. The S-CSCF uses information included in the service profile to decide when and, in particular, which application server(s) is contacted when a user sends a SIP request or receives a request from somebody. Moreover, the service profile may contain further instructions about what kind of media policy the S-CSCF needs to apply – for example, it may indicate that a user is only allowed to use audio and application media components but not video media components.

The S-CSCF is responsible for key routing decisions as it receives all UE-originated and UE-terminated sessions and transactions. When the S-CSCF receives a UE-originating request via the P-CSCF it needs to decide if application servers are contacted prior to sending the request further on. After possible application server(s) interaction the S-CSCF either continues a session in IMS or breaks to other domains (CS or another IP network). Moreover, if the UE uses a Mobile Station ISDN (MSISDN) number to address a called party then the S-CSCF converts the MSISDN number (i.e., a tel URL) to SIP Universal Resource Identifier (URI) format prior to sending the request further, as the IMS does not route requests based on MSISDN numbers. Similarly, the S-CSCF receives all requests which will be terminated at the UE. Although, the S-CSCF knows the IP address of the UE from the registration it routes all requests via the P-CSCF, as the P-CSCF takes care of SIP compression and security functions. Prior to sending a request to the P-CSCF, the S-CSCF may route the request to an application server(s), for instance, checking possible redirection instructions. Figure 2.7 illustrates the S-CSCF's role in routing decisions.

Figure 2.7   S-CSCF routing and basic IMS session setup.

In addition, the S-CSCF is able to send accounting-related information to the Online Charging System for online charging purposes (i.e., supporting pre-paid subscribers).

2.2.2 Databases

There are two main databases in the IMS architecture: Home Subscriber Server (HSS) and Subscription Locator Function (SLF).

The HSS is the main data storage for all subscriber and service-related data of the IMS. The main data stored in the HSS include user identities, registration information, access parameters and service-triggering information [3GPP TS 23.002]. User identities consist of two types: private and public user identities (see Sections 3.5.1.1 and 3.5.1.2). The private user identity is a user identity that is assigned by the home network operator and is used for such purposes as registration and authorization, while the public user identity is the identity that other users can use for requesting communication with the end-user. IMS access parameters are used to set up sessions and include parameters like user authentication, roaming authorization and allocated S-CSCF names. Service-triggering information enables SIP service execution. The HSS also provides user-specific requirements for S-CSCF capabilities. This information is used by the I-CSCF to select the most suitable S-CSCF for a user (see Section 3.9). In addition to functions related to IMS functionality, the HSS contains the subset of Home Location Register and Authentication Center (HLR/AUC) functionality required by the Packet-Switched (PS) domain and the Circuit-Switched (CS) domain. The structure of the HSS is shown in Figure 2.8. Communication between different HSS functions is not standardized.

HLR functionality is required to provide support to PS domain entities, such as SGSN and GGSN. This enables subscriber access to PS domain services. In similar fashion the HLR provides support for CS domain entities, like MSC/MSC servers. This enables subscriber access to CS domain services and supports roaming to Global System for Mobile Communications (GSM)/UMTS CS domain networks. The AUC stores a secret key for each mobile subscriber, which is used to generate dynamic security data for each mobile subscriber. Data are used for mutual authentication of the International Mobile Subscriber Identity (IMSI) and the network. Security data are also used to provide integrity protection and ciphering of the communication over the radio path between the UE and the network. There may be more than one HSS in a home network, depending on the number of mobile subscribers, the capacity of the

Figure 2.8 Structure of HSS.

equipment and the organization of the network. There are multiple reference points between the HSS and other network entities.

The SLF is used as a resolution mechanism that enables the I-CSCF, the S-CSCF and the AS to find the address of the HSS that holds the subscriber data for a given user identity when multiple and separately addressable HSSs have been deployed by the network operator (see Figure 2.12).

2.2.3 Service functions

Three functions in this book are categorized as IMS service-related functions – namely, Multimedia Resource Function Controller (MRFC), Multimedia Resource Function Processor (MRFP) and Application Server (AS).

Keeping in mind the layered design, ASs are not pure IMS entities; rather, they are functions on top of IMS. However, ASs are described here as part of IMS functions because ASs are entities that provide value-added multimedia services in the IMS, such as presence and Push to talk Over Cellular. An AS resides in the user's home network or in a third-party location. The third party here means a network or a standalone AS. The main functions of the AS are:

- The possibility to process and impact an incoming SIP session received from the IMS.
- The capability to originate SIP requests.
- The capability to send accounting information to the charging functions.

The services offered are not limited purely to SIP-based services since an operator is able to offer access to services based on the Customized Applications for Mobile network Enhanced Logic (CAMEL) Service Environment (CSE) and the Open Service Architecture (OSA) for its IMS subscribers [3GPP TS 23.228]. Therefore, AS is the term used generically to capture the behaviour of the SIP AS, OSA Service Capability Server (SCS) and CAMEL IP Multimedia Service Switching Function (IM-SSF).

Using the OSA an operator may utilize such service capability features as call control, user interaction, user status, data session control, terminal capabilities, account management, charging and policy management for developing services [3GPP TS 29.198]. An additional benefit of the OSA framework is that it can be used as a standardized mechanism for providing third-party ASs in a secure manner to the IMS, as the OSA itself contains initial access, authentication, authorization, registration and discovery features (the S-CSCF does not provide authentication and security functionality for secure direct third-party access to the IMS). As the support of OSA services is down to operator choice, it is not architecturally sound to support OSA protocols and features in multiple entities. Therefore, OSA SCS is used to terminate SIP signalling from the S-CSCF. The OSA SCS uses an OSA Application Program Interface (API) to communicate with an actual OSA application server.

The IM-SSF was introduced in the IMS architecture to support legacy services that are developed in the CSE. It hosts CAMEL network features (trigger detection points,

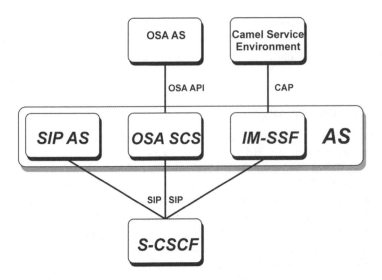

Figure 2.9 Relationship between different application server types.

CAMEL Service Switching Finite State Machine, etc.) and interworks with the CAMEL Application Part (CAP) interface.

A SIP AS is a SIP-based server that hosts a wide range of value-added multimedia services. A SIP AS could be used to provide presence, messaging, Push to talk Over Cellular and conferencing services. The different functions of SIP servers are described in more detail in Sections 12.3 and 3.13.4, as part of service provisioning.

Figure 2.9 shows how different functions are connected. From the perspective of the S-CSCF SIP AS, the OSA service capability server and the IM-SSF exhibit the same reference point behaviour.

An AS may be dedicated to a single service and a user may have more than one service, therefore there may be one or more ASs per subscriber. Additionally, there may be one or more ASs involved in a single session. For example, an operator could have one AS to control terminating traffic to a user based on user preferences (e.g., redirecting all incoming multimedia sessions to an answer machine between 5 p.m. and 7 a.m.) and another AS to adapt the content of instant messages according to the capabilities of the UE (screen size, number of colours, etc.).

MRFC and MRFP together provide mechanisms for bearer-related services such as conferencing, announcements to a user or bearer transcoding in the IMS architecture. The MRFC is tasked to handle SIP communication to and from the S-CSCF and to control the MRFP. The MRFP in turn provides user-plane resources that are requested and instructed by the MRFC. The MRFP performs the following functions:

- Mixing of incoming media streams (e.g., for multiple parties).
- Media stream source (for multimedia announcements).
- Media stream processing (e.g., audio transcoding, media analysis) [3GPP TS23.228, TS 23.002].

Currently, the role of MRFC and MRFP in the IMS architecture is minor, as in IMS conferencing work [3GPP TS 24.147] the MRFC is co-located with an AS and the reference point between the MRFC and MRFP is not yet well defined.

2.2.4 Interworking functions

This section introduces four interworking functions, which are needed for exchanging signalling and media between IMS and the CS CN.

Section 2.2.1.3 explained that the S-CSCF decides when to break to the CS CN. For breaking out the S-CSCF sends a SIP session request to the Breakout Gateway Control Function (BGCF); it further chooses where a breakout to the CS domain occurs. The outcome of a selection process can be either a breakout in the same network in which the BGCF is located or another network. If the breakout happens in the same network, then the BGCF selects a Media Gateway Control Function (MGCF) to handle the session further. If the breakout takes place in another network, then the BGCF forwards the session to another BGCF in a selected network [3GPP TS 23.228]. The latter option allows routing of signalling and media over IP near to the called user.

When a SIP session request hits the MGCF it performs protocol conversion between SIP protocols and the ISDN User Part (ISUP), or the Bearer Independent Call Control (BICC) and sends a converted request via the Signalling Gateway (SGW) to the CS CN. The SGW performs signalling conversion (both ways) at the transport level between the IP-based transport of signalling (i.e., between Sigtran SCTP/IP and SS7 MTP) and the Signalling System No. 7 (SS7) based transport of signalling. The SGW does not interpret application layer (e.g., BICC, ISUP) messages, as is shown in Figure 2.10. The MGCF also controls the IMS Media Gateway (IMS-MGW). The IMS-MGW provides the user-plane link between CS CN networks and the IMS. It terminates the bearer channels from the CS network and media streams from the backbone network (e.g., RTP streams in an IP network or AAL2/ATM connections in an ATM backbone), executes the conversion between these terminations and performs transcoding and signal processing for the user plane when needed. In addition, the IMS-MGW is able to provide tones and announcements to CS users.

Similarly, all incoming call control signalling from a CS user to an IMS user is destined to the MGCF that performs the necessary protocol conversion and sends a SIP session request to the I-CSCF for a session termination. At the same time, the MGCF interacts with the IMS-MGW and reserves necessary IMS-MGW resources at

Figure 2.10   Signalling conversion in the SGW.

the user plane. The overall IMS–CS interworking concept is covered in Section 3.14 which includes two example figures.

2.2.5 Support functions

The PDF is responsible for making policy decisions based on session and media-related information obtained from the P-CSCF. It acts as a policy decision point for SBLP control (see Section 3.10).

Session establishment in the IMS involves an end-to-end message exchange using SIP and SDP. During the message exchange UEs negotiate a set of media characteristics (e.g., common codec(s)). If an operator applies the SBLP, then the P-CSCF will forward the relevant SDP information to the PDF together with an indication of the originator. Correspondingly, the PDF allocates and returns an authorization token which the P-CSCF will pass on to the UE. The PDF notes and authorizes the IP flows of the chosen media components by mapping from SDP parameters to authorized IP QoS parameters for transfer to the access network – i.e., GGSN in case of UMTS/ GPRS access (GPRS access is assumed hereafter) – via the Go interface. On receiving the PDP context activation or modification, the GGSN asks for authorization information from the PDF. The PDF compares the received binding information with the stored authorization information and returns an authorization decision. If the binding information is validated as correct, then the PDF communicates the media authorization details in the decision to the GGSN.

In addition to a bearer authorization decision the PDF receives information as to when an SBLP-governed PDP context is released or the UE has lost/recovered its radio bearer(s) or as to when an SBLP-governed PDP is using streaming or conversational traffic class. Based on this information the PDF is able to inform the P-CSCF about the occurred event. This allows the P-CSCF to effect charging, and it may even start releasing an IMS session on behalf of the user. Moreover, the PDF is able to request GGSN to deactivate a particular SBLP-governed PDP context.

The Security Gateway (SEG) has the function of protecting control-plane traffic between security domains. The security domain refers to a network that is managed by a single administrative authority. Typically, this coincides with operator borders. The SEG is placed at the border of the security domain and it enforces the security policy of a security domain toward other SEGs in the destination security domain. In the IMS all traffic within the IMS is routed via SEGs, especially when the traffic is inter-domain, meaning that it originates from a different security domain from the one where it is received. When protecting inter-domain IMS traffic, both confidentiality as well as data integrity and authentication are mandated [3GPP TS 33.203]. The concept behind a security domain is described more thoroughly in Section 3.19.3.2.

THIG functionality could be used to hide the configuration, capacity and topology of the network from outside an operator's network. If an operator wants to use hiding functionality then the operator must place a THIG function in the routing path when receiving requests or responses from other IMS networks. Similarly, the THIG must be placed in the routing path when sending requests or responses to other IMS networks. The THIG performs the encryption and decryption of all headers which reveal topology information about the operator's IMS network.

2.2.6 Charging entities

Different charging entities and corresponding reference points will be described separately in Section 3.11.

2.2.7 GPRS entities

2.2.7.1 Serving GPRS Support Node (SGSN)

The Serving GPRS Support Node (SGSN) links the RAN to the packet core network. It is responsible for performing both control and traffic-handling functions for the PS domain. The control part contains two main functions: mobility management and session management. Mobility management deals with the location and state of the UE and authenticates both the subscriber and the UE. The control part of session management deals with connection admission control and any changes in the existing data connections. It also supervises 3G network services and resources. Traffic handling is the part of session management that is executed. The SGSN acts as a gateway for user data tunnelling: in other words, it relays user traffic between the UE and the GGSN. As a part of this function, the SGSN also ensures that connections receive the appropriate QoS. In addition, the SGSN generates charging information.

2.2.7.2 Gateway GPRS Support Node (GGSN)

The Gateway GPRS Support Node (GGSN) provides interworking with external packet data networks. The prime function of the GGSN is to connect the UE to external data networks, where IP-based applications and services reside. The external data network could be the IMS or the Internet, for instance. In other words, the GGSN routes IP packets containing SIP signalling from the UE to the P-CSCF and vice versa. Additionally, the GGSN takes care of routing IMS media IP packets toward the destination network (e.g., to GGSN in the terminating network). The interworking service provided is realized as access points that relate to the different networks the subscriber wants to connect. In most cases the IMS has its own access point. When the UE activates a bearer (PDP context) toward an access point (IMS), the GGSN allocates a dynamic IP address to the UE. This allocated IP address is used in IMS registration and when the UE initiates a session as a contact address of the UE. Additionally, the GGSN polices and supervises the PDP context usage for IMS media traffic and generates charging information.

2.3 IMS reference points

This section explains how the previously described network entities are connected to each other and what protocol is used; moreover, the IMS architecture is depicted (Figure 2.11). You will also find an overview of SIP-based reference points (i.e., where SIP is used and the main procedures involved). However, you will realize that the level of description of SIP-based reference points is not so deep as with Diameter-based reference points. The reason for this division is that several chapters

Figure 2.11   IMS architecture.

in this book are dedicated to SIP and SDP procedures where such descriptions are given in detail.

For the sake of clarity, it is impossible to include everything in one figure; so, please note the following:

- Figure 2.11 does not show charging-related functions or reference points (see Section 3.11. for more details).
- The figure does not show different types of ASs (see Section 2.2.3 for more details).
- The figure does not show the user-plane connections between different IMS networks and the AS.
- The figure does not show the SEG at the Mm, Mk, Mw reference points.
- The dotted line between the entities indicates a direct link.
- IMS Service Control (ISC), Cx, Dx, Mm, Mw terminate at both the S-CSCF and the I-CSCF.

2.3.1   Gm reference point

The Gm reference point connects the UE to the IMS. It is used to transport all SIP signalling messages between the UE and the IMS. The IMS counterpart is the P-CSCF. Procedures in the Gm reference point can be divided into three main categories: registration, session control and transactions:

- In the registration procedure the UE uses the Gm reference point to send a registration request with an indication of supported security mechanisms to the P-CSCF. During the registration process the UE exchanges the necessary parameters for authenticating both itself and the network, gets implicit registered user identities, negotiates the necessary parameters for an SA with the P-CSCF and possibly starts SIP compression. In addition, the Gm reference point is used to inform the UE if network-initiated de-registration or network-initiated re-authentication occurs.

- Session control procedures contain mechanisms for both mobile-originated sessions and mobile-terminated sessions. In mobile-originated sessions the Gm reference point is used to forward requests from the UE to the P-CSCF. In mobile-terminated sessions the Gm reference point is used to forward requests from the P-CSCF to the UE.
- Transaction procedures are used to send standalone requests (e.g., MESSAGE) and to receive all responses (e.g., 200 OK) to that request via the Gm reference point. The difference between transaction procedures and session control procedures is that a dialog is not created.

2.3.2   Mw reference point

The Gm reference point links the UE to the IMS (namely, to the P-CSCF). Next, a SIP-based reference point between different CSCFs is needed. This reference point is called Mw. The procedures in the Mw reference point can be divided into three main categories – registration, session control and transaction:

- In the registration procedure the P-CSCF uses the Mw reference point to forward a registration request from the UE to the I-CSCF. The I-CSCF then uses the Mw reference point to pass the request to the S-CSCF. Finally, the response from the S-CSCF traverses back via the Mw reference point. In addition, the S-CSCF uses the Mw reference point in network-initiated de-registration procedures to inform the UE about network-initiated de-registration and network-initiated re-authentication to inform the P-CSCF that it should release resources regarding a particular user.
- Session control procedures contain mechanisms for both mobile-originated sessions and mobile-terminated sessions. In mobile-originated sessions the Mw reference point is used to forward requests both from the P-CSCF to the S-CSCF and from the S-CSCF to the I-CSCF. In mobile-terminated sessions the Mw reference point is used to forward requests both from the I-CSCF to the S-CSCF and from the S-CSCF to the P-CSCF. This reference point is also used for network-initiated session releases: for example, the P-CSCF could initiate a session release toward the S-CSCF if it receives an indication from the PDF that media bearer(s) are lost. In addition, charging-related information is conveyed via the Mw reference point.
- Transaction procedures are used to pass a standalone request (e.g., MESSAGE) and to receive all responses (e.g., 200 OK) to that request via the Mw reference point. As already stated, the difference between transaction procedures and session control procedures is that a dialog is not created.

2.3.3   IMS Service Control (ISC) reference point

In the IMS architecture, ASs are entities that host and execute services, such as presence, messaging and session forwarding. Therefore, there has to be a reference point for sending and receiving SIP messages between the CSCF and an AS. This reference point is called the IMS Service Control (ISC) reference point and the selected protocol is SIP. ISC procedures can be divided into two main categories – routing the initial SIP request to an AS and AS-initiated SIP requests:

- When the S-CSCF receives an initial SIP request it will analyse it. Based on the analysis the S-CSCF may decide to route the request to an AS for further processing. The AS may terminate, redirect or proxy the request from the S-CSCF.
- An AS may initiate a request (e.g., on behalf of a user).
- The concept of service control is thoroughly described in Section 3.13.

2.3.4   Cx reference point

Subscriber and service data are permanently stored in the HSS. These centralized data need to be utilized by the I-CSCF and the S-CSCF when the user registers or receives sessions. Therefore, there has to be a reference point between the HSS and the CSCF. This reference point is called the Cx reference point and the selected protocol is Diameter. The procedures can be divided into three main categories: location management, user data handling and user authentication. Table 2.1 summarizes the available Cx commands.

2.3.4.1   Location management procedures

Location management procedures can be further divided in two groups: registration and de-registration, and location retrieval.

When the I-CSCF receives a SIP REGISTER request from the P-CSCF via the Mw reference point it will invoke a user registration status query or, as it is known in the standards, a User-Authorization-Request (UAR) command. After receiving the UAR command the HSS sends a User-Authorization-Answer (UAA) command. It contains the S-CSCF Name and/or S-CSCF Capabilities (if the UAR command does not fail due, say, to the private and public identities received in the request not belonging to the same user) depending on the user's current registration status. S-CSCF capabilities are returned if the user does not have an S-CSCF Name assigned yet in the HSS or if the I-CSCF explicitly requests S-CSCF Capabilities. Otherwise, the S-CSCF name is returned. When capabilities are returned the I-CSCF needs to perform S-CSCF selection as described in Section 3.9.

We explained above how the I-CSCF finds an S-CSCF that will serve the user. Having done this, the I-CSCF forwards a SIP REGISTER request to the S-CSCF. When the S-CSCF receives the SIP REGISTER request from the I-CSCF it uses a Server-Assignment-Request (SAR) command to communicate with the HSS. The SAR command is used to inform the HSS about which S-CSCF will be serving the user when the expires value is not equal to zero. Similarly, if the expires value equals zero, then the SAR command is used to announce the fact that the S-CSCF is no longer serving a user. A precondition for sending the SAR command is that the user has been successfully authenticated by the S-CSCF. After receiving the SAR command the HSS will respond with a Server-Assignment-Answer (SAA) command. It contains the User

Table 2.1 Cx commands.

Command-Name	Purpose	Abbreviation	Source	Destination
User-Authorization-Request/Answer	User-Authorization-Request/Answer (UAR/UAA) commands are used between the I-CSCF and the HSS during SIP registration for retrieving S-CSCF name or S-CSCF capabilities for S-CSCF selection and during SIP deregistration for retrieving S-CSCF name when the SIP method is REGISTER	UAR UAA	I-CSCF HSS	HSS I-CSCF
Server-Assignment-Request/Answer	Server-Assignment-Request/Answer (SAR/SAA) commands are used between the S-CSCF and the HSS to update the S-CSCF name to the HSS and to download the user profile data to the S-CSCF	SAR SAA	S-CSCF HSS	HSS S-CSCF
Location-Info-Request/Answer	Location-Info-Request/Answer (LIR/LIA) commands are used between the I-CSCF and the HSS during the SIP session set-up to obtain the name of the S-CSCF that is serving the user or S-CSCF capabilities for S-CSCF selection	LIR LIA	I-CSCF HSS	HSS I-CSCF
Multimedia-Auth-Request/Answer	Multimedia-Auth-Request/Answer (MAR/MAA) commands are used between the S-CSCF and the HSS to exchange information to support the authentication between the end user and the home IMS network	MAR MAA	S-CSCF HSS	HSS S-CSCF
Registration-Termination-Request/Answer	Registration-Termination-Request/Answer (RTR/RTA) commands are used between the S-CSCF and the HSS when the HSS administratively de-registers one or more of the user's public identities	RTR RTA	HSS S-CSCF	S-CSCF HSS
Push-Profile-Request/Answer	Push-Profile-Request/Answer (PPR/PPA) commands are used between the HSS and the S-CSCF when user profile data are changed by a management operation in HSS and the data need to be updated to the S-CSCF	PPR PPA	HSS S-CSCF	S-CSCF HSS

Profile based on the set values in the SAR request and, optionally, the addresses of the charging functions.

So far we have described how user-initiated registration and de-registration (user initiated or S-CSCF initiated) procedures are handled over the Cx reference point. There is still the need for additional operations to bring about network-initiated de-registration (e.g., due to a stolen UE or when a subscription is terminated). In this case it is the HSS that starts network-initiated de-registration by using a command called Registration-Termination-Request (RTR). The RTR command is acknowledged by a Registration-Termination-Answer (RTA) command, which simply indicates the result of the operation.

Previously, we have described how the I-CSCF uses a user registration status query (UAR command) to find the S-CSCF when it receives a SIP REGISTER request. Correspondingly, there has to be a procedure to find the S-CSCF when a SIP method is different from REGISTER. The required procedure is to make use of a Location-Info-Request (LIR) command. The HSS responds with a Location-Info-Answer (LIA) command. The response contains the S-CSCF Name or S-CSCF Capabilities – the latter are returned if the user does not have the S-CSCF Name assigned, otherwise the SIP URI of the S-CSCF is returned.

2.3.4.2   User data handling procedures

During the registration process, user and service-related data will be downloaded from the HSS to the S-CSCF via the Cx reference point using SAR and SAA commands as described earlier. However, it is possible for these data to be changed later when the S-CSCF is still serving a user. To update the data in the S-CSCF the HSS initiates a Push-Profile-Request (PPR) command. Update takes place immediately after the change with one exception: when the S-CSCF is serving an unregistered user or the S-CSCF is kept for an unregistered user as described in Section 3.9.5 and there is a change in the registered part of user profile, then the HSS will not send a PPR command. The PPR command is acknowledged by a Push-Profile-Answer (PPA) command, which simply indicates the result of the operation.

2.3.4.3   Authentication procedures

IMS user authentication relies on a pre-configured shared secret. Shared secrets and sequence numbers are stored in the IP Multimedia Services Identity Module (ISIM) in the UE and in the HSS in the network. Because S-CSCF takes care of user authorization, there exists the need to transfer security data over the Cx reference point. When the S-CSCF needs to authenticate a user it sends a Multimedia-Auth-Request (MAR) command to the HSS. The HSS responds with a Multimedia-Auth-Answer (MAA) command. The answer contains among other information Authentication Data. It includes an authentication vector, which is comprised of an Authentication Scheme (e.g., Digest-AKAv1-MD5), Authentication Information (authentication challenge RAND and the token AUTN), Authorization Information (expected response, or XRES), Integrity Key and, optionally, a Confidentiality Key. Additionally, it contains an Item Number, which indicates the order in which the authentication vectors are to be consumed when multiple vectors are returned. The usage of these elements is further described in Section 10.6.

2.3.5   Dx reference point

When multiple and separately addressable HSSs have been deployed in a network, neither the I-CSCF nor the S-CSCF know which HSS they need to contact. However, they need to contact the SLF first. For this purpose the Dx reference point has been introduced. The Dx reference point is always used in conjunction with the Cx reference point. The protocol used in this reference point is based on Diameter. Its functionality is implemented by means of the routing mechanism provided by an enhanced Diameter redirect agent.

To get an HSS address the I-CSCF or the S-CSCF sends to the SLF the Cx requests aimed for the HSS. On receipt of the HSS address from the SLF, the I-CSCF or the S-CSCF will send the Cx requests to the HSS. Figure 2.12 shows how the SLF is used to find a correct HSS when the I-CSCF receives an INVITE request and three HSSs have been deployed.

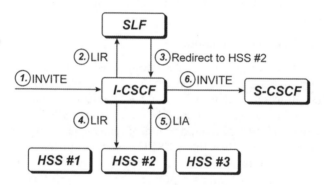

Figure 2.12   HSS resolution using the SLF.

2.3.6   Sh reference point

An AS (SIP AS or OSA SCS) may need user data or need to know to which S-CSCF to send a SIP request. This type of information is stored in the HSS. Therefore, there has to be a reference point between the HSS and the AS. This reference point is called the Sh reference point and the protocol is Diameter. Procedures are divided into two main categories: data handling and subscription/notification. Table 2.2 summarizes the available Sh commands. The HSS maintains a list of ASs that are allowed to obtain or store data.

2.3.6.1   Data handling

Data-handling procedures contain the possibility to retrieve user data from the HSS. Such user data can contain service-related data (transparent or non-transparent), registration information, identities, initial filter criteria, the S-CSCF Name serving

Table 2.2 Sh commands.

Command-Name	Purpose	Abbreviation	Source	Destination
User-Data-Request/Answer	User-Data-Request/Answer (UDR/UDA) commands are	UDR	AS	HSS
	used to deliver the user data of a particular user	UDA	HSS	AS
Profile-Update-Request/Answer	Profile-Update-Request/Answer (PUR/PUA) commands are	PUR	AS	HSS
	used to update transparent data in the HSS	PUA	HSS	AS
Subscribe-Notifications-Request/Answer	Subscribe-Notifications-Request/Answer commands are	SNR	AS	HSS
	used to make a subscription/ cancel a subscription to user's data on which notifications of change are required	SNA	HSS	AS
Push-Notification-Request/Answer	Push-Notification-Request/ Answer commands are used to	PNR	HSS	AS
	send the changed data to the AS	PNA	AS	HSS

the user, addresses of the charging functions and even location information from the CS and PS domains. Transparent data are understood syntactically but not semantically by the HSS. They are data that an AS may store in the HSS to support its service logic. On the contrary, non-transparent data are understood both syntactically and semantically by the HSS. The AS uses the User-Data-Request (UDR) command to request data. The request contains information about the request data. The HSS responds with the User-Data-Answer (UDA).

The AS can update transparent data in the HSS using the Profile-Update-Request (PUR) command, which contains data to be updated. The PUR command is acknowledged by a Profile-Update-Answer (PUA) command, which simply indicates the result of the operation.

2.3.6.2 Subscription/Notification

Subscription/Notification procedures allow the AS to get a notification when particular data for a specific user is updated in the HSS. The AS sends a Subscribe-Notifications-Request (SNR) command to receive a notification of when a user's data indicated in the SNR command are changed in the HSS. The HSS acknowledges the subscription request by a Subscribe-Notifications-Answer (SNA) command, which simply indicates the result of the operation.

If the AS has sent the SNR command and requested a notification with subscription request type, then the HSS sends a Push-Notification-Request (PNR) command to the AS when particular data have changed giving details of the changed data. The PNR command is acknowledged by a Push-Notification-Answer (PNA) command, which simply indicates the result of the operation.

2.3.7   Si reference point

When the AS is a CAMEL AS (IM-SSF) it uses the Si reference point to communicate with the HSS. The Si reference point is used to transport CAMEL subscription information including triggers from the HSS to the IM-SSF. The used protocol is Mobile Application Part (MAP).

2.3.8   Dh reference point

When multiple and separately addressable HSSs have been deployed in the network, the AS cannot know which HSS it needs to contact. However, the AS needs to contact the SLF first. For this purpose the Dh reference point was introduced in Release 6. In Release 5 the correct HSS is discovered by using proprietary means. The Dh reference point is always used in conjunction with the Sh reference point. The protocol used in this reference point is based on Diameter. Its functionality is implemented by means of the routing mechanism provided by an enhanced Diameter re-direct agent.

To get an HSS address, the AS sends to the SLF the Sh request aimed for the HSS. On receipt of the HSS address from the SLF, the AS will send the Sh request to the HSS.

2.3.9   Mm reference point

For communicating with other multimedia IP networks, a reference point between the IMS and other multimedia IP networks is needed. The Mm reference point allows the I-CSCF to receive a session request from another SIP server or terminal. Similarly, the S-CSCF uses the Mm reference point to forward IMS UE-originated requests to other multimedia networks. At the time of writing, a detailed specification of the Mm reference point has not been provided. However, it is very likely that the protocol would be SIP.

2.3.10   Mg reference point

The Mg reference point links the CS edge function, MGCF, to IMS (namely, to the I-CSCF). This reference point allows MGCF to forward incoming session signalling from the CS domain to the I-CSCF. The protocol used for the Mg reference point is SIP. MGCF is responsible for converting incoming ISUP signalling to SIP.

2.3.11   Mi reference point

When the S-CSCF discovers that a session needs to be routed to the CS domain it uses the Mi reference point to forward the session to BGCF. The protocol used for the Mi reference point is SIP. Section 3.14 contains further details about IMS–CS interworking.

2.3.12   Mj reference point

When BGCF receives session signalling via the Mi reference point it selects the CS domain in which breakout is to occur. If breakout occurs in the same network, then it forwards the session to MGCF via the Mj reference point. The protocol used for the Mj reference point is SIP. Section 3.14 contains further details about IMS–CS interworking.

2.3.13   Mk reference point

When BGCF receives session signalling via the Mi reference point it selects the CS domain in which breakout is to occur. If the breakout occurs in another network, then it forwards the session to BGCF in the other network via the Mk reference point. The protocol used for the Mk reference point is SIP. Section 3.14 contains further details about IMS–CS interworking.

2.3.14   Mn reference point

The Mn interface is the control reference point between the MGCF and IMS-MGW. The Mn interface controls the user plane between IP access and IMS-MGW (Mb reference point). Also, it controls the user plane between CS access (Nb and TDM interfaces) and IMS-MGS. The Mn interface is based on H.248 and is equivalent to the usage (encoding, decoding, etc.) of the Mc interface specified to control the CS-MGW. The difference between these two interfaces is that the Mn interface introduces new H.248 procedures for handling IP access end termination and also some additional procedures for CS end termination handling. The H.248 is primarily used to perform the following tasks:

- reserve and connect terminations;
- connect or release echo canceller to terminations;
- connect or release tones and announcements to terminations;
- send/receiver DTMF tones.

2.3.15   Ut reference point

The Ut reference point is the reference point between the UE and the AS. It enables users to securely manage and configure their network services related information hosted on an AS. Users can use the Ut reference point to create Public Service Identities (PSIs), such as a resource list, and manage the authorization policies that are used by the service. Examples of services that utilize the Ut reference point such as presence, Push to talk Over Cellular and conferencing. The AS may need to provide security for the Ut reference point.

Hypertext Transfer Protocol (HTTP) is the chosen data protocol for the Ut reference point. Any protocol chosen for an application that makes use of the Ut reference point needs to be based on HTTP. This reference point is standardized in Release 6. Usage of the Ut reference point is described in more detail in Chapter 8.

2.3.16 Mr reference point

When the S-CSCF needs to activate bearer-related services it passes SIP signalling to the MRFC via the Mr reference point. The functionality of the Mr reference point is not fully standardized: for example, it is not specified how the S-CSCF informs the MRFC to play a specific announcement. The used protocol in the Mr reference point is SIP.

2.3.17 Mp reference point

When the MRFC needs to control media streams (e.g., to create connections for conference media or to stop media in the MRFP) it uses the Mp reference point. This reference point is fully compliant with H.248 standards. However, IMS services may require extensions. This reference point is not standardized in Release 5 nor in Release 6.

2.3.18 Go reference point

It is in operators' interests to ensure that the QoS and source and destination addresses of the intended IMS media traffic matches the negotiated values at the IMS level. This requires communication between the IMS (control plane) and the GPRS network (user plane). The Go reference point was originally defined for this purpose. Later on, the charging correlation was added as an additional functionality. The protocol used is the Common Open Policy Service (COPS) protocol. Go procedures can be divided into two main categories:

- Media authorization – as far as access is concerned, the Policy Enforcement Point (PEP) (e.g., GGSN) uses the Go reference point to ask whether a requested bearer activation can be accepted from the PDF that acts as a Policy Decision Point (PDP). The PEP also uses the Go reference point to notify the PDP about necessary bearer modification and bearer releases (e.g., PDP context). As far as the IMS is concerned, the PDF uses the Go reference point to explicitly indicate when a bearer can or cannot be used; it may also request the PEP to initiate a bearer release. Media authorization is thoroughly explained in the context of the SBLP in Section 3.10.
- Charging correlation – via the Go reference point the IMS is able to pass an IMS Charging Identifier (ICID) to the GPRS network (user plane). In similar manner, the GPRS network is able to pass a GPRS charging identifier to the IMS. With this procedure it is possible to later merge GPRS charging and IMS charging information in a billing system. This concept is further explained in Section 3.11.6.

2.3.19 Gq reference point

When a standalone PDF is deployed the Gq reference point is used to transport policy setup information between the application function and the PDF. The term "application function" is used because it is intended that a PDF could authorize traffic other than IMS traffic. In the IMS case the P-CSCF plays the role of an application function. This reference point was standardized in Release 6 and the selected protocol is Diameter.

The P-CSCF sends policy information to the PDF about every SIP message that includes an SDP payload. This ensures that the PDF passes the proper information to perform media authorization for all possible IMS session setup scenarios. The media authorization (SBLP) concept is thoroughly explained in Section 3.10. The P-CSCF provides the following policy-related information to the PDF to be used in SBLP [3GPP TS 23.207, 29.207, 29.209]:

- Information about the media stream. It includes, for instance, the following:
  - direction of traffic (bi-directional, uplink/downlink);
  - source/destination IP address and port number;
  - transport protocol;
  - maximum requested bandwidth uplink/downlink;
  - status of each media component (enabled/disabled per uplink/downlink direction);
  - information about the grouping rules of media components;
  - type of media (audio, video, data, application, control, text, message, other).
- Resource Reservation Policy. It contains information as to whether the P-CSCF wants to be contacted in each bearer authorization or whether the PDF can use the available information to make the decision itself.
- Indication Forwarding Policy. This is used to inform the PDF if the P-CSCF is interested in receiving indications of loss of bearer, recovery of bearer or release of bearer.
- IMS Charging Identifier (ICID). This information is delivered by the PDF to the access network in order to enable charging correlation as described in Section 3.11.6.
- Information about the used application. This information may be used by the PDF to differentiate QoS for different application services.
- Information about SIP forking. This information is needed so that the PDF is able to calculate the proper authorization, as the PDF shall authorize the maximum bandwidth required by any of the SIP dialogs but not the sum of the bandwidths required by all SIP dialogs (reserving some of the bandwidth would decrease capacity and performance in the access networks).

Moreover, the P-CSCF is able to request the PDF to remove previously authorized resources. The PDF uses the Gq reference point to deliver an authorization token (see Section 3.10.1.1), GPRS charging identifier (see Section 3.10.7), the GGSN's IP address and fulfill other requests from the P-CSCF, as listed above.

2.3.20 Charging reference points

Charging-related reference points Rf, Ro, Rx are described in Sections 3.11.5.1, 3.11.5.2 and 3.11.5.3.

Table 2.3 Summary of reference points.

Name of reference point	Involved entities	Purpose	Protocol
Gm	UE, P-CSCF	This reference point is used to exchange messages between UE and CSCFs	SIP
Mw	P-CSCF, I-CSCF, S-CSCF	This reference point is used to exchange messages between CSCFs	SIP
ISC	S-CSCF, I-CSCF, AS	This reference point is used to exchange messages between CSCF and AS	SIP
Cx	I-CSCF, S-CSCF, HSS	This reference point is used to communicate between I-CSCF/S-CSCF and HSS	Diameter
Dx	I-CSCF, S-CSCF, SLF	This reference point is used by I-CSCF/S-CSCF to find a correct HSS in a multi-HSS environment	Diameter
Sh	SIP AS, OSA SCS, HSS	This reference point is used to exchange information between SIP AS/OSA SCS and HSS	Diameter
Si	IM-SSF, HSS	This reference point is used to exchange information between IM-SSF and HSS	MAP
Dh	SIP AS, OSA, SCF, IM-SSF, HSS	This reference point is used by AS to find a correct HSS in a multi-HSS environment	Diameter
Mm	I-CSCF, S-CSCF, external IP network	This reference point will be used for exchanging messages between IMS and external IP networks	Not specified
Mg	MGCF → I-CSCF	MGCF converts ISUP signalling to SIP signalling and forwards SIP signalling to I-CSCF	SIP
Mi	S-CSCF → BGCF	This reference point is used to exchange messages between S-CSCF and BGCF	SIP

Reference point	Entities	Description	Protocol
Mj	BGCF → MGCF	This reference point is used to exchange messages between BGCF and MGCF in the same IMS network	SIP
Mk	BGCF → BGCF	This reference point is used to exchange messages between BGCFs in different IMS networks	SIP
Mr	S-CSCF, MRFC	This reference point is used to exchange messages between S-CSCF and MRFC	SIP
Mp	MRFC, MRFP	This reference point is used to exchange messages between MRFC and MRFP	H.248
Mn	MGCF, IMS-MGW	This reference point allows control of user-plane resources	H.248
Ut	UE, AS (SIP AS, OSA SCS, IM-SSF)	This reference point enables UE to manage information related to his services	HTTP
Go	PDF, GGSN	This reference point allows operators to control QoS in a user plane and exchange charging correlation information between IMS and GPRS network	COPS
Gq	P-CSCF, PDF	This reference point is used to exchange policy decisions-related information between P-CSCF and PDF	Diameter
Ro	AS, MRCF, S-CSCF, OCS	This reference point is used by AS/MRFC/S-CSCF for online charging towards OCS. Note: there might exist an interworking function between the S-CSCF and OCS.	Diameter
Rf	P-CSCF, S-CSCF, I-CSCF, BGCF, MGCF, AS, MRFC, CDF	This reference point is used by IMS entities for offline charging towards CDF.	Diameter
Rx	P-CSCF, AS, Charging Rule Function	This reference point allows dynamic charging-related service information to be exchanged between Charging Rules Function (CRF) and IMS entities. This information is used by the CRF for the selection and completion of charging rules.	Diameter

3

IMS concepts

3.1 Overview

This chapter begins with a first-glance description of IP Multimedia Subsystem (IMS) registration and session establishment. It depicts the IMS entities that are involved. The intention is not to show a full-blown solution; rather, it is to give an overview and help the reader to understand the different IMS concepts explained in this chapter. Detailed registration and session establishment flows will be shown and explained later in the book.

Prior to IMS registration the User Equipment (UE) must discover the IMS entity to which it will send a REGISTER request. This concept is called a Proxy-Call Session Control Function (P-CSCF) discovery and is described in Section 3.8. In addition, before a registration process the UE needs to fetch user identities from identity modules. Identity modules are covered in Section 3.6 and identities are presented in Section 3.5. During the registration a Serving-CSCF (S-CSCF) will be assigned (Section 3.9), authentication will be performed and corresponding security associations will be established (Section 3.19), a user profile (Section 3.12) will be downloaded to the assigned S-CSCF, Session Initiation Protocol (SIP) compression will be initialized (Section 3.16) and implicitly registered public user identities will be delivered (Section 3.3). The concept of sharing a single user identity between multiple terminals is covered in Section 3.7.

Section 3.10 explains how Internet Protocol (IP) policy control is applied when a user is establishing a session, and Section 3.13 shows how services can be provisioned. Section 3.11 shows how an operator is able to charge a user. Interworking with the Circuit Switched (CS) network is briefly described in Section 3.14. Simultaneous usage of CS and Packet Switched (PS) components are covered in Section 3.18. In addition, fixed mobile convergence (Section 3.15) and IP version interworking (Section 3.17) are covered.

3.2 Registration

Prior to IMS registration, which allows the UE to use IMS services, the UE must obtain an IP connectivity bearer and discover an IMS entry point (i.e., the P-CSCF): for example, in the case of General Packet Radio Service (GPRS) access the UE

The IMS: IP Multimedia Concepts and Services Second Edition
Miikka Poikselkä, Georg Mayer, Hisham Khartabil and Aki Niemi © 2006 John Wiley & Sons, Ltd

Figure 3.1 High-level IMS registration.

performs the GPRS attach procedure and activates a Packet Data Protocol (PDP) context for SIP signalling. Chapter 17 gives a short overview of the PDP context, and P-CSCF discovery is explained in Section 3.8. This book does not describe the GPRS attach procedure (for further information see [3GPP TS 23.060]).

IMS registration contains two phases: the leftmost part of Figure 3.1 shows the first phase – how the network challenges the UE. The rightmost part of Figure 3.1 shows the second phase – how the UE responds to the challenge and completes the registration.

First, the UE sends a SIP REGISTER request to the discovered P-CSCF. This request would contain, say, an identity to be registered and a home domain name (address of the Interrogating-CSCF, or I-CSCF). The P-CSCF processes the REGISTER request and uses the provided home domain name to resolve the IP address of the I-CSCF. The I-CSCF, in turn, will contact the Home Subscriber Server (HSS) to fetch the required capabilities for S-CSCF selection. After S-CSCF selection the I-CSCF forwards the REGISTER request to the S-CSCF. The S-CSCF realizes that the user is not authorized and, therefore, retrieves authentication data from the HSS and challenges the user with a 401 Unauthorized response. Second, the UE will calculate a response to the challenge and send another REGISTER request to the P-CSCF. Again the P-CSCF finds the I-CSCF and the I-CSCF, in turn, will find the S-CSCF. Finally, the S-CSCF checks the response and, if it is correct, downloads a user profile from the HSS and accepts the registration with a 200 OK response. Once the UE is successfully authorized, the UE is able to initiate and receive sessions. During the registration procedure both the UE and the P-CSCF learn which S-CSCF in the network will be serving the UE. Table 3.1 summarizes information storage before, during and after the registration process.

It is the UE's responsibility to keep its registration active by periodically refreshing its registration. If the UE does not refresh its registration, then the S-CSCF will silently remove the registration when the registration timer lapses. When the UE wants to de-register from the IMS it sets a registration timer to 0 and sends a REGISTER request. Sections 10.5 and 10.14 contain more detailed descriptions of IMS registration and de-registration.

Table 3.1 Information storage before, during and after the registration process.

Node	Before registration	During registration	After registration
UE	P-CSCF address, home domain name, credentials, public user identity, private user identity	P-CSCF address, home domain name, credentials, public user identity, private user identity, security association	P-CSCF address, home domain name, credentials, public user identity (and implicitly registered public user identities), private user identity, security association, service route information (S-CSCF)
P-CSCF	No state information	Initial network entry point, UE IP address, public and private user IDs, security association	Final network entry point (S-CSCF), UE address, registered public user identity (and implicitly registered public user identities), private user ID, security association, address of CDF
I-CSCF	HSS or SLF address	HSS or SLF entry, P-CSCF address, S-CSCF address	HSS or SLF address
S-CSCF	HSS or SLF address	HSS address/name, user profile (limited—as per network scenario), proxy address/name, public/private user ID, UE IP address	HSS address/name, user profile (limited—as per network scenario), proxy address/name, public/private user ID, UE IP address
HSS	User profile, S-CSCF selection parameters	User profile, S-CSCF selection parameters, visited network information if user is roaming	User profile, S-CSCF selection parameters, information about which user identities are registered, name of the S-CSCF allocated to the user

3.3 Mechanism to register multiple user identities at a go

SIP allows one public user identity to be registered at a time; so, if a user has more than one public user identity, then she has to register every public user identity individually. This may be frustrating and time consuming from the end-user perspective. Obviously, registering four public user identities would consume four times as much radio resource in the case of Universal Mobile Telecommunications System (UMTS) than registering one public user identity. It was for these reasons that Third Generation Partnership Project (3GPP) developed a mechanism to register more than one public user identity at a time. This concept is called "implicit registration".

Figure 3.2   Example of implicit registration sets.

An implicit registration set is a group of public user identities that are registered via a single registration request. When one of the public user identities within the set is registered, all public user identities associated with the implicit registration set are registered at the same time. Similarly, when one of the public user identities within the set is de-registered, all public user identities that have been implicitly registered are de-registered at the same time. Public user identities belonging to an implicit registration set may point to different service profiles. Some of these public user identities may point to the same service profile [3GPP TS 23.228].

To get implicitly registered public user identities the UE must send a SUBSCRIBE request for a registration event package to the S-CSCF. When the S-CSCF receives the SUBSCRIBE request it will return the implicitly registered public user identity with a NOTIFY request. For example, a user has four public user identities that are grouped in two implicit registration sets (Figure 3.2). The first set contains *joe.smith@brandnewcar.com* and *tel:+358501234567*. The second set contains *joe.smith@ims.example.com* and *tel:+358503334444*. When Joe sends a REGISTER request containing *joe.smith@brandnewcar.com* as an identity to be registered, the allocated S-CSCF performs a normal registration procedure and, after successful authorization, the S-CSCF downloads the service profiles that are associated with the public user identities belonging to the implicit registration set (service profile 1). To obtain the implicitly registered public user identities, Joe's UE sends a SUBSCRIBE request to the S-CSCF. When the S-CSCF receives the SUBSCRIBE request it will return the implicitly registered public user identity, *tel:+358501234567*, within NOTIFY.

3.4   Session initiation

When User A wants to have a session with User B, UE A generates a SIP INVITE request and sends it via the Gm reference point to the P-CSCF. The P-CSCF processes the request: for example, it decompresses the request and verifies the originating user's identity before forwarding the request via the Mw reference point to the S-CSCF. The S-CSCF processes the request, executes service control which may include interactions with Application Servers (ASs) and eventually determines the entry point of the home operator of User B based on User B's identity in the SIP INVITE request. The I-CSCF receives the request via the Mw reference point and contacts the HSS over the Cx reference point to find the S-CSCF that is serving User B. The request is passed to the S-CSCF via the Mw reference point. The S-CSCF takes charge of processing the terminating session, which may include interactions with ASs and eventually delivers the request to the P-CSCF over the Mw reference point. After further processing (e.g., compression and privacy checking), the P-CSCF uses the Gm reference point to deliver the SIP INVITE request to UE B. UE B generates a response – 183 Session Progress – which traverses back to UE A following the route that was created on the way from UE A (i.e., UE B → P-CSCF → S-CSCF → I-CSCF → S-CSCF → P-CSCF → UE A) (Figure 3.3). After a few more round trips, both sets of UE complete session establishment and are able to start the actual application (e.g., a game of chess). During session establishment an operator may control the usage of bearers intended for media traffic. Section 3.10 explains how this can be done. Chapter 11 contains a more detailed description of IMS session initiation.

Just to give a taste of what is coming in the book, the high-level content of a SIP INVITE request is given in Table 3.2. Each column gives the information elements that are inserted, removed or modified. The meaning of each information element is covered later in the book.

Figure 3.3 High-level IMS session establishment flow.

Table 3.2 The high-level content of a SIP INVITE request during session establishment.

UE (A)	P-CSCF (A)	S-CSCF (A)
User A identity	*Inserted information*	*Inserted information*
User B identity	One piece of routing	Interoperator identifier
Contact address	information (Record-Route	
Access information	header)	
Routing information (Via	IMS charging information	
and Route headers)	Verified A party identity	
Support of reliable		
responses	*Removed information*	*Removed information*
Support of preconditions	Security information	One piece of routing
Security information	Proposed A party identity	information (Route
Privacy indication		header)
Compression indication		Access information
SDP payload reflecting user's		
terminal capabilities and	*Modified information*	*Modified information*
user preferences for the	Routing information (Route,	Routing information
session, MIME subtype	Via)	(Record-Route and Via)
"telephone-event",		Verified A party identity,
bandwidth information		also includes Tel-URL type
		of identity from now on

I-CSCF (B)	S-CSCF (B)	P-CSCF (B)
Inserted information	*Inserted information*	*Inserted information*
One piece of routing	None	Authorization token
information (Route		
header)		
Removed information	*Removed information*	*Removed information*
None	Interoperator identifier	IMS charging information
		One piece of routing
		information (Route
		header)
		A party identity is removed
		if privacy is required
Modified information	*Modified information*	*Modified information*
Routing information (Via)	Routing information (R-URI,	Routing information (Via,
	Route, Via, Record route)	Record route)

3.5   Identification

3.5.1   Identification of users

3.5.1.1   Private user identity

The private user identity is a unique global identity defined by the home network operator, which may be used within the home network to uniquely identify the user from a network perspective [3GPP TS 23.228]. It does not identify the user herself; on the contrary, it identifies the user's subscription. Therefore, it is mainly used for authentication purposes. It is possible to utilize private user identities for accounting and administration purposes as well. The IMS architecture imposes the following requirements for private user identity [3GPP TS 23.228, TS 23.003]:

- The private user identity will take the form of a Network Access Identifier (NAI) defined in [RFC2486].
- The private user identity will be contained in all registration requests passed from the UE to the home network.
- The private user identity will be authenticated only during registration of the user (including re-registration and de-registration).
- The S-CSCF will need to obtain and store the private user identity on registration and on unregistered termination.
- The private user identity will not be used for routing of SIP messages.
- The private user identity will be permanently allocated to a user and securely stored in an IMS Identity Module (ISIM) application. The private user identity will be valid for the duration of the user's subscription within the home network.
- It will not be possible for the UE to modify the private user identity.
- The HSS will need to store the private user identity.
- The private user identity will optionally be present in charging records based on operator policies.

Example of NAI	*form_user@realm*

3.5.1.2 Public user identity

User identities in IMS networks are called public user identities. They are the identities used for requesting communication with other users. Public identities can be published (e.g., in phone books, Web pages, business cards).

As stated earlier in the book, IMS users will be able to initiate sessions and receive sessions from many different networks, such as GSM networks and the Internet. To be reachable from the CS side, the public user identity must conform to telecom numbering (e.g., *+358501234567*). In similar manner, requesting communication with Internet clients, the public user identity must conform to Internet naming (e.g., *joe.doe@example.com*).

The IMS architecture imposes the following requirements for public user identity [3GPP TS 23.228, TS 23.003]:

- The public user identity/identities will take the form of either a SIP Uniform Resource Identifier (URI) or a telephone Uniform Resource Locator (tel URL) format.
- At least one public user identity will be securely stored in an ISIM application.

- It will not be possible for the UE to modify the public user identity.
- A public user identity will be registered before the identity can be used to originate IMS sessions and IMS session-unrelated procedures (e.g., MESSAGE, SUBSCRIBE, NOTIFY).
- A public user identity will be registered before terminating IMS sessions, and terminating IMS session-unrelated procedures will be delivered to the UE of the user that the public user identity belongs to. This does not prevent the execution of services in the network by unregistered users.
- It will be possible to register multiple public user identities through one single UE request. This is described further in Section 3.3.
- The network will not authenticate public user identities during registration.

The tel URL scheme is used to express traditional E.164 numbers in URL syntax. The tel URL is described in [RFC2806], and the SIP URI is described in [RFC3261] and [RFC2396]. Examples of public user identities are given below. A more detailed description of SIP URI and tel URL syntaxes can be found in Sections 12.5 and 12.6.

Example of SIP URI	*sip:joe.doe@ims.example.com*

Example of tel URL	*tel:+358 50 1234567*

3.5.1.3 Derived public user identity and private user identity

In Sections 3.5.1.1 and 3.5.1.2 the concepts of private user identity and public user identity have been explained. It was stated that these identities are stored in an ISIM application. When the IMS is deployed there will be a lot of UE in the market place that does not support the ISIM application; therefore, a mechanism to access the IMS without the ISIM was developed.

In this model, private user identity, public user identity and home domain name are derived from an International Mobile Subscriber Identifier (IMSI). This mechanism is suitable for UE that has a Universal Subscriber Identity Module (USIM) application.

Private user identity
The private user identity derived from the IMSI is built according to the following steps [3GPP TS 23.003]:

1. The user part of the private user identity is replaced with the whole string of digits from IMSI.
2. The domain part of the private user identity is composed of the MCC and MNC values of IMSI and has a pre-defined domain name, *IMSI.3gppnetwork.org*. These three parts are merged together and separated by dots in the following order: Mobile Network Code (MNC, a digit or a combination of digits uniquely identifying the

public land mobile network), Mobile Country Code (MCC, code uniquely identifying the country of domicile of the mobile subscriber) and pre-defined domain name.
For example:

```
IMSI in use: 234150999999999; where:
MCC: 234;
MNC: 15;
MSIN: 0999999999; and
Private user identity is:
234150999999999@234.15.IMSI.3gppnetwork.org
```

Temporary public user identity

If there is no ISIM application to host the public user identity, a temporary public user
identity will be derived, based on the IMSI. The temporary public user identity will take
the form of a SIP URI, "*sip:user@domain*. The user and domain part are derived
similarly to the method used for private user identity [3GPP TS 23.003]. Following
our earlier example a corresponding temporary public user identity would be:

```
sip:234150999999999@234.15.IMSI.3gppnetwork.org
```

The IMS architecture imposes the following requirements for a temporary public user
identity [3GPP TS 23.228]:

- It is strongly recommended that the temporary public user identity is set to "barred"
  for IMS non-registration procedures so that it cannot be used for IMS communication. The following additional requirements apply if the temporary public user
  identity is "barred":
    - the temporary public user identity will not be displayed to the user and will not be
      used for public usage (e.g., displayed on a business card);
    - the temporary public user identity will only be used during the registration to
      obtain implicitly registered public user identities (the concept of implicitly regis-
      tered public user identities is explained in Section 3.3).
- Implicitly registered public user identities will be used for session handling, in other
  SIP messages and at subsequent registration processes.
- After the initial registration, only the UE will use the implicitly registered public user
  identity(s).
- The temporary public user identity will only be available to CSCF and HSS nodes.

3.5.1.4   Relationship between private and public user identities

Here a basic example shows how different identities are linked to each other. Joe is
working for a car sales company and is using a single terminal for both his work life and
his personal life. To handle work-related matters he has two public user identities:
sip:joe.smith@brandnewcar.com and *tel:+358501234567*. When he is off-duty he uses
two additional public user identities to manage his personal life: *sip:joe.smith@ims.
example.com* and *tel:+358503334444*. By having two sets of public user identities he

Figure 3.4 Relationship between user identities.

could have a totally different treatment for incoming sessions: for example, he is able to direct all incoming work-related sessions to a messaging system after 5 p.m. and during weekends and holidays.

Joe's user and service-related data are maintained in two different service profiles. One service profile contains information about his work life identities and is downloaded to the S-CSCF from the HSS when needed: that is, when Joe registers a work life public user identity or when the S-CSCF needs to execute unregistered services for a work life public user identity. Similarly, another service profile contains information about his personal life identities and is downloaded to the S-CSCF from the HSS when needed. The concept of service profile is explained in Section 3.12.1.

Figure 3.4 shows how Joe's private user identity, public user identities and service profiles are linked together.

3.5.2   Identification of services (public service identities)

With the introduction of standardized presence, messaging, conferencing and group service capabilities it became evident that there must be identities to identify services and groups that are hosted by ASs. Identities for these purposes are also created on the fly: that is, they may be created by the user on an as-needed basis in the AS and are not registered prior to usage. Ordinary public user identities were simply not good enough; so, Release 6 introduced a new type of identity, the public service identity. Public service identities take the form of a SIP URI or are in tel URL format: for example, in messaging services there is a public service identity for the messaging list service (e.g., *sip:messaginglist_joe@ims.example.com*) to which the users send messages and then the messages are distributed to other members on the messaging list by the messaging list server. The same applies to conferencing services (i.e., audio/video and messaging sessions), where a URI for the conferencing service is created.

3.5.3   Identification of network entities

In addition to users, network nodes that handle SIP routing need to be identifiable using a valid SIP URI. These SIP URIs would be used when identifying these nodes in the header fields of SIP messages. However, this does not require that these URIs be globally published in a Domain Name System (DNS) [3GPP TS 23.228]. An operator could name its S-CSCF as follows:

| Example of network entity naming | *sip:finland.scscf1@ims.example.com* |

3.6   Identity modules

3.6.1   *IP Multimedia Services Identity Module (ISIM)*

An IP Multimedia Services Identity Module (ISIM) is an application residing on the Universal Integrated Circuit Card (UICC), which is a physically secure device that can be inserted and removed from UE. There may be one or more applications in the UICC. The ISIM itself stores IMS-specific subscriber data mainly provisioned by an IMS operator. The stored data can be divided into six groups as shown in Figure 3.5. Most of the data are needed when a user performs an IMS registration [3GPP TS 31.103]:

- Security keys consist of integrity keys, ciphering keys and key set identifiers. Integrity keys are used to prove integrity protection of SIP signalling. Ciphering keys are used to provide confidential protection of SIP signalling. Confidential protection is not used in Release 5; however, in Release 6 it should be possible to use confidential protection. At the time of writing there is a need for key set identifiers.

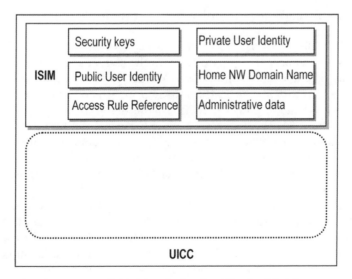

Figure 3.5   IP Multimedia Services Identity Module.

- The private user identity simply contains the private user identity of the user. It is used in a registration request to identify the user's subscription (see Section 3.5.1.1 for further information).
- The public user identity contains one or more public user identities of the user. It is used in a registration request to identify an identity to be registered and is used to request communication with other users (see Section 3.5.1.2 for further information).
- The home network domain name consists of the name of the entry point of the home network. It is used in a registration request to route the request to the user's home network.
- Administrative data include various data, which could be used, say, by IMS subscribers for IMS operations or by manufacturers to execute proprietary auto-tests.
- Access rule reference is used to store information about which personal identification number needs to be verified in order to get access to the application.

3.6.2   Universal Subscriber Identity Module (USIM)

The Universal Subscriber Identity Module (USIM) is required for accessing the PS domain (GPRS) and unambiguously identifies a particular subscriber. Similarly to the ISIM, the USIM application resides on the UICC as a storage area for subscription and subscriber-related information. Additionally, it may contain applications that use the features defined in the USIM Application Toolkit.

The USIM contains such data as the following: security parameters for accessing the PS domain, IMSI, list of allowed access point names, Multimedia Message Service (MMS) related information [3GPP TS 31.102, TS 22.101, TS 21.111]. Section 3.5.1.3 describes how UE with the USIM could derive the necessary information to access the IMS.

3.7   Sharing a single user identity between multiple devices

Traditionally, in the CS every single user has her own Mobile Station International ISDN (MSISDN) number that is used to reach the user. It is not possible for a single user to use multiple terminals with the same MSISDN number simultaneously. Having two mobile stations with identical MSISDN numbers would cause significant conflicts in the network. Nowadays, users may have more than one item of UE with totally different capabilities: big/small screen, camera/no camera, full keyboard and so forth. Different UE may serve different purposes (e.g., one for gaming, another for ordinary voice and video sessions). From the user's point of view, the user should be reachable via the same identity regardless of the number of UEs that she is using simultaneously. The IMS makes this feature possible.

Release 6 IMS allows users to register the same public user identity from a number of items of UE. In addition, a user is able to indicate her preferences regarding a single UE at the registration phase. Different registrations can be differentiated by means of the private user identity and the used IP address. Figure 3.6 shows an example in which a user has two items of UE: one for video sessions and another for chat and gaming applications. When someone is calling the user – say, Joe – it is his S-CSCF that makes

UE #1

Private User Identity: 543210abcd@ims.example.com
Public User Identity: sip:joe.doe@ims.example.com
Contact: sip:[5555::a:b:c:d]; media="application"

S-CSCF

sip: joe.doe@ims.example.com
Contact: sip:[5555::a:b:c:d]; media="application"
Contact: sip:[5555::b:b:b:b]; media="video"

UE #2

Private User Identity: 60000eeee@ims.example.com
Public User Identity: sip:joe.doe@ims.example.com
Contact: sip:[5555::b:b:b:b]; media="video"

Figure 3.6   Sharing a single user identity between multiple devices.

the decision as to which UE is going to be contacted in the first place. This decision can be done based on the preferences given at the registration phase: for example, if the incoming session contains a video component, then the S-CSCF could select UE #2, which is Joe's primary preference for video sessions. In addition to preference-based routing, the S-CSCF may perform forking. There are two types of forking:

- sequential forking;
- parallel forking.

Sequential forking means that different items of UE are contacted one by one: for example, the S-CSCF first sends the request to UE #2 and, if Joe fails to respond, within a certain time limit the S-CSCF then tries to reach Joe through UE #1.

Parallel forking means that different items of UE are contacted at the same time: for example, when two items of UE are ringing, Joe can decide which UE to use for the incoming session; however, in the end the session can only be connected to a single item of UE.

3.8   Discovering the IMS entry point

In order to communicate with the IMS, an item of UE has to know at least one IP address of the P-CSCF. The mechanism by which the UE retrieves these addresses is called "P-CSCF discovery". Two dynamic mechanisms for P-CSCF discovery have been standardized in 3GPP: the Dynamic Host Configuration Protocol's (DHCP) DNS procedure and the GPRS procedure. Additionally, it is possible to configure either the P-CSCF name or the IP address of the P-CSCF in the UE.

In the GPRS procedure (Figure 3.7), the UE includes the P-CSCF address request flag in the PDP context activation request (or secondary PDP context activation request) and receives the IP address(es) of the P-CSCF in the response. This information is transported in the protocol configuration options information element [3GPP TS 24.008]. The mechanism the Gateway GPRS Support Node (GGSN) used to get the IP

Figure 3.7 A GPRS-specific mechanism for discovering the P-CSCF.

Figure 3.8 A generic mechanism for discovering the P-CSCF.

address(es) of the P-CSCF(s) is not standardized. This mechanism does not work with pre-Release 5 GGSNs.

In the DHCP DNS procedure (Figure 3.8), the UE sends a DHCP query to the IP connectivity access network (e.g., GPRS), which relays the request to a DHCP server. According to [RFC3319] and [RFC3315], the UE could request either a list of the SIP server domain names of the P-CSCF(s) or a list of the SIP server IPv6 addresses of the P-CSCF(s). When domain names are returned, the UE needs to perform a DNS query (NAPTR/SRV) to find an IP address of the P-CSCF. The DHCP DNS mechanism is an access-independent way of discovering the P-CSCF.

3.9 S-CSCF assignment

Section 3.8 explained how the UE discovers the IMS entry point – i.e., the P-CSCF. The next entity on a session signalling path is the S-CSCF. There are three cases when the S-CSCF is assigned:

- when a user registers with the network;
- when an unregistered user receives a SIP request;
- when a previously assigned S-CSCF is not responding.

3.9.1 S-CSCF assignment during registration

When a user is registering with a network the UE sends a REGISTER request to the discovered P-CSCF, which finds the user's home network entity – i.e., the I-CSCF – as described in Section 3.2. Then the I-CSCF exchanges messages with the HSS (UAR and UAA) as described in Section 2.3.4. As a result, the I-CSCF receives S-CSCF capabilities, as long as there is no previously assigned S-CSCF. Based on the received capabilities the I-CSCF selects a suitable S-CSCF. Capability information is transferred between the HSS and the I-CSCF within the Server-Capabilities Attribute Value Pair (AVP). The Server-Capabilities AVP contains [3GPP TS 29.228 and TS 29.229]:

- Mandatory-Capability AVP – the type of this AVP is unsigned and contains the mandatory capabilities of the S-CSCF. Each mandatory capability available in an individual operator's network will be allocated a unique value.
- Optional-Capability AVP – the type of this AVP is unsigned and contains the optional capabilities of the S-CSCF. Each optional capability available in an individual operator's network will be allocated a unique value.
- Server-Name AVP – this AVP contains a SIP URI used to identify a SIP server.

Based on the mandatory and optional capability AVPs, an operator is able to distribute users between S-CSCFs, depending on the different capabilities (required capabilities for user services, operator preference on a per-user basis, etc.) that each S-CSCF may have. It is the operator's responsibility to define (possibly based on the functionality offered by each S-CSCF installed in the network) the exact meaning of the mandatory and optional capabilities. As a first choice, the I-CSCF will select the S-CSCF that has all the mandatory and optional capabilities for the user. If that is not possible, then the I-CSCF applies a "best-fit"algorithm. None of the selection algorithms is standardized (i.e., solutions are implementation-dependent). Figure 3.9 gives an example.

Figure 3.9 Example of an S-CSCF assignment.

Using the Server-Name AVP, an operator has the possibility to steer users to certain
S-CSCFs; for example, having a dedicated S-CSCF for the same company/group to
implement a VPN service or just making S-CSCF assignment very simple.

3.9.2   S-CSCF assignment for an unregistered user

Section 3.4 and Figure 3.3 explained at a high level how a session is routed from UE A
to UE B. It can be seen from the figure that the I-CSCF is a contact point within an
operator's network. In Section 2.3.4.1 location retrieval procedures were explained
(i.e., an incoming SIP request will trigger LIR/LIA commands to find out which
S-CSCF is serving User B). If the HSS does not have knowledge of a previously
assigned S-CSCF, then it returns S-CSCF capability information and the S-CSCF
assignment procedure will take place in the I-CSCF as described in Section 3.9.1.

3.9.3   S-CSCF assignment in error cases

3GPP standards allow S-CSCF re-assignment during registration when the assigned
S-CSCF is not responding: that is, when the I-CSCF realizes that it cannot reach the
assigned S-CSCF it sends the UAR command to the HSS and explicitly sets the type of
authorization information element to the value registration_and_capabilities. After
receiving S-CSCF capabilities, the I-CSCF performs S-CSCF assignment as described
in Section 3.9.1.

3.9.4   S-CSCF de-assignment

The S-CSCF is de-assigned when a user de-registers from the network or the network
decides to de-register the user (e.g., because registration has timed out or the subscription has expired). It is the responsibility of the S-CSCF to clear the stored S-CSCF
name from the HSS.

3.9.5   Maintaining S-CSCF assignment

When a user de-registers from the network or a registration timer expires in the S-CSCF
an operator may decide to keep the same S-CSCF assigned for the unregistered user. It
is the responsibility of the S-CSCF to inform the HSS that the user has been de-
registered; however, the S-CSCF could indicate that it is willing to maintain the user
profile. This optimizes the load of the Cx reference point because there is no need to
transfer the user profile once the user registers again or receives sessions while she has a
services-related unregistered state.

3.10   Mechanism for controlling bearer traffic

Separation of the control plane and the user plane was maybe one of the most important issues of IMS design. Full independence of the layers is not feasible because,
without interaction between the user plane and the control plane, operators are not able

Figure 3.10   SBLP entities.

to control Quality of Service (QoS), source/destination of IMS media traffic, and when the media starts and stops. Therefore, a mechanism to authorize and control the usage of the bearer traffic intended for IMS media traffic was created; it was based on the Session Description Protocol (SDP) parameters negotiated at the IMS session. This overall interaction between the GPRS and the IMS is called a Service-Based Local Policy (SBLP) control. Later on, charging correlation was specified as an additional capability. Figure 3.10 shows the functional entities involved in the SBLP. The figure depicts a Release 6 based architecture model where a standalone Policy Decision Function (PDF) and the Gq reference point are used. In the Release 5 architecture, PDF is integrated with the P-CSCF. The example in this section does not explain how the P-CSCF maps SIP/SDP information to Diameter information elements nor how the information is transferred to PDF using Diameter commands over the Gq reference point. On the contrary, the example focuses on the essentials: what information is needed from SIP/SDP session signalling and how it is used at the PDF. The utilization of the Gq reference point is covered in detail in Section 3.10.8:

- IP Bearer Service (BS) manager – manages the IP BS using a standard IP mechanism. It resides in the GGSN and optionally in the UE.
- Translation/Mapping function – provides the interworking between the mechanism and parameters used within the UMTS BS and those used within the IP BS. It resides in the GGSN and optionally in the UE.
- UMTS BS manager – handles resource reservation requests from the UE. It resides in the GGSN and in the UE.
- Policy enforcement point – is a logical entity that enforces policy decisions made by the PDF. It resides in the IP BS manager of the GGSN.
- Policy decision function – is a logical policy decision element that uses standard IP mechanisms to implement SBLP in the IP media layer. In Release 5 it resides in the

P-CSCF. In Release 6 it is a standalone entity. The PDF is effectively a policy decision point according to [RFC2753] that defines a framework for policy-based admission control.

There are seven SBLP functions. These seven functions are described in the following subsections:

- bearer authorization;
- approval of QoS commit;
- removal of QoS commit;
- indication of bearer release;
- indication of bearer loss/recovery;
- revoke authorization;
- exchange of charging identifiers.

3.10.1 Bearer authorization

Session establishment and modification in the IMS involves an end-to-end message exchange using SIP and SDP. During the message exchange, UEs negotiate a set of media characteristics (e.g., common codec(s)). If an operator applies the SBLP, then the P-CSCF will send the relevant SDP information to the PDF together with an indication of the originator. The PDF notes and authorizes the IP flows of the chosen media components by mapping from SDP parameters to authorized IP QoS parameters for transfer to the GGSN via the Go interface.

When the UE is activating or modifying a PDP context for media it has to perform its own mapping, from SDP parameters and application demands to some UMTS QoS parameters. PDP context activation or modification will also contain the received authorization token and flow identifiers as the binding information.

On receiving the PDP context activation or modification, the GGSN asks for authorization information from the PDF. The PDF compares the received binding information with the stored authorization information and returns an authorization decision. If the binding information is validated as correct, then the PDF communicates details of the media authorization in the decision to the GGSN. These details contain IP QoS parameters and packet classifiers related to the PDP context.

The GGSN maps the authorized IP QoS parameters to authorized UTMS QoS parameters and, finally, the GGSN compares the UMTS QoS parameters against the authorized UMTS QoS parameters of the PDP context. If the UMTS QoS parameters from the PDP context request lie within the limits authorized by the PDF, then the PDP context activation or modification will be accepted. Figure 3.10 shows this functionality and the PDF is shown as a part of the P-CSCF for simplicity. When a standalone PDF exists, then the P-CSCF needs to map SIP/SDP signalling information to appropriate Diameter information elements and send an appropriate Diameter request to the PDF via the Gq reference point. This functionality is further described in Section 3.10.8.

From Figure 3.11 we can find two different phases: authorize QoS resources (Steps 1–6) and resource reservation (Steps 7–14). Next, we take a deeper look at both steps

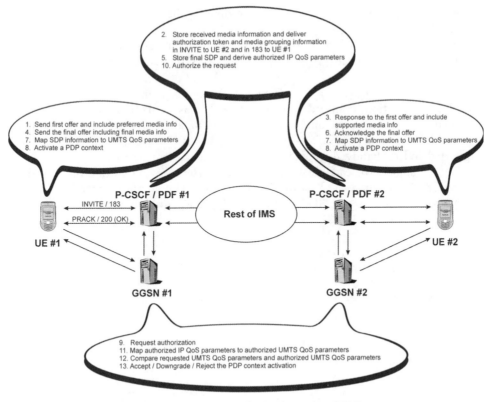

Figure 3.11 Bearer authorization using SBLP.

and, then, the final step of bearer authorization – i.e., approval of QoS commit – is described.

3.10.1.1   Authorize QoS resources

Steps 2 and 5 in Figure 3.11 correspond to authorization of the QoS resources procedure. During session setup the PDF collects IP QoS authorization data. These data comprise:

- Flow identifier – used to identify the IP flows that are described within a media component associated with a SIP session. A flow identifier consists of two parts: (1) the ordinal number of the position of the media component description in the SDP session description information and (2) the ordinal number of the IP flow(s) within the media component description assigned (in order of increasing port numbers).
- Data rate – this information is derived from SDP bandwidth parameters. The data rate will include all the overheads coming from the IP layer and the layers above (e.g., UDP, RTP or RTCP). If multiple codecs per media are agreed to be used in a session,

then the authorized data rate is set according to the codec requiring the highest bandwidth.
- QoS class – the QoS class information represents the highest class that can be used for the media component. It is derived from the SDP media description.

Let's say that Tobias (UE #1 in Figure 3.11) wants to talk to his sister Theresa (UE #2). In addition to an ordinary voice call, Tobias wants to activate bidirectional and unidirectional video streams. Therefore, his terminal builds a SIP INVITE containing an SDP that reflects Tobias's preferences and his UE capabilities. SDP contains supported codecs, bandwidth requirements (plus characteristics of each) and assigned local port numbers for each possible media flow. Here we concentrate only on those parameters that are necessary for the SBLP. Chapter 11 contains a general description for the whole session setup. SDP sent from UE #1 would look like this:

```
v=0
o=- 3262464865 3262464868 IN IP6 5555::1:2:3:
t=3262377600 3262809600
m=video 50230 RTP/AVP 31
c=IN IP6 5555::1:2:3:4
b=AS:35
b=RS:700
b=RR:700
m=video 50240 RTP/AVP 31
c=IN IP6 5555::1:2:3:4
b=AS:32
b=RS:640
b=RR:640
a=sendonly
m=audio 3456 RTP/AVP 97 96
c=IN IP6 5555::1:2:3:4
b=AS:25.4
b=RS:500
b=RR:500
```

When PDF #1 in Figure 3.11 receives information based on this, it is able to formulate the authorization data for the downlink direction (from GGSN #1 to UE #1). When Theresa's UE responds, PDF #1 is able to formulate the authorization data for the uplink direction (from UE #1 to GGSN #1). Note that Theresa is not willing to receive unidirectional video, therefore the corresponding port number is set to zero:

```
v=0
o=- 3262464865 3262464868 IN IP6 5555::1:2:3:4
t=3262377600 3262809600
m=video 60230 RTP/AVP 31
c=IN IP6 5555::5:6:7:8
b=AS:35
```

```
b=RS:700
b=RR:700
m=video 0 RTP/AVP 31
c=IN IP6 5555::5:6:7:8
b=AS:32
b=RS:640
b=RR:640
a=recvonly
m=audio 3550 RTP/AVP 0
c=IN IP6 5555::5:6:7:8
b=AS:25.4
b=RS:500
b=RR:500
```

From this information, PDF #1 and PDF #2 are able to construct the necessary flow identifiers. Table 3.3 shows the flow identifiers in PDF #1.

Data rates The PDF derives the data rate value for the media IP flow(s) from the "b = AS" SDP parameter. For possibly associated Real-time Transport Control Protocol (RTCP) IP flows, the PDF will use SDP "b = AS", "b = RR" and "b = RS" parameters, if present. When SDP "b = RR" or "b = RS" are missing the data rate for RTCP IP flows is derived from the available parameters as described in [3GPP TS 29.208]. Description of RTCP bandwidth:

- If b = RS and b = RR exist, then the RTCP bandwidth for uplink (UL) and downlink (DL) = (bRS + bRR)/1,000].
- If either b = RS or b = RR is missing, then the RTCP bandwidth for UL and DL = MAX[0.05 * bAS, bRS/1,000 or bRR/1,000].
- If both b = RS and b = RR are missing, then the RTCP bandwidth for UL and DL = 0.05 * bAS.

Table 3.5 shows maximum data rates per flow identifier as calculated in PDF #1.

Table 3.3 Flow identifier information in PDF #1.

Order of "m =" line	Type of IP flow	Destination IP address	Port number of the IP flows	Flow identifier
1	RTP (video) DL	5555::1:2:3:4	50230	<1, 1>
1	RTP (video) UL	5555::5:6:7:8	60230	<1, 1>
1	RTCP DL	5555::1:2:3:4	50231	<1, 2>
1	RTCP UL	5555::5:6:7:8	60231	<1, 2>
3	RTP (audio) DL	5555::1:2:3:4	3456	<3, 1>
3	RTP (audio) UL	5555::6:7:8:9	3550	<3, 1>
3	RTCP DL	5555::1:2:3:4	3457	<3, 2>
3	RTCP UL	5555::6:7:8:9	3551	<3, 2>

Table 3.4 The maximum data rates per media type.

Media type (m-line in the SDP)	Maximum authorized QoS class
Bidirectional audio or video	A
Undirectional audio or video	B
Application	A
Data	E
Control	C
Others	F

Table 3.5 The maximum data rates and QoS class per flow identifier in PDF #1.

	Flow identifier			
	< 1, 1 >	< 1, 2 >	< 3, 1 >	< 3, 2 >
Maximum data rate downlink (kbps)	35	0.7	25.4	0.5
Maximum data rate uplink (kbps)	35	0.7	25.4	0.5
Maximum QoS class	A	A	A	A

QoS class   The PDF maps media-type information to the highest QoS class that can be used for the media. The PDF will use an equal QoS class for both uplink and downlink directions when both directions are used [3GPP TS 29.207]. [3GPP TS 29.208] contains detailed derivation rules (a summary is presented in Table 3.4). Table 3.5 shows how the information in Table 3.4 is utilized in our example. The maximum authorized QoS class for an RTCP IP flow is the same as for the corresponding RTP media IP flow.

The authorized IP QoS was created and stored in the PDFs during Steps 2 and 5 in Figure 3.11 (bearer authorization). The authorized IP QoS comprises the QoS class and data rate. At the same time the PDFs created the flow identifiers that will be used to create packet classifiers in the GGSNs later on. Table 3.5 summarizes things so far.

Authorization token   In Figure 3.11, Step 2 states: "deliver authorization token to UE #1 and UE #2." But, what is an authorization token?:

- It is a unique identifier across all PDP contexts associated with an access point name.
- It is created in the PDF when the authorization data are created.
- It consists of the IMS session identifier and the PDF identifier.
- Its syntax conforms to [RFC3520].
- It is delivered to the UE by means of [RFC3313].
- The UE includes it in a PDP context activation/modification request.
- The GGSN uses a PDF identifier within the authorization token to find the PDF that holds the authorized IP QoS information.
- The PDF uses the authorization token to find the right authorized data when receiving requests from the GGSN.

Media grouping SIP and the IMS allow multimedia sessions to be set up which may comprise a number of different components, such as audio and video. Any participating party may add or drop a media component from an ongoing session. As stated in Section 2.1.6, all components should be individually identifiable for charging purposes, and it must be possible to charge for each of these components separately in a session.

Unfortunately, Release 5 GGSN is able to produce only one GGSN Call Detail Record (CDR) for a PDP context. Therefore, it is impossible to separate traffic for each media component within the same PDP context. As the current model for charging data generation and correlation does not allow multiplexing media flows in the same secondary PDP context, there must be a mechanism on the IMS level to force the UE to open separate PDP contexts for each media component. For this purpose a keep-it-separate indication was defined.

When the P-CSCF receives an initial INVITE request for a terminating session setup or a 183 (Session Progress) response to an INVITE request for an originating session setup, the P-CSCF may modify SDP according to [RFC3524] to indicate to the UE that a particular media stream(s) is grouped according to a local policy [3GPP TS 24.229].

[RFC3524] defines the Single Reservation Flow (SRF) group type (a = group : SRF). SRF groups are used in the following way:

- If a network wants to carry particular media in the same PDP context, then the P-CSCF sets the same SRF value for these media components.
- If a network wants to carry particular media in different PDP contexts, then the P-CSCF sets different SRF values for each media component.
- If a network does not set the SRF indication, then the UE is free to group media streams as it likes.

The following further restrictions and guidelines are given in standards [3GPP TS 23.228] and [3GPP TS 24.229]:

- The P-CSCF will apply and maintain the same policy throughout the complete SIP session.
- If a media stream is added and a grouping of media stream(s) was indicated in the initial INVITE or 183 (Session Progress) response, the P-CSCF will modify SDP according to [RFC3524] to indicate to the UE that the added media stream(s) will be grouped into either a new group or into one of the existing groups.
- The P-CSCF will not apply [RFC3524] to SDP for additional media stream(s), if grouping of media stream(s) was not indicated in the initial INVITE request or 183 (Session Progress) response.
- The P-CSCF will not indicate re-grouping of media stream(s) within SDP.
- All associated IP flows (e.g., RTP/RTCP) used by the UE to support a single media component are assumed to be carried within the same PDP context.
- In Release 5 it is assumed that media components from different IMS sessions are not carried within the same PDP context. This requirement does not apply in Release 6 as Release 6 adds a capability to charge on an IP flow basis.

Following our example, P-CSCF #1 forces a separate PDP context for all media types
in the 183 (Session Progress) response as follows (183 toward UE #1):

```
v=0
o=- 3262464865 3262464868 IN IP6 5555::1:2:3:4
t=3262377600 3262809600
a=group:SRF 1
a=group:SRF 2
a=group:SRF 3
m=video 60230 RTP/AVP 31
a=mid: 1
c=IN IP6 5555::5:6:7:8
b=AS:35
b=RS:700
b=RR:700
m=video 0 RTP/AVP 31
a=mid: 2
c=IN IP6 5555::5:6:7:8
b=AS:32
b=RS:640
b=RR:640
a=recvonly
m=audio 3550 RTP/AVP 0
a=mid: 3
c=IN IP6 5555::5:6:7:8
b=AS:25.4
b=RS:500
b=RR:500
```

Forking issues   When the P-CSCF receives a forked response it will pass the information listed in Section 2.3.19 to the PDF. As the PDF receives the forking indication it
also assigns the previously allocated authorization token to the forked response. Additionally, the PDF authorizes any additional media components and any increased
QoS requirements for previously authorized media components, as requested by the
forked response. Thus, the QoS authorized for a media component equals the highest
QoS requested for that media component by any of the forked responses [3GPP TS
29.207].

3.10.1.2   Resource reservation

UE functions
When the UE receives an authorization token within the end-to-end message exchange
it knows that the SBLP is applied in the network. Therefore, it has to generate the
requested QoS parameters and flow identifiers for a PDP context activation (modification) request. The requested QoS parameters include the information listed in Table 3.6.
From the SBLP point of view the first three rows in Table 3.6 present values that are

Table 3.6 Requested QoS parameters.

Traffic class	Maximum bit rate for downlink
Guaranteed bit rate for downlink	Maximum bit rate for uplink
Guaranteed bit rate for uplink	
SDU format information	Maximum SDU size
SDU error ratio	Residual BER
Delivery of erroneous SDUs	Traffic-handling priority
Transfer delay	Allocation/Retention priority
Source statistics descriptor	Delivery order

interesting. Interested readers can find detailed descriptions of other QoS parameters in [3GPP TS 23.107]. Here the traffic class, guaranteed bit rate and maximum bit rate are described:

- Traffic class – the four different traffic classes defined for UMTS are conversational, streaming, interactive and background. By including the traffic class, UMTS can make assumptions about the traffic source and optimize the transport for that traffic type.
- Guaranteed Bit Rate (GBR) – describes the bit rate the UMTS bearer service will guarantee to the user or application.
- Maximum Bit Rate (MBR) – describes the upper limit a user or application can accept or provide. This allows different rates to be used for operation (e.g., between GBR and MBR).

The traffic class values – GBR for downlink/uplink and MBR for downlink/uplink – should not exceed the derived values of maximum authorized bandwidth and maximum authorized traffic class per flow identifier. The maximum authorized bandwidth in the UE is derived from SDP in the same way as was done in the PDF. The maximum authorized traffic class is derived according to Table 3.7. The exact derivation rules for both parameters are described in [3GPP TS 29.208].

[3GPP TS 26.236] gives recommendations on how other requested QoS parameters for conversational codec applications could be set. Correspondingly, [3GPP TS 26.234] gives recommendations on how other requested QoS parameters for streaming codec applications could be set.

Table 3.7 The maximum authorized traffic class per media type in the UE.

Media type (m-line in SDP)	UMTS traffic class
Bidirectional audio or video	Conversational
Unidirectional audio or video	Streaming
Application	Conversational
Data	Interactive
Control	Interactive
Others	Background

Table 3.8 The values of the maximum authorized UMTS QoS parameters per flow identifier as calculated by UE #1 (Tobias) from the example.

	Flow identifier			
	< 1, 1 >	< 1, 2 >	< 3, 1 >	< 3, 2 >
Maximum data rate DL (kbps)	35	0.7	25.4	0.5
Maximum data rate UL (kbps)	35	0.7	25.4	0.5
Maximum QoS class	Conversational	Conversational	Conversational	Conversational

Flow identifiers are derived in the UE in the same manner as in the PDF. Table 3.8 shows the maximum authorized UMTS QoS parameters per flow identifier as calculated by the UE.

Next, the UE needs to decide how many PDP contexts are needed. The key factors are the nature of media streams (i.e., required traffic class) and the received grouping indication from the P-CSCF. In our example there are two different types of bidirectional media: video and audio. Both media would require high QoS (low delay and preserved time relation); therefore, a single conversational traffic class PDP context would be suitable. However, the P-CSCF has indicated that a separate PDP context is required for each IMS media component. Therefore, UE #1 should activate two different PDP contexts. Otherwise, PDP context activation would fail due to the SBLP decision enforced by the PDF. When the P-CSCF has not indicated a need for separate PDP contexts then the UE can optimize the number of PDP contexts. In our example it could open one new PDP context for video and audio. Moreover, the UE could even transport new components with an existing PDP if a suitable PDP context existed. Finally, Table 3.9 presents the maximum authorized UMTS QoS parameters per PDP context as calculated by UE #1.

The UE has now completed Step 7 in Figure 3.11. After deriving and choosing the suitable, requested QoS parameters, the UE activates the necessary PDP contexts. The authorization token and flow identifiers are inserted within the traffic flow template information element. A detailed description of how the authorization token and flow identifiers are carried in the traffic flow template information element is provided in

Table 3.9 The values of the maximum authorized UMTS QoS parameters per PDP context as calculated by UE #1 from the example.

	PDP context #	
	1	2
Maximum authorized bandwidth DL (kbps)	35.7	25.9
Maximum authorized bandwidth UL (kbps)	35.7	25.9
Maximum authorized traffic class	Conversational	Conversational

[3GPP TS 24.008]. The requested QoS parameters are inserted within the QoS information element. A detailed description of how the requested QoS parameters are carried in the QoS information element is provided in [3GPP TS 24.008].

GGSN functions

When a GGSN receives a secondary PDP context activation request to an Access Point Name (APN) for which the Go reference point is enabled, the GGSN:

- Identifies the correct PDF by extracting the PDF identify from the provided authorization token. If an authorization token is missing, then the GGSN may either reject the request or accept it within the limit imposed by a locally stored QoS policy. The Release 6 standard requires that the GGSN accept at least one secondary PDP context activation request without binding information and SBLP authorization if the requested PDP context is non-real time (UMTS traffic class "background" or "interactive") and such a PDP context is not yet assigned to the UE for the same APN [3GPP TS 29.207]. This kind of functionality is needed when an operator offers a single IMS APN for services that are not imposed on SBLP (e.g., Push to talk Over Cellular or session-based messaging) and services that are imposed on SBLP (e.g., voice and video sessions).
- Requests authorization information from the PDF for the IP flows carried by a PDP context. This request is a Common Open Policy Service (COPS) request and contains the provided authorization token and the provided flow identifiers. The exact content of the request is described in Chapter 21.
- Enforces the decision after receiving an authorization decision. The authorization decision is given as a COPS authorization_decision message. The exact content of the decision is described in Chapter 21. The main components of the decision are:
  ○ direction indication – uplink, downlink;
  ○ authorized IP QoS – data rate, maximum authorized QoS class;
  ○ packet classifiers (also called a gate description) – source IP address and port number(s), destination IP address and port number(s), protocol ID.
- Maps the authorized IP QoS to the authorized UMTS QoS.
- Compares the requested QoS parameters with the authorized UTMS QoS. If all the requested parameters lie within the limits, then PDP context activation will be accepted. In other words, if the following criteria are fulfilled [3GPP TS 29.208]:
  ○ the requested GBR DL/UL (if the requested traffic class is conversational or streaming) or MBR DL/UL (if the requested traffic class is interactive or background) is less than or equal to the maximum authorized data rate DL/UL; and
  ○ the requested traffic class is less than or equal to the maximum authorized traffic class.
  If the requested QoS exceeds the authorized UMTS QoS, then the requested UMTS QoS information is downgraded to the authorized UMTS QoS information.
- Constructs a gate description based on the received packet classifier. The gate description allows a gate function to be performed. The gate function enables or disables the forwarding of IP packets. If the gate is closed, then all the packets of the related IP flows are dropped. If the gate is open, then the packets of the related IP flows are allowed to be forwarded. The opening of the gate may be part of the

authorization decision event or may be a standalone decision as described in Section 3.10.8.2. The closing of the gate may be part of the revoke authorization decision.
- Stores the binding information.
- May cache the policy decision data of the PDF decisions.

During the secondary PDP context modification the GGSN may use previously cached information for a local policy decision in case the modification request does not exceed the previously authorized QoS. If the GGSN does not have cached information, then it performs the above-described functions.

PDF functions
When a PDF receives a COPS request the PDF validates that:

- The authorization token is valid.
- The corresponding SIP session exists. If media components from separate sessions are multiplexed in the same PDP context, there is more than one authorization token, and the PDF identifies one IMS session per each token. This capability was added in Release 6.
- The binding information contains valid flow identifier(s).
- The authorization token has not changed in an authorization modification request.
- The UE follows the grouping indication.
- If validation is successful, then the PDF will determine and communicate the authorized IP QoS, packet classifiers and the gate status to be applied to the GGSN. When valid binding information consists of more than one flow identifier, the information sent back to the GGSN will include the aggregated QoS for all the IP flows and suitable packet filter(s) for these IP flows.

In our example, UE #1 needs to activate two PDP contexts. When a secondary PDP context activation for the first PDP context (bidirectional video) arrives at GGSN #1, which extracts the authorization token and flow identifiers (1, 1 and 1, 2) from the traffic flow template and sends them to PDF #1, PDF #1 uses the authorization token to identify the corresponding IMS session. PDF #1 verifies that the session exists and returns the authorized IP QoS parameters and packet classifiers corresponding to the flow identifiers (1, 1 and 1, 2). GGSN #1 maps the authorized IP QoS to the authorized UMTS QoS; it compares the values and realizes that everything is OK. Finally, GGSN #1 accepts the request and installs the gate, based on the received packet classifiers. The same is applied to other, related PDP contexts.

Additionally, the PDF is able to send a new standalone decision to the GGSN when it receives modified SDP information from the P-CSCF. This may be needed, for example, in the case of forking.

3.10.2 Approval of the QoS commit function

During the resource reservation procedure a PDF sends packet classifiers to a GGSN. Based on the packet classifiers, the GGSN formulates a gate to policy-control incoming and outgoing traffic. It is the PDF's decision when to open the gate. When the gate is open, the GGSN allows traffic to pass through the GGSN. Opening the gate could be sent as a response to an initial authorization request from the GGSN or the decision can be sent as a standalone decision. If a standalone decision is used, then an operator can ensure that user-plane resources are not used before the IMS session is finally accepted (i.e., when a SIP 200 OK message is received). In this case end-users will lose all announcements which are to be delivered before completing the session, as the GGSN will drop all incoming user-plane IP packets.

3.10.3 *Removal of the QoS commit function*

This function closes a gate in the GGSN when a PDF does not allow traffic to traverse through the GGSN. This function is used, for example, when a media component of a session is put on hold due to media re-negotiation.

3.10.4 *Indication of bearer release function*

When the GGSN receives a delete PDP context request and the PDP context has been previously authorized via the Go reference point, the GGSN informs the PDF of the bearer release related to the SIP session by sending a COPS delete request-state message. The PDF removes the authorization for the corresponding media component(s). When the PDF receives a report that a bearer has been released, it could request the P-CSCF to release the session(s) and revoke all the related media authorization with the procedure described in Section 3.10.6.

3.10.5 *Indication of bearer loss/recovery*

When the MBR value equals 0 kbit/s in an update PDP context request, the GGSN needs to send a COPS report message to the PDF. Similarly, when the MBR is modified from 0 kbit/s, the GGSN sends a COPS report message to the PDF after receiving an update from the Serving GPRS Support Node (SGSN).

Using this mechanism the IMS is able to learn that the UE has lost/recovered its radio bearer(s) when a streaming or conversational traffic class is in use in the GPRS system. [3GPP TS 23.060] states that the SGSN needs to send an update PDP context request to the GGSN when the Radio Network Controller (RNC) informs the SGSN about Iu release or radio access bearer release. When the PDF receives a report that the MBR equals 0 kbit/s, it could request the P-CSCF to release the session(s) and revoke all the related media authorization with the procedure described in Section 3.10.6.

3.10.6 *Revoke function*

This function is used to force the release of previously authorized bearer resources in a GPRS network. With this mechanism the PDF is able, for example, to ensure that the UE releases a PDP context when a SIP session is ended or that the UE modifies the PDP context when a media component bound to a PDP context is removed from the session. If the UE fails to do so within a pre-defined time set by an operator, then the PDF revokes the resources.

3.10.7   Charging identifiers exchange function

The Go reference point is the link between the IMS and the GPRS networks. For charging correlation to be carried out as described in Section 3.11.6, the IMS layer needs to know the corresponding GPRS layer charging identifier and vice versa. These charging identifiers are exchanged during the bearer authorization phase. An IMS charging identifier is delivered to the GGSN within the authorization_decision message, while a GPRS charging identifier is delivered to the PDF as part of the authorization report.

3.10.8   Usage of Gq reference point

Using the Gq reference point PDF can be deployed as a standalone network entity. When a standalone PDF is used, the P-CSCF needs to pass necessary information derived from SIP/SDP session signalling to the PDF. Similarly, the PDF uses the Gq to send replies and independent notifications to the P-CSCF. For passing the information, the P-CSCF and PDF use Diameter commands as defined in RFC3588, NASREQ Diameter application and 3GPP TS 29.209. This section gives a brief summary of these Diameter commands and when they are used between the P-CSCF and PDF to support different SBLP functions.

Four request and answer pairs are used in the Gq reference point:

- AA-Request/AA-Answer (AAR&AAA);
- Re-Auth-Request/Re-Auth-Answer (RAR&RAA);
- Session-Termination-Request/Session-Termination-Answer (STR&STA);
- Abort-Session-Request/Abort-Session-Answer (ASR&ASA).

3.10.8.1   Bearer authorization

When SBLP is used in the network the P-CSCF needs to send relevant information obtained from SIP/SDP session signalling to the PDF. This information is used in the PDF to make an authorization decision about the request received via the Go reference point. Moreover, the P-CSCF needs to get an authorization token and GPRS charging identifier for charging correlation purposes. For this purpose AAR&AAA and RAR&RAA Diameter command pairs are utilized.

After receiving a SIP request or response containing SDP information, the P-CSCF sends an AAR to the PDF. The AAR contains, among other basic elements, information about the media stream, resource reservation policy, indication forwarding policy, IMS charging identifier, information about the used application and information about SIP forking. Following our example, P-CSCF #1 in Figure 3.11 will first issue an AAR command when it receives a SIP INVITE request from Tobias's UE. This command will carry the necessary information to construct downlink information (flow identifiers, calculate maximum bandwidth and derive maximum authorized QoS class) at the PDF. P-CSCF #1 secondly issues an AAR command when it receives a 183 Session Progress response from Teresa's UE. The second AAR command will carry the necessary information to construct uplink information. The PDF is now ready to authorize the whole bearer activation request.

Within an initial AAR command the P-CSCF can also indicate whether the P-CSCF wants to be contacted in each bearer authorization or whether the PDF can use available information to make the decision itself (resource reservation policy). Moreover, the P-CSCF may indicate that it is interested in receiving indications of loss of bearer, recovery of bearer or release of bearer (indication forwarding policy). In these cases, the PDF sends an RAR command to the P-CSCF after receiving an appropriate message from GGSN via the Go reference point. The RAR command is acknowledged with an RAA command. The AAR command is acknowledged with an AAA command. An authorization token and/or GPRS charging identifier and/or GGSN's IP address to the P-CSCF are delivered within the AAA command.

3.10.8.2 Approval and removal of the QoS commit

The approval and removal of QoS commit – i.e., opening/closing/removing appropriate gate(s) in GGSN – is given as part of an AAR command. For this purpose a 3GPP-specific Diameter AVP, called "Flow Status AVP", was developed [3GPP TS 29.209]. This AVP may have five different values: enable uplink, enable downlink, enabled (both directions), disabled and removed. Different "enabled" values are used to open gates (i.e., letting user-plane traffic to pass). The P-CSCF may either instruct the PDF to open appropriate gates in an initial AAR command or in an AAR command that is triggered at the reception of SIP 200 OK in the P-CSCF. The P-CSCF uses a "disabled" value, for example, when a SIP session or one of its media components is put on hold. A "removed" value is used to remove authorization of the IP flow(s). This could occur, for example, when a certain component is removed from a SIP session.

3.10.8.3 Indication of bearer release function

After receiving a bearer release notification from GGSN via the Go reference point, the PDF needs to notify the P-CSCF, if the P-CSCF has requested this type of notification. The PDF uses an ASR command if all IP flows within a SIP session are affected. Within the ASR command 3GPP has defined a 3GPP-specific AVP, called "Abort Cause AVP", to indicate the reason for bearer release. An RAR command is used instead if not all IP flows within a SIP session are affected. Within the RAR command 3GPP has defined a 3GPP-specific AVP, called "Specific-Action AVP", to indicate the reason for bearer release. This could occur, for example, when only the video component of a multimedia session is dropped.

3.10.8.4 Indication of bearer loss/recovery

When GGSN reports bearer loss or bearer recovery via the Go reference point, the PDF needs to notify the P-CSCF, if the P-CSCF has requested this type of notification. In order to do so, the PDF issues an RAR command including a Specific-Action AVP, to indicate either the loss of bearer or recovery of bearer.

Table 3.10 Gq commands.

Command-Name	Purpose	Abbreviation	Source	Destination
AA-Request	P-CSCF uses AAR to push SIP session information and IMS charging correlation identifier towards PDF.	AAR	P-CSCF	PDF
AA-Answer	AAA delivers acknowledgement to AAR, authorization token and GPRS charging correlation identifier to P-CSCF.	AAA	PDF	P-CSCF
Re-Auth-Request	RAR delivers reports (bearer loss/recovery, some component(s) of session released) and contains a request to send SIP session information.	RAR	PDF	P-CSCF
Re-Auth-Answer	RAA is used to acknowledge the RAR command. It may contain, for example, SIP session information.	RAA	P-CSCF	PDF
Session-Termination-Request	STR is used to ensure that bearer resources are released together with a SIP session release.	STR	P-CSCF	PDF
Session-Termination-Answer	STA is used to acknowledge the STR command.	STA	PDF	P-CSCF
Abort-Session-Request	ASR is used to inform P-CSCF that all bearers related to a particular SIP session have been released.	ASR	PDF	P-CSCF
Abort-Session-Answer	ASA is used to acknowledge the ASR command	ASR	P-CSCF	PDF

3.10.8.5 Revoke function

In order to trigger the release of PDP context(s) associated with a SIP session, the P-CSCF sends an STR command. This could occur, for instance, when a SIP session is released. When the PDF receives this command it will send another message via the Go reference point to trigger GGSN-initiated PDP context deactivation to prevent bearer misuse after SIP session termination.

3.11 Charging

Flat-free and volume-based charging schemas are traditional charging models in current IP-based communication networks. The IMS enables new charging models which, in turn, enables different business models for IMS operators. The capability to charge based on session or event or service is one of the key advantages that the IMS enables for operators. There are foreseen benefits for end-users as well. For instance, an operator is able to offer peer-to-peer games as a pre-paid service (i.e., a user needs to have money in her account before consuming services) and other multimedia sessions as a post-paid service (i.e., a user pays for services periodically such as once per month), or instant messages could be available as a flat-free service and session based-messaging charged differently (e.g., based on duration of the session or based on transferred bytes).

In order to offer a post-paid service, the IMS needs to support a mechanism for offline charging. Offline charging is a charging process where charging information is mainly collected after the session and the charging system does not affect in real time the service being used. In this model a user typically receives a bill on a monthly basis, which shows the chargeable items during a particular period. A pre-paid service requires online charging support. This means that IMS network entities need to consult the Online Charging System (OCS) before allowing users to use services. The OCS is responsible for interacting in real time with the user's account and for controlling or monitoring the charges related to service usage.

IMS network entities are configured to detect when a chargeable trigger condition is met. After detection, the entity collects the necessary information from a SIP request and either requests permission from a charging system (online charging) to continue processing the SIP request or sends relevant information to a charging system for creating a CDR for post-processing (offline charging), and allows the SIP request to continue. A chargeable trigger could be session initiation, session modification, session termination request (session-based charging) or it could be on any SIP transaction – e.g., MESSAGE, PUBLISH, SUBSCRIBE requests (event-based charging). Moreover, a trigger could be the presence of some SIP header or SDP information. Based on the received information the charging system either takes credit from the user's account (online charging) or transfers CDR(s) to the billing system. This sounds rather easy, but standardizing a charging solution has been very painful. Solutions defined in 3GPP Release 5 have changed in Release 6, and Release 6 work is still ongoing at the time of writing. This section is based on Release 6 charging solutions as they were in June 2005.

3.11.1 Charging architecture

Due to the different nature of charging models different architecture solutions for offline and online are defined. Figure 3.12 shows the high-level IMS charging architecture. The left side of the figure depicts offline charging and the right side shows online charging. Impacts of flow-based charging are not shown in the figure.

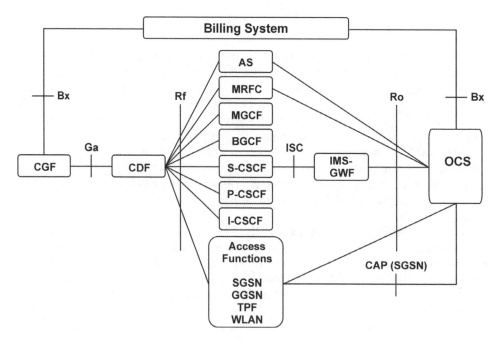

Figure 3.12 IMS charging architecture.

From the figure you can see that all IMS entities handling SIP signalling are able to communicate with the offline charging entity – i.e., Charging Data Function (CDF) – using a single Diameter-based Rf reference point [3GPP TS 32.299]. The CDF receives a Diameter request also from access network entities and based on the information provided from various entities it creates CDRs which are delivered to the Charging Gateway Function (CGF) via the Ga reference point [3GPP TS 32.295]. Finally, the CGF processes the received CDRs and transfers the final CDR(s) to the billing system using the Bx reference point [3GPP TS 32.240].

In contrast, only three IMS entities (AS, MRFC and S-CSCF) are involved in online charging. Moreover, the S-CSCF cannot communicate directly with the OCS due to bad design in the Release 5 timeframe. An IMS-Gateway Function (IMS-GWF) is used to perform the necessary protocol conversion. The OCS supports two reference points from other network entities. SGSN uses the CAMEL Application Part (CAP) and the rest of the entities use the Diameter-based Ro reference point. Like CGF in offline charging the OCS is also able to create CDRs in addition to credit control handling (approving resources in real time).

3.11.2 Offline charging

IMS signalling traverses through various IMS entities and, as stated earlier, all entities are able to generate offline charging information. In fact, each offline charging-capable entity contains an integrated function called a Charging Trigger Function (CTF). The CTF is aware of charging triggers (such as the beginning of IMS sessions, IMS session modification, IMS session termination, sending of message, subscribing to an event, publishing presence information) and is able to decide when it needs to contact the CDF, the central point in the offline charging system. When a trigger condition is met the CTF collects charging information from the signalling message and sends the offline charging information to the CDF using Diameter *Accounting Requests* (ACRs) via the Rf interface. The request contains much information about the event that launched the trigger (e.g., type of request INVITE/MESSAGE/SUBSCRIBE, calling party address, called party address, time stamps). The CDF uses a Diameter *Accounting Answer* (ACA) to acknowledge the received request. In case of an IMS session, at least two ACR/ACA pairs are sent (at the start of the session and at the end of the session). Further ACR(s) may be used as well if the session properties are changed (e.g., media components added or removed, the codecs of the media component and bandwidth have changed, session is placed on hold). In case of a single end-user to network transaction (e.g., sending an instant message) a single ACR/ACA is sufficient. The usage of Diameter requests is further described in Section 3.11.5.1 (Rf reference point).

So far we have described how charging information is delivered from an IMS entity to the CDF. From Figure 3.12 we can see that there are further steps ahead before a billing system can send an actual bill to a user. The next step is to transfer CDRs from the CDF on towards the CGF. The CGF is needed as there can be multiple CDFs involved in a single session/non-session event because different IMS entities may send charging information to different CDFs (e.g., due to roaming or configuration reasons). The CGF validates, consolidates, pre-processes the incoming CDR (e.g., filters unnecessary fields and adds operator-specific information), and may correlate different CDRs before passing them to the billing system. Table 3.11 summarizes the key procedures supported by different offline charging functions.

Table 3.11 Summary of offline charging functions

Offline charging function	Key procedures
Charging Triggering Function (CTF)	Monitors SIP signalling Detects trigger condition Extracts information from SIP signalling and assembles charging information Sends charging information to CDF
Charging Data Function (CDF)	Constructs CDRs Delivers CDRs to CGF
Charging Gateway Function (CGF)	Correlates, consolidates, filters unnecessary fields and adds operator-specific information to the received account information CDR error handling and storage Delivers CDRs to billing system Pre-processes CDRs.
Billing system	Creates the actual bill

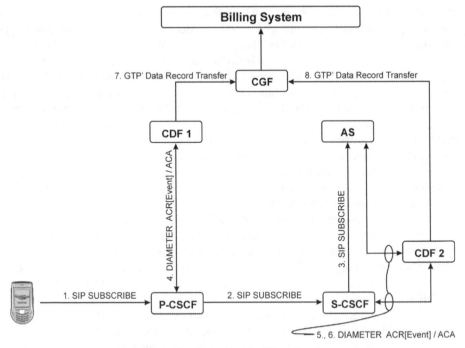

Figure 3.13   Example of offline charging.

   In Figure 3.13 an offline charging example is given. In this example a user sends a subscription request to an AS. It is assumed that the S-CSCF and the AS are using the same CDF (CDF #2 in the figure) and the P-CSCF is using a different CDF (CDF #1 in the figure). The subscription request could, for example, be a request to discover who are the members of a particular Push to talk Over Cellular group. In Steps 1–3 a SUBSCRIBE request traverses from the UE to the AS. In Steps 4–6 CTFs inside the IMS entities detect a chargeable event, create an ACR and send it to the CDF. In Steps 7–8 CDFs send appropriate CDRs to the CGF which, in turn, delivers the CDR(s) to the billing system. CDRs from the CDF to the CGF are transferred using a Data Record Transfer request in GPRS Tunnelling Protocol which includes functions for charging (GTP′) [3GPP TS 32.295]. Please note that it is possible for CDF #2 to create a single CDR from the charging information received. Moreover, it is possible for a CGF to consolidate the received CDRs and send a single CDR to the billing system.
   Although in this example it was assumed that multiple CDFs are in use, it is possible that a single CDF address is distributed to all IMS entities involved in a SIP session or transaction. This enables sending charging information to a single CDF. How this can be done is explained in Section 11.8.

3.11.3   Online charging

The purpose of online charging is to perform credit control before usage of IMS services/resources. Two different models exist: direct debiting and unit reservation. In direct debiting, an IMS network entity contacts the OCS and asks permission to grant the usage of services/resources. The OCS uses an internal rating function to find the appropriate tariff for an event based on the received information, if the cost was not given in the request. After resolving the tariff and the price, the OCS checks whether the user has enough credits in her account. If so, the OCS deducts a suitable amount of money from the user's account and grants the request from the IMS entity. In the unit reservation model, the OCS receives a credit control request from an IMS entity and uses an internal rating function to determine the price of the desired service according to the service-specific information provided by the IMS entity, if the cost was not given in the request. Then the OCS reserves a suitable amount of money from the user's account and returns the corresponding number of resources to the requesting IMS entity. Among the number of resources could be, for example, time or allowed data volume. When resources granted to the user have been consumed or the service has been successfully delivered or terminated, the IMS entity informs the OCS of the number of resources consumed. Finally, the OCS deducts the used amount from the user's account [3GPP TS 32.240, TS 32.296]. It is also possible for the OCS to receive subsequent requests from the IMS entity during service execution if all granted resources are consumed. In this case the OCS needs to perform a new credit authorization.

The direct debiting model is appropriate when the IMS entity knows that it could deliver the requested service to the user itself. For example, a game AS could send a credit control request and inform the OCS of the service (say, a game of rally) and the number of items (say, 2) to be delivered. Then the OCS uses the rating function to resolve the tariff (€0.3) and to calculate the price based on the number of delivered units (€0.6). Finally, €0.6 is deducted from the user's account and the OCS informs the game AS that 2 units have been granted within the credit control answer. 3GPP's definition for this online charging model is Immediate Event Charging [3GPP TS 32.240].

The unit reservation is suitable when the IMS entity is unable to determine before-hand whether the service could be delivered or when the required number of resources are not known prior to the use of a specific service (e.g., duration of multimedia session). The unit reservation model is usually applied to sessions (3GPP's term for this is Session Charging with Unit Reservation) but it is also possible to apply non-session related requests (3GPP's term for this is Event charging with unit reservation).

In this book we do not probe the OCS further but readers who are interested to learn more about the internal functions of the OCS are recommended to read 3GPP's specification: Online Charging System (OCS) applications and interfaces [3GPP TS 32.296].

3.11.4 Flow-based charging

Current standardized charging solutions enable user-plane charging at the APN and PDP level. 3GPP Release 6 and Release 7 will introduce the ability to charge in a more granular manner. The flow-based charging model introduces the ability to charge for service data flows identified by service flow filters according to certain charging rules. Charging rules contain information that allows the filtering of traffic to identify the packets belonging to a particular service data flow (e.g., IMS, FTP, browsing), and

Figure 3.14 Flow-based charging architecture.

allows definition of how the service data flow is to be charged. Charging rules are normally requested by a Traffic Plane Function (TPF) at the bearer establishment, upon a specified trigger event and upon bearer termination. This request is made using the Gx reference point towards a Charging Rule Function (CRF). The CRF is further connected to an application function via the Rx reference point (in the IMS case it is usually the P-CSCF or AS). The Rx reference point enables transport of information (e.g., dynamic media stream information) from the application function to the CRF. An example of such information would be filter information to identify an IMS session and its connection parameters (e.g. end points, media description). The architecture is shown in Figure 3.14.

There is ongoing Release 7 work to harmonize and merge policy control (SBLP) and flow-based charging architecture and procedures. As the architecture is subject to further changes it is not further described in this edition. Interested readers are advised to see 3GPP TS 23.808 (Evolution of policy control and charging) and 3GPP TS 23.125 (Overall high-level functionality and architecture impacts of flow-based charging).

3.11.5 *Charging reference points*

For charging purposes three IMS-related reference points, Ro (online charging), Rf (offline charging), and Rx exist. All reference points are based on the Diameter protocol developped by Internet Engineering Task Force (IETF).

3.11.5.1 Rf reference point (offline charging)

At the beginning of this chapter we explained that the CTF inside the IMS entity is responsible for detecting when it is necessary to report a chargeable event to the CDF (charging system in general). This critical task is achieved by sending a Diameter ACR via the Rf reference point to the CDF. The CDF replies back using another Diameter command, an Accounting Answer (ACA).

The basic Diameter functionality is described in [RFC3588] and forms the base of the Rf reference point (see Chapter 19 for a basic introduction to the Diameter protocol). In addition to the Diameter base protocol, 3GPP has defined its own extensions as a 3GPP Diameter accounting AVP to meet the charging requirements of 3GPP. A 3GPP-specific AVP contains information which is considered valuable especially in the 3GPP environment but not necessarily in broad Internet. For instance, it includes media description of the session (audio, video, message, chat), authorized Quality of Service, involved ASs. The ACR used in 3GPP contains suitable Diameter protocol AVPs and 3GPP Diameter accounting AVPs. The use of AVPs is specified per IMS entity and ACR type: for example, ACRs generated by the S-CSCF could contain information about the contacted ASs and ACRs generated by the P-CSCF could contain authorized QoS information [3GPP TS 32.299].

The offline charging system needs to support both session- and event-based charging. IMS entities know the request type that need to be indicated to the CDF. This is achieved using a suitable value in the Accounting-Record-Type AVP ("event" for events and "start", "interim", "stop" for sessions) in the ACR. IMS session-related ACRs are called start, interim and stop and are sent at the start, during and at the end of a session, as the name implies. Non-session related ACRs are called event ACRs. Event ACRs cause the CDF to generate corresponding CDRs, while session ACRs cause the CDF to open, update and close corresponding CDRs. Let's take two examples on usage of the Rf reference point, IMS session and instant message (non-session related), to explain the usage of Diameter ACR.

In an IMS session three different phases can be detected (session initiation, session modification and session release). At beginning of a session (200 OK, acknowledging an INVITE is received), the CTF inside an IMS entity monitors the signalling traffic and detects a triggering point defined for reception of 200 OK, acknowledging an INVITE. When the triggering point is met the CDF collects information from the signalling message (e.g., calling party address, called party address, time stamps, "audio" SDP media compoment), assembles the charging information that matches the detected chargeable event and forwards the charging information towards the CDF via the Rf reference point using an ACR[Start] request. Usage of an ACR[Start] prompts the CDF to open a CDR for this session. When the same session is modified (RE-INVITE or UPDATE is received) – e.g., a video component is added – the CTF could again trigger this event and collect the necessary information again (e.g., calling party address, called party address, time stamps, "audio + video" SDP media component). This changed charging information is sent again to the CDF, but this time an ACR[Interim] request is used. Finally, when the session ends (BYE is received) the CTF constructs the ACR[Stop] request to indicate session termination. Based on these three charging events, the CDF can create a single CDR including total session time, audio session time and video session time.

After receiving a non-session related request (here MESSAGE) a trigger point can be met again. The CTF collects the necessary information from the request (e.g., calling party address, called party address, time stamps, content length) and this time it constructs an ACR[Event] request to indicate event-based charging. As a result of this ACR[Event] the CDF knows to apply event-based charging, generates a CDR immediately and passes it on to the CGF.

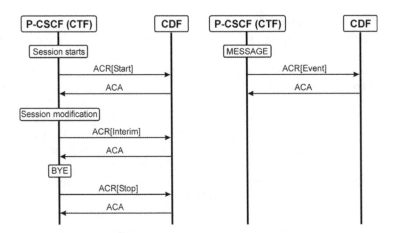

Figure 3.15   Session- and event-based offline charging example.

It is the operator's choice which SIP method, ISDN User Part (ISUP) or Bearer Independent Call Control (BICC) message triggers sending of the ACR. However, two mandatory items have been defined:

- Whenever SIP 200 OK, acknowledging an initial SIP INVITE, is received or the MGCF receives an ISUP/BICC answer, ACR[Start] will be sent to the CDF.
- Whenever SIP BYE is received or the Media Gateway Control Function (MGCF) receives an ISUP/BICC release, ACR[Stop] will be sent to the CDF.

3.11.5.2   Ro reference point (online charging)

In order to perform online charging the OCS needs to get the necessary information from the requesting IMS entity. For this purpose the Ro reference point is defined. It transfers credit control requests and answers between the OCS and three different IMS entities which are able to perform online charging (AS, MRFC and S-CSCF via the IMS-GWF). *Credit Control Request* and *Credit Control Answer* commands from [Draft-ietf-aaa-diameter-cc] are used for this purpose. In addition, 3GPP has defined 3GPP Credit Control AVPs to enhance IETF's solution [3GPP TS 32.299] to meet the charging requirements of 3GPP.

To enable direct debiting (see Section 3.11.3 for details) the IMS entity sends a Credit Control Request to the OCS and uses the value "EVENT_REQUEST" in CC-Request-Type AVP and the value "DIRECT_DEBITING" in Requested-Action AVP. For example, a messaging AS may receive a request from a user to send an instant message to somebody (1). The messaging server knows that the user is a pre-paid user and, therefore, it needs to seek permission from the OCS. It constructs a *Credit Control Request*, sets the *CC-Request-Type AVP*, *Requested_Action AVP* and any other required AVP correctly, and then sends the request to the OCS (2). The OCS gets the request and, if it does not contain information about the price of service, the OCS uses a rating function before consulting the user's account. This example is depicted in Figure

3.15. When the user has enough credits in the account the OCS grants the request with a *Credit Control Answer* (3). Finally, the messaging server allows the service and sends the instant message towards the destination (4). This example is depicted in the left-part of Figure 3.16.

To enable session charging with unit reservation (see Section 3.11.3 for definition) the IMS entity sends a *Credit Control Request* to the OCS and then uses values "INITIAL_REQUEST", "UPDATE_REQUEST" and "TERMINATION_RE-QUEST" in the *CC-Request-Type AVP* as follows:

- "INITIAL_REQUEST" value is used when the IMS entity receives the first service delivery request.
- "UPDATE_REQUEST" value is used when the IMS entity request reports the number of units used and indicates a request for additional units.
- "TERMINATION_REQUEST" value is used when the IMS entity reports that the content/service delivery is complete or the final allocated units have been consumed.

To enable event charging with unit reservation (see Section 3.11.3 for definition) the IMS entity sends a *Credit Control Request* to the OCS and then uses values "INITIAL_REQUEST" and "TERMINATION_REQUEST" in the *CC-Request-Type AVP* as follows:

- "INITIAL_REQUEST" value is used when the IMS entity receives the first service delivery request.
- "TERMINATION_REQUEST" value is used when the IMS entity reports that content/service delivery is complete.

For example, an operator who offers a pre-paid service for its customers needs to route all signalling traffic via an AS or IMS-GWF in order to apply online charging. In this example, the AS-based approach is assumed. In Step (1) in Figure 3.16 the AS receives a SIP session request (INVITE). The SIP INVITE triggers a *Credit Control Request* including the value "INITIAL_REQUEST" in the *CC-Request-Type AVP* as this is the first request in this session. The OCS receives the request (2) and uses the information provided to decide whether or not to grant this request. The *Credit Control Answer* (3) contains the number of service units granted and based on this the AS is able to allow the SIP session to continue (4). When the number of units granted are used up or there is a need for additional units (e.g., due to addition of a media component) the AS sends a new *Credit Control Request* but this time the value in the *CC-Request-Type AVP* is different (6). Again the OCS makes a credit control decision and communicates it, based on which the SIP session continues (7, 8). When the session is terminated or all units are used up the AS sends a third *Credit Control Request* indicating termination of the session and uses an appropriate value in the *CC-Request-Type AVP*. This example is depicted in the right-hand side of Figure 3.16.

3.11.5.3   Rx reference point

When flow-based charging is used in the network the IMS entities (in practice, the P-CSCF and AS) assist the CRF by transferring media information of the IMS

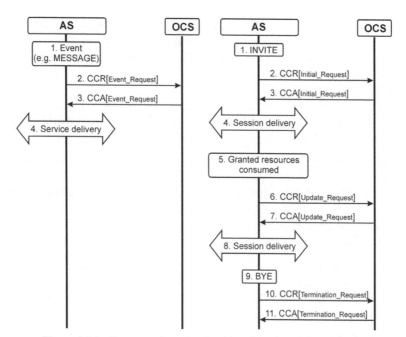

Figure 3.16 Event- and session-based online charging example.

session to the CRF. The CRF uses the provided information to generate dynamic charging rules which are communicated to the access network (TPF). Four different procedures exist on the Rx reference point: information transfer from IMS entities to the CRF, CRF-initiated information request, announcing IMS session release and announcing bearer release.

Table 3.12 shows which Diameter requests are used in each procedure. The reader is recommended to compare this table with the Gq command table (the commands used are exactly the same).

3.11.6 Charging information correlation

Due to the layering design (see Section 2.1.11), IMS entities are not aware of the user-plane traffic volumes related to IMS sessions, and IP connectivity network entities (e.g., SGSN and GGSN) are not aware of the status of control-plane signalling (i.e., the status of IMS sessions). From the operator's perspective, it is desirable to have a possibility to correlate charging information created at the user plane and the control plane. Exchanging charging identifiers – the IMS Charging Identifier (ICID) and the GPRS Charging Identifier (GCID) – through the Go reference point enables charging correlation between the IMS and the GPRS networks.

During the session establishment phase the UE activates the necessary secondary PDP context(s). During the PDP context authorization process the GGSN and the PDF exchange charging identifiers as follows:

1. The PDF passes the ICID to the GGSN in the authorization decision.
2. The GGSN passes the GCID to the PDF in a report about the authorization decision.

The PDF also passes the GCID to the P-CSCF, which forwards the GCID to the IMS entities in its own network where it is included on IMS CDRs. The GGSN includes the ICID on its G-CDR (i.e., a GGSN Charging Data Record), but does not pass the ICID to the SGSN. When a single IMS session requires several secondary PDP contexts, one or more GCIDs are mapped to one ICID. In addition, the GGSN is responsible for updating GCID information at the IMS level when secondary PDP context or media flows are removed or added during the session. As a last link, the SGSN creates an S-CDR (i.e., an SGSN Charging Data Record) that includes GCID and GGSN

Table 3.12   Rx commands.

Command-Name	Purpose	Abbreviation	Source	Destination
AA-Request	When a new SIP session is being established and media information for this session is available at the P-CSCF or the session is modified, the P-CSCF shall send the corresponding Service Information (e.g., media information) to the CRF by sending the AA-Request message.	AAR	P-CSCF	CRF
AA-Answer	AA-Answer delivers an acknowledgement to the AAR.	AAA	CRF	P-CSCF
Re-Auth-Request	Re-Auth-Request delivers reports (some component(s) of session released) and may contain a request to send SIP session information.	RAR	CRF	P-CSCF
Re-Auth-Answer	This command is used to acknowledge the STR command.	RAA	P-CSCF	CRF
Session-Termination-Request	This command is used to announce that the SIP session is released.	STR	P-CSCF	PDF
Session-Termination-Answer	This command is used to acknowledge the STR command.	STA	PDF	P-CSCF
Abort-Session-Request	This command is used to inform the P-CSCF that all bearers related to a particular SIP session have been released.	ASR	PDF	P-CSCF
Abort-Session-Answer	This command is used to acknowledge the ASR command.	ASR	P-CSCF	PDF

addresses. This is a unique identifier for each PDP context. Figure 3.17 shows an example of an IMS session that contains two media components which are transported in separate PDP contexts.

As seen from the example above, the 3GPP IMS architecture defines the ICID and GCID for charging data correlation and a mechanism for exchanging the identifiers between the IMS and the PS domain.

3.11.7 Charging information distribution

Section 3.11.6 explained how charging information is correlated. The present section gives an overview of how charging information is distributed between different IMS entities. An example of SIP details is given in Section 11.8.

The first IMS entity within the SIP signalling path generates an ICID. This ICID is passed along the SIP signalling path to all entities involved, except the UE: that is, the P-CSCF in the terminating network will remove the ICID. The ICID is used for correlating charging data between IMS components. The ICID applies for the duration of the event with which it is associated: for example, an ICID assigned for session establishment is valid until session termination, etc. We can see from Figure 3.18 that IMS and GPRS charging identifiers are exchanged when the bearer is author-ized. In addition, Figure 3.18 indicates when accounting requests are sent to the CDF. The address of the CDF is distributed during registration or, alternatively, it is config-ured in IMS entities.

3.12   User profile

When a user obtains an IMS subscription from an operator, the operator needs to assign a user profile. The user profile contains at least one private user identity and single service profile. Figure 3.19 depicts the general structure of a user profile [3GPP TS 29.228]. The private user identity was described in Section 3.5.1.1, but it should be understood that a user profile may contain more than one private user identity, if a user is using a shared public user identity as described in Section 3.7. Figure 3.4 shows that a single IMS subscription may contain multiple service profiles; this allows different treatment for different public user identities as explained in Section 3.5.1.4.

3.12.1   Service profile

A service profile is a collection of user-specific information that is permanently stored in the HSS. It is transferred from the HSS to an assigned S-CSCF in two user data-handling operations – Server-Assignment-Answer (SAA) and Push-Profile-Request (PPR) – as described in Sections 2.3.4.1 and 2.3.4.2. The service profile is carried in one Diameter AVP, where it is included as an Extensible Markup Language (XML) document. The service profile is further divided into three parts:

- Public Identification;
- Core Network Service Authorization;
- Initial Filter Criteria.

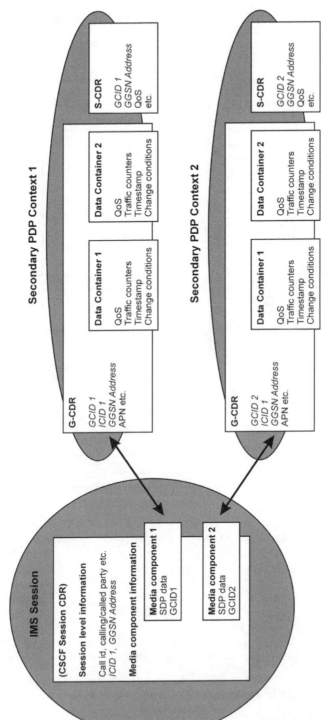

Figure 3.17 IMS charging correlation.

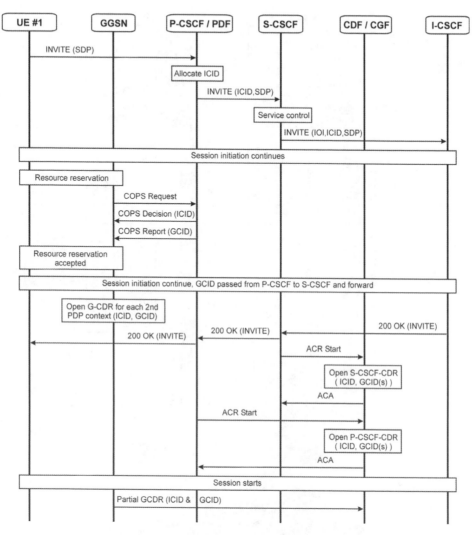

Figure 3.18 Distribution of charging information.

3.12.1.1 Public Identification

Public Identification comprises those user public identities that are associated with a service profile. Identities can be either SIP URIs or tel URIs. Each public user identity contains an associated barring indication. If the barring indication is set, then the S-CSCF will prevent that public identity (e.g., a temporary public user identity) from being used in any IMS communication other than registrations and re-registrations.

3.12.1.2 Media policy information

Media policy information is carried in the Core Network Service Authorization. It contains an integer that identifies a subscribed media profile in the S-CSCF

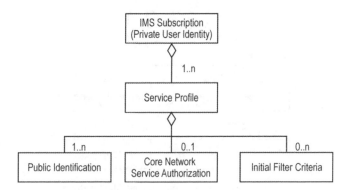

Figure 3.19 Structure of IMS user profile.

(e.g., allowed SDP parameters). This information allows operators to define different subscriber profiles in their IMS networks. They may define different customer classes, such as gold, silver and bronze. Gold could mean that a user is able to make video calls and all ordinary calls. Silver could mean that a user is able to use wideband Adaptive Multi-Rate (AMR) as a speech codec, but she is not allowed to make video calls and so on. Transferring just the integer value between the HSS and the S-CSCF saves the storage space in the HSS and optimizes the usage of the Cx reference point.

The S-CSCF needs to have a static database that contains the mapping between the integer value and the subscribed media profile. The meaning of the integer value is not standardized (i.e., it is operator specific). Figure 3.20 gives an illustrative example.

Figure 3.20 Media authorization in the S-CSCF.

3.12.1.3 Service-triggering information

Service-triggering information is presented in the form of initial filter criteria. Initial filter criteria describe when an incoming SIP message is further routed to a specific AS. Figure 3.21 shows that initial filter criteria are composed of either zero or one instance of a trigger point and one instance of an AS [3GPP TS 29.228]. Each initial filter criterion within the service profile has a unique priority value (integer) that is utilized in the S-CSCF. When multiple initial filter criteria are assigned the S-CSCF assesses them in numerical order: that is, an initial filter criterion with a higher priority number will be assessed after one with a smaller priority number.

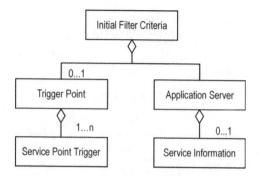

Figure 3.21   Structure of initial filter criteria.

Trigger point

The trigger point describes conditions that should be checked to discover whether the indicated AS should be contacted. The absence of a trigger point will indicate unconditional triggering to an AS. Each trigger point contains one to multiple instances of the Service Point Trigger. Service Point Triggers may be linked by means of logical expressions (AND, OR, NOT). Section 3.13 will give a more detailed explanation of how trigger points are used.

Application Server (AS)

The Application Server (AS) defines the AS that is contacted if the trigger points are met. The AS may contain information about the default handling of the session if contact with the AS fails. Default handling will either terminate the session or let the session continue based on the information in the initial filter criteria. In addition, the AS contains zero or one instance of the service information. Service information enables provisioning of information that is to be transferred transparently via the S-CSCF to an AS when the conditions of initial filter criteria are satisfied during registration.

3.13   Service provision

3.13.1   Introduction

The IMS is not a service in itself; on the contrary, it is a SIP-based architecture for enabling an advanced IP service and application on top of the PS network. IMS provides the necessary means for invoking services; this functionality is called "service provision". IMS service provisioning contains three fundamental steps:

1. Define possible service or service sets.
2. Create user-specific service data in the format of initial filter criteria when a user orders/modifies a subscription.
3. Pass an incoming initial request to an AS.

Item (1) is not addressed in this book because it is up to operators and service providers to define what kind of services they are willing to offer their subscribers. The other two steps are described next.

3.13.2 Creation of filter criteria

Whenever a user obtains an IMS subscription and her subscription contains some value-added services or an operator is willing to utilize ASs as part of its IMS infrastructure, they need to create service-specific data. These service-specific data are part of the user's user profile. More precisely, service-specific data are represented as initial filter criteria. Hereafter, we only concentrate on initial filter criteria. Section 3.12 describes how initial filter criteria fit into a user profile. When constructing initial filter criteria an operator needs to consider these questions:

- What is a trigger point?
- What is the correct AS when the trigger point is met?
- What is the priority of an initial filter criterion?
- What should be done if the AS is not responding?

The trigger point is used to decide whether an AS is contacted. It contains one to multiple instances of a Service Point Trigger [3GPP TS 29.228]. The Service Point Trigger comprises the items shown in Figure 3.22:

- Request-URI – identifies a resource that the request is addressed to (e.g., *sportnews@ims.example.com*).
- SIP Method – indicates the type of request (e.g., INVITE or MESSAGE).
- SIP Header – contains information related to the request. A Service Point Trigger could be based on the presence or absence of any SIP header or the content of any SIP header. The value of the content is a string that is interpreted as a regular expression. A regular expression could be as simple as a proper noun (e.g., John) in the FROM header that indicates the initiator of the request.
- Session Case – can be any one of three possible values, Originating, Terminating or Terminating_Unregistered, that indicate whether the filter should be used by the S-CSCF that is handling the originating service, terminating service or terminating for an unregistered end-user service. An originating case refers to when the S-CSCF is serving the calling user. A terminating case refers to when the S-CSCF is serving the called user.

Figure 3.22 Structure of Service Point Trigger.

- Session Description – defines a Service Point Trigger for the content of any SDP field within the body of a SIP Method. Regular expressions can be used to match the trigger.

Based on the above an operator could build, for example, initial filter criteria to handle unregistered users, such as an IMS user who has not registered any of her public user identities. The following initial filter criterion routes an incoming session to a voice mail server (*sip:vmail@ims.example.com*) when the user is not registered. To make this happen the operator has to set a SIP Method to match INVITE and a session case to match the value of Terminating_Unregistered. If the voice mail server cannot be contacted, then the default handling should be that the session is terminated. Initial filter criteria are coded in XML, as shown below (see [3GPP TS 29.228] for the exact coding rules of initial filter criteria):

```
Method="INVITE" AND SessionCase="2"
<?xml version="1.0" encoding="UTF-8"?>
<testDatatype xmlns:xsi="http://www.w3.org/2001/XMLSchema-instance"
xsi:noNamespaceSchemaLocation="D:\ CxDataType.xsd">
   <IMSSubscription>
      <PrivateID>privatexzyjoe@ims.example.com </Identity>
      <ServiceProfile>
         <PublicIdentity>
            <Identity>sip: joe.doe@ims.example.com </Identity>
         </PublicIdentity>
         <PublicIdentity>
            <Identity>tel:+358503334444</Identity>
         </PublicIdentity>
      <InitialFilterCriteria>
         <Priority>0</Priority>
         <TriggerPoint>
            <ConditionNegated>0</ConditionTypeCNF>
            <SPT>
               <ConditionNegated>0</ConditionNegated>
               <Group>0</Group>
               <Method>INVITE</Method>
            </SPT>

            <SPT>
               <ConditionNegated>0</ConditionNegated>
               <Group>0</Group>
               <SessionCase>2</SessionCase>
            </SPT>
         </TriggerPoint>
            <ApplicationServer>
               <ServerName>sip:vmail@ims.example.com</ServerName>
               <DefaultHandling>1</DefaultHandling>
```

```
          </ApplicationServer>
        </InitialFilterCriteria>
      </ServiceProfile>
   <IMSSubscription>
  </testDatatype>
```

3.13.3 Selection of AS

Initial filter criteria are downloaded to the S-CSCF on user registration or on a terminating initial request for an unregistered user. After downloading the user profile
from the HSS, the S-CSCF assesses the filter criteria for the initial request alone,
according to the following steps [3GPP TS 24.229]:

1. Check whether the public user identity is barred; if not, then proceed.
2. Check whether this request is an originating request or a terminating request.
3. Select the initial filter criteria for a session case (originating, terminating, or terminating for an unregistered end-user).
4. Check whether this request matches the initial filter criterion that has the highest
   priority for that user by comparing the service profile with the public user identity
   that was used to place this request:
   o if this request matches the initial filter criterion, then the S-CSCF will forward this
      request to that AS, check to see whether it matches the next following filter
      criterion of lower priority and apply the filter criteria on the SIP Method
      received from the previously contacted AS;
   o if this request does not match the highest priority initial filter criterion, then check
      to see whether it matches the following filter criterion's priorities until one does
      match;
   o if no more (or none) of the initial filter criteria apply, then the S-CSCF will
      forward this request based on the route decision.

There exists one clear difference in how the S-CSCF handles originating and terminating initial filter criteria. When the S-CSCF realizes that an AS has changed the Request-
URI in the case of terminating initial filter criteria, it stops checking and routes the
request based on the changed value of the Request-URI. In an originating case the
S-CSCF will continue to evaluate initial filter criteria until all of them have been
evaluated.

If the contacted AS does not respond, then the S-CSCF follows the default-handling
procedure associated with initial filter criteria: that is, either terminate the session or let
the session continue based on the information in the filter criteria. If the initial filter
criteria do not contain instructions to the S-CSCF regarding the failure to contact the
AS, then the S-CSCF will let the call continue as the default behaviour [3GPP TS
24.229].

According to our initial filter criteria example, incoming INVITE requests will be
routed to a voice mail server, *vmail@ims.example.com*, when Joe is not registered in the
network. In exceptional cases, when the voice mail server is not responding, the
S-CSCF is instructed to release a session attempt.

3.13.4   AS behaviour

Section 3.13.3 described how the request is routed to an AS. After receiving the request the AS initiates the actual service. To carry the service out the AS may act in three different modes:

- Terminating User Agent (UA) – the AS acts as the UE. This mode could be used to provide a voice mail service.
- Redirect server – the AS informs the originator about the user's new location or about alternative services that might be able to satisfy the session. This mode could be used for redirecting the originator to a particular Web page.
- SIP proxy – the AS processes the request and then proxies the request back to the S-CSCF. While processing, the AS may add, remove or modify the header contents contained in the SIP request according to the proxy rules specified in [RFC3261].
- Third-party call control/back-to-back UA – the AS generates a new SIP request for a different SIP dialog, which it sends to the S-CSCF.

These modes are described in more detail in Section 12.3. In addition to these modes, an AS can act as an originating UA. When the application is acting as an originating UA it is able to send requests to the users: for example, a conferencing server may send SIP INVITE requests to a pre-defined number of people at 9 a.m. for setting up a conference call. Another example could be a news server sending a SIP MESSAGE to a soccer fan to let him know that his favourite team has scored a goal.

3.14   Connectivity between traditional CS users and IMS users

For the time being, most users utilize traditional CS UE: that is, fixed line telephones and all kinds of cellular terminals. Therefore, it is desirable for the IMS to interwork with legacy CS networks to support basic voice calls between IMS users and CS network users. This requires interworking both at the user plane and the control plane because the used protocols are different in both planes. Control-plane interworking is tasked to the MGCF. It performs mapping from SIP signalling to the BICC or ISUP used in CS legacy networks, and vice versa. The IMS Media Gateway (IMS-MGW), in turn, translates protocols at the user plane. It terminates the bearer channels from the CS (PSTN/ISDN/GSM) networks as well as media streams from IP or ATM-based PS networks and provides the translation between these terminations. Additional functions, such as codec interworking, echo cancellation and continuity check, can also be provided. The terminations are controlled by the MGCF. Network configurations for handling both IMS and CS-originated calls are explained next.

3.14.1   IMS-originated session toward a user in the CS core network

When an IMS user initiates a session she does not need to bother about whether the called user is an IMS user or a CS user. She simply makes a call and the IMS takes care of finding the called party. The session request from the calling user will always arrive at

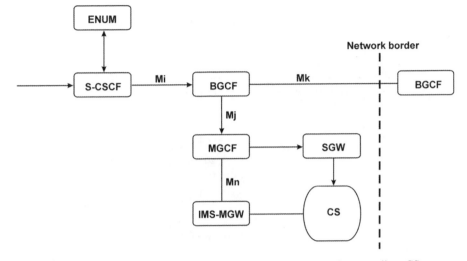

Figure 3.23 IMS–CS interworking configuration when an IMS user calls a CS user.

the S-CSCF serving the calling user, based on a route learned during IMS registration. When the S-CSCF receives a session request using a tel URL type of user identity (*tel:+358501234567*), it has to perform an ENUM query to convert the tel URL to a SIP URI, as IMS routing principles do not allow routing with tel URLs. If the S-CSCF is able to convert the identity to SIP URI format it will route the session further to the target IMS network, and when this conversion fails the S-CSCF will try to reach the user in the CS network. To break out to the CS network, the S-CSCF routes the session request further to the Breakout Gateway Control Function (BGCF) in the same network. The selected BGCF has two options: either selecting the breakout point in the same network or selecting another network to break out to the CS network. In the former case the BGCF selects an MGCF in the same network in order to convert SIP signalling to ISUP/BICC signalling and control the IMS-MGW. In the latter case the BGCF selects another BGCF in a different IMS network to select an MGCF in its network for handling breakout. The MGCF acts as an end point for SIP signalling; so, it negotiates media parameters together with the IMS UE and, similarly, negotiates media parameters together with the CS entity (e.g., with an MSC server). Figure 3.23 visualizes the interworking concept when an IMS-originated session is terminated in the CS network. The arrows in the figure show how the first signalling message traverses from the S-CSCF to the CS network.

3.14.2 *CS-originated session toward a user in IMS*

When a CS user dials an E.164 number that belongs to an IMS user, it will be handled in the CS network like any other E.164 number; however, after routing analysis it will be sent to an MGCF in the IMS user's home network. After receiving the ISUP/BICC signalling message, the MGCF interacts with the IMS-MGW to create a user-plane connection, converts ISUP/BICC signalling to SIP signalling and sends a SIP INVITE

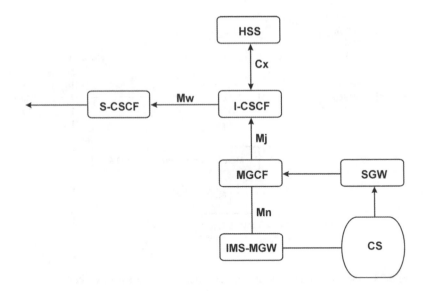

Figure 3.24 IMS–CS interworking configuration when a CS user calls an IMS user.

to the I-CSCF, which finds the S-CSCF for the called user with the help of the HSS (as described in Section 2.3.4.1). Then the S-CSCF takes the necessary action to pass the SIP INVITE to the UE. Thereafter, the MGCF continues communication with the UE and the CS network to set the call up. Figure 3.24 shows how the functions interwork when a CS-originated call is terminated by the IMS network. The arrows in the figure show how the first signalling message traverses from the CS to the IMS user.

3.15 Fixed and mobile convergence

Since both wireless and wireline networks can be integrated using the IMS architecture, the IMS becomes an inexpensive medium for Fixed to Mobile Convergence (FMC). FMC (aka wireless/wireline integration) is a way of connecting mobile phones to the fixed line infrastructure. It is currently one of the crucial strategic issues in the tele-communications industry.

The ultimate goal of this convergence is to deliver a seamless end-user experience across multiple locations, devices and services. With the convergence between the fixed and mobile line networks, telecommunications operators can provide services to users irrespective of their location, access technology or the type of device they are using.

Convergence can be viewed from three separate angles:

- convergence of services;
- convergence of networks;
- convergence of devices.

Wireless/wireline integration creates a new telecom services paradigm where the system is user driven as opposed to being a network- or device-driven paradigm. This integration also presents the user with the best of both worlds, fixed and mobile, by providing the user with services in terms of the convenience and availability of mobile services and the reliability and quality of fixed services.

Operator benefits

FMC offers business opportunities for all types of operators. Lifestyle changes, efficient pricing models and personalization of services will encourage users to opt for operators with an FMC solution.

Both consumers and businesses consider mobility an important feature today. Fixed line operators can increase revenue by increasing their service coverage to include those offered by mobile operators.

As the mobile market is now almost saturated, adding new subscribers is no longer the type of growth that mobile operators want to persue. Wireless operators need to seek new and innovative ways of increasing usage of their current customer base. This can be achieved by offering new services to their users. Wireless operators can give their subscribers a new way of being reached: one number, one handset, always on.

Personal mobility

Personal mobility provides end-users with much greater choice, convenience, productivity and cost savings than is available today with mobility. It enables carriers and service providers to rejuvenate their businesses.

Many users today have multiple phones (mobile phone, office phone and home phone), each having a different number and using separate access networks.

Personal mobility would allow a user to have one phone that lets her talk over WiFi and seamlessly switch her call to a cellular network when she is out of WiFi coverage, and vice versa. It also allows the user to unify services such as voice mail into one and makes it accessible, thus boosting a user's productivity. It also opens the door for more advanced services like group management (see Chapter 8) to be more effective.

Personal mobility is enabled by convergence; specifically, the convergence of fixed, wireless and IP data networks. It enables a user to take advantage of the enormous potential that combining these networks provides. Users have a choice of access to both fixed and mobile phones and can choose which one to use depending on cost and convenience. They can choose between making a call on a fixed or a mobile phone if they are in the workplace or from home where both are available.

The seamless handoff between WiFi networks and cellular networks with minimal or no visibility to the user, as well as a set of convergence applications that take advantage of such a seamless handoff, enables a service provider to deliver personal mobility and deliver even more value to users.

3.16 SIP compression

The IMS supports multimedia services using the SIP call control mechanism. SIP is a client server, text-based signalling protocol used to create and control multimedia sessions with two or more participants. The messages also contain a large number of headers and header parameters, including extensions and security-related information. Setting up a SIP session is a tedious process involving codec and extension negotiations as well as QoS interworking notifications. In general, this provides a flexible framework that allows sessions with differing requirements to be set up. However, the drawback is the large number of bytes and the many messages exchanged over the radio interface. The increased message size means that:

- Call setup procedures using SIP will take much more time to be completed compared with those using existing cellular-specific signalling, which means that the end-user will experience a delay in call establishment that will be unexpected and likely unacceptable.
- Intra-call signalling will, in some way or another, adversely affect voice quality/ system performance.

Therefore, support for real-time multimedia applications requires particular attention when SIP call control is used. To speed up session establishment, 3GPP has mandated the support of SIP compression by both the UE and the P-CSCF [3GPP TS 23.221]. Although the support of compression is mandatory, 3GPP was not happy to mandate its usage because in the future Wireless Local Area Network (WLAN) terminals may not need to use SIP compression at all. At the time of writing, it is required that the UE and the P-CSCF implement compression functionalities as defined in [RFC3320], [RFC3485], [RFC3486] and [3GPP TS 24.229]. The first-mentioned RFC gives an overall solution to how SIP messages between two entities can be compressed. The second RFC defines a SIP/SDP-specific static dictionary that a signalling compression solution may use in order to achieve higher efficiency. The third RFC explains how the UE could signal that compression is desired for one or more SIP messages (see Section 12.13.15 and Chapter 23 for more details).

3.17   Interworking between IPv4 and IPv6 in the IMS

3.17.1   Introduction

In the Internet at large, deployment of IPv6 has not progressed as fast as was hoped or expected when the initial concept of the IMS first came about. Many factors have contributed and continue to contribute to this, but in general any change in the core routing infrastructure of the Internet is likely to take a much longer time than one adopted at the edges of the network.

Network Address Translators (NATs) constitute a change that falls into the latter category. NATs are widely acknowledged as detrimental to new services as well as existing ones (e.g., SIP-based VoIP), but in terms of investment they offer a very affordable and easily deployable option to extending the fast-depleting address space of IPv4,[1] thus slowing down deployment of the long-term solution in the form of IPv6.

---

[1] Actually, there are many other use-cases for deploying NATs, which have to do with security (firewall functionality), naming (e.g., private naming of hosts in a home-grown network), as well as addressing, just to name a few.

In many ways, what is happening in the Internet at large is descriptive of what is happening in the mobile domain. While IPv6 is seen as technically superior to IPv4 for the IMS, mandating IPv6 support represents quite a big barrier to deployment of any system. Using IPv6 exclusively also requires that roaming partners and service providers migrate to IPv6. This is why many early deployments have chosen to work with existing GPRS access networks, as well as a security infrastructure (more on this in section 3.8.6), and often this existing network infrastructure does not support IPv6. In addition, the 3GPP2 IP multimedia system, called Multimedia Domain (MMD), operates both over IPv4 and IPv6. This means that any interworking between the IMS and the MMD naturally involves IPv4–IPv6 interworking as well.

Hence, it has become apparent that some deployment of the IMS based on IPv4 is inevitable, although originally specified as supporting IPv6 only. Usually, these IMS deployments based on IPv4 are called "early IMS implementations", and either the UE, the IMS core network, or both exclusively use IPv4.

There are two main consequences:

- IPv4 implementations will need to deal with having to use private address space, since not all mobile devices can be afforded an address from the public address space.
- Once IPv6-capable implementations are introduced alongside IPv4 implementations, they need the ability to work together in a seamless way.

This chapter illustrates the issues and solutions for IPv4 and IPv6 interworking in IMS. In Release 6, 3GPP has worked to provide guidelines to the implementers of early IMS systems in the form of a technical report [TR 23.981]. This means that, while early IMS implementation is still not endorsed, at least some form of guidelines is given to address some of the most pressing concerns.

Address translation

The basic component for doing address translation is called a Network Address Translator (NAT). NATs allow a single public IP address to be shared between a number of hosts. The hosts behind the NAT device are given IP addresses that are in a private address space (i.e., in 192.168.0.0/16 or 10.0.0.0/24), and therefore not routable in the Internet. The NAT device then creates temporary bindings between the public and private address space, on a per-connection basis. A binding is simply a mapping between the public IP address and port to a private address and port associated with a specific transport (UDP or TCP). The binding lasts only as long as the session lasts, or until the binding times out. Usually, these timeouts are quite aggressive in order to save resources.

Another type of NAT is specifically designed to translate between protocol versions. A Network Address Translator–Protocol Translator (NAT-PT) translates between IPv6 and IPv4. In effect, it takes the IP datagram and replaces the IPv6 header with an IPv4 header and vice versa.

In the IMS, the existence of such translation devices alone is not enough. SIP and SDP both contain plain IP addresses, such as in the SIP Contact header field, or in the SDP media lines. Those addresses will neither be understood nor reachable to a peer using the wrong IP version. Therefore, there needs to be an Application Level Gateway

(ALG) device that inspects SIP messages and replaces any occurrence of the original IP address with the translated one. NATs, in general, do not allow inbound traffic until outbound traffic is generated (i.e., the binding is created), which means that the ALG must also actively create bindings based on the media information contained in SDP.

Often the ALG device is co-located with the NAT device, simply because it needs to be aware of the translation bindings, as well as install them. A Session Border Controller (SBC) is one type of such translator, although SBCs usually perform many other functions as well.

As may be expected, address translators have several drawbacks:

- They are a single point of failure; if the translation binding is lost, the entire session becomes unusable.
- ALGs either brake security, or do not operate correctly in the presence of an end-to-end security mechanism. Modification of messages in transit is, in effect, an active Man-in-the-Middle (MITM) attack.
- ALGs are not feature transparent; a new version of SDP, or another header field containing an IP address will not be recognized or translated correctly.
- Scalability problems; ALG functionality is resource intensive, and NATs require both resources for storing bindings as well as bandwidth.

It is no surprise that IPv4–IPv6 interworking in the IMS attempts to avoid using NATs and ALGs as much as possible. However, they are suggested for certain interconnecting scenarios, which are further discussed in Section 3.17.3.

3.17.2   IPv6-only versus dual stack

One key method for successful transition to IPv6 is for hosts (and routers) to support a dual IP layer approach, also known as a "dual stack operation". Dual stack is a technique for providing support for both versions of IP – i.e., IPv6 as well as IPv4. A dual stack host can send and receive both versions of IP and can therefore communicate with both kinds of nodes. A dual stack node is also able to be configured with both address versions, although the configuration mechanisms may be different. For example, the IPv4 address could be configured using DHCP, and the IPv6 address using stateless address autoconfiguration.

There are also other functions that a dual stack node needs to support. For example, it needs to support both IPv4 "A" and IPv6 "AAAA" DNS records, irrespective of over which version of IP datagrams such results are received. For example, a DNS query made over IPv4 for a domain might return both A and AAAA records, or even AAAA records only. Then, the order in which the IP addresses are used depends on the application, but may also be influenced by the order in which the DNS server returns the respective records.

A node may also disable either protocol version from its stack. However, in practice, many IMS nodes will need to keep IPv4 functioning, as there are other applications – such as the WWW – that require IPv4 to be fully usable.

In terms of migration, dual stack UEs and IMS core offer more options than IPv4-only implementations. Even with a specific IP version disabled in the stack, later on when IPv6 becomes more prevalent, it can be enabled. The main benefit in dual stack hosts is that they minimize the need for NAT devices in the network. IPv6-only hosts don't appear to be feasible well into the future, since the majority of services will still remain IPv4-only for a long time.

3.17.3 Interworking scenarios

The main problem with IPv4–IPv6 interworking is, of course, the different IP version. The existence of nodes that only understand one version make it necessary for some form of translation. This applies to both signalling traffic as well as user-plane or media traffic. There are also many ways that the translation can happen, depending on whether the IMS core supports IPv6.

We split the discussion in this section into intra-domain and inter-domain scenarios. Intra-domain scenarios deal with interworking between a UE and the IMS core of a single IMS provider, whereas the inter-domain scenarios deal with the interconnection of different IMS providers' networks, and the end-to-end considerations between UEs belonging to different IMS operators' networks.

3.17.4 Intra-domain scenarios

The intra-domain scenarios are very straightforward; the assumption is that between the UE and the P-CSCF, the IP version stays constant:

- if the IMS core supports IPv4 only – only dual stack and IPv4 clients are supported;
- if the IMS core supports dual stack – all variants of clients are supported;
- if the IMS core is IPv6 only – only dual stack and IPv6 clients are supported.

3.17.5 Inter-domain scenarios

The inter-domain scenarios are much more complicated. The complexity is brought on by the fact that there are many more variables to consider. In addition to the sender and receiver UE and IMS core, there is also the transit network between the two domains, which can also exhibit support for one of either IPv4, dual stack or IPv6 behaviour. In addition, the IMS core may also be using IPv4 addresses from a private address space, requiring a NAT at the edge of the network in order to interconnect with other domains.

A dual-stack IMS core may need an additional NAT device that is associated with the S-CSCF. The reason for this is that the interconnection may use IPv4, which the I-CSCF and the S-CSCF support, but the UE doesn't. In that case, a NAT-PT device is needed to translate the IP versions. This is illustrated in Figure 3.25, showing the possible placement of the various NAT devices.

3.17.6 Configuration and bootstrapping

Usually, after establishing L2 connectivity the UE receives an IP address using a standard configuration, such as DHCP, or some other mechanism.

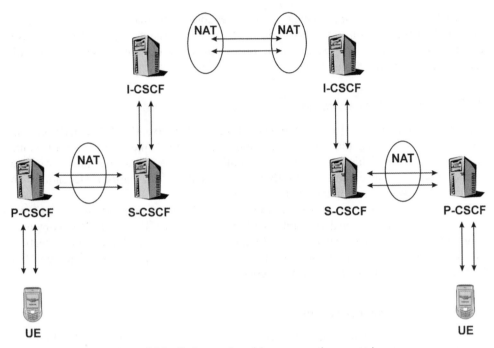

Figure 3.25   End-to-end and interconnection scenarios.

The next step for the UE is to find its contact point to the IMS – namely, the address of the P-CSCF. In IMS, the existing discovery mechanisms for the P-CSCF address are either IPv6 specific, or use a Release 5 (or later) based GPRS. For deployments of IMS based on IPv4 other mechanisms may be needed. There are a few alternative options to discovering the P-CSCF address:

- Discovery using the GPRS could be altered so that it returns both IPv4 and IPv6 addresses upon PDP context establishment.
- Using the DHCP option for configuring the P-CSCF address.
- Out-of-band provisioning mechanisms, such as one using the Short Messaging Service (SMS), Over-the-Air (OTA) or Open Mobile Alliance (OMA) device management framework.
- Pre-configuration.

Using an out-of-band mechanism requires no changes to the existing network infrastructure and is, therefore, probably preferred by most early IMS implementers. Pre-configuring the address is also an option, but it is inflexible regarding later changes in naming or topology.

In addition to configuring the outbound proxy (i.e., the P-CSCF) address, there is additional address information carried in the SIP header, which needs interworking considerations as well. For example, any route entries or addresses of charging entities need to be represented in both IPv4 and IPv6.

3.17.7 IPv4-only access networks

Even though not strictly an IMS implementation issue, due to functionality in GPRS, the access network is a consideration in IPv4–IPv6 interworking. The reason for this is that the bearers are dependent on the IP version; an SGSN/GGSN incapable of supporting IPv6 would not allow an IPv6 PDP context to be established. Older, existing access networks might simply not support IPv6. Therefore, the first interworking issue is with the GPRS access network, both in roaming and non-roaming scenarios. If the GPRS SGSN or the GGSN do not support IPv6, the UE is unable to use native IPv6, regardless of whether any of the other IMS elements are supporting it. The only available option for using IPv6 in these scenarios is for the UE and IMS core to use a tunnelling mechanism, where each IPv6 packet is encapsulated in an IPv4 packet, and de-capsulated at an exit point or router – in effect, treating the IPv4 connectivity as L2 transport. This is illustrated in Figure 3.26.

Encapsulation needs to be applied between the host and a network-based node in order to be able to transparently communicate with a IPv6-only host. So, in effect, the tunnel end point will route the packets to the correct host in the network, after de-capsulation of the packet. Tunnels are addressed and set up using a tunneling mechanism, such as the Intra-Site Automatic Tunnel Addressing Protocol (ISATAP) [Draft.ietf-ngtrans-isatap].

In ISATAP, interface addresses are created using a static mapping from the IPv4 link address (i.e., the last 32 bits of the interface address). They are accompanied by a routing table, so that packets can then be routed to the proper interface. However, there are several downsides to the tunnelling approach:

- it breaks SBLP by using the Go interface;
- the home network needs to be installed with a tunnelling server that serves as a tunnel end point;
- such a server also needs to be configured or discoverable by the UE;
- the tunnel end point is a single point of failure;
- using a tunnel in roaming scenarios is likely to be inefficient, as all traffic is routed via the home network – this causes triangular routing which will increase latency;

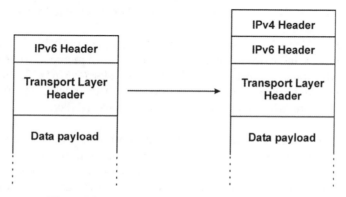

Figure 3.26 IPv6 to IPv4 tunnelling mechanism.

- it requires a dual stack host in any case, so using native IPv4 may be a better approach.

3.18   Combination of CS and IMS services – combinational services

3.18.1   Introduction

Whilst the IMS is designed as a standalone system to provide multimedia services to its users, it can also be used to provide additional functionalities within existing telephony networks. During the initial phase of IMS deployment mobile network users will make use mainly of the CS domain of the mobile networks for audio sessions. In such CS calls the IMS is not required for provision of the basic service – i.e., the signalling and media connection between the two users, as well as the QoS.

But also such CS calls should benefit from the flexible and broad service provisioning capabilities of IMS. New work has been started by 3GPP recently, to enable such so-called "combinational services". As the work is still in progress at the time of writing, only the basic principles can be shown here.

In order to enable such a service combination, the two UEs first need to become aware of the supported capabilities at the other end. Section 3.18.2 shows how IMS-related capabilities can be exchanged after a CS call has been established between two users. In Section 3.18.3 it is shown how, based on that capability exchange, an IMS service can be invoked that enhances the ongoing CS call.

3.18.2   Capability exchange

In the following example we will assume that two users are connected over the CS domain – i.e., they have a CS connection between them. In this example Tobias called his sister via her international phone number. In order to allow the combinational services to work, several conditions have to be fulfilled:

- within CS signalling, the calling network needs to provide Tobias's phone number (the MSISDN) in full international format to Theresa's UE – e.g., +44123456789;
- Tobias's MSISDN also needs to be registered in Tobias's home IMS network as a tel URL – e.g. tel:+44123456789;
- also, Theresa's MSISDN must be sent via CS signalling to Tobias's UE. Theresa's MSISDN must be indicated in international number format and must be registered in Theresa's IMS home network as a tel URL – e.g. *tel:+36987654321*; and
- both Tobias and Theresa need to be registered with their tel URLs from the very same UEs they are using to establish CS calls with each other.

In order to simplify the example, we assume that both users, Tobias and Theresa, are each registered only from one UE. This is done in order to avoid forking cases (see Section 3.7). Theresa and Tobias both have an IMS-based application installed on their UEs that allows them to share videostreams in real time in addition to an existing CS voice call. This peer-to-peer application will only work if the application is present on

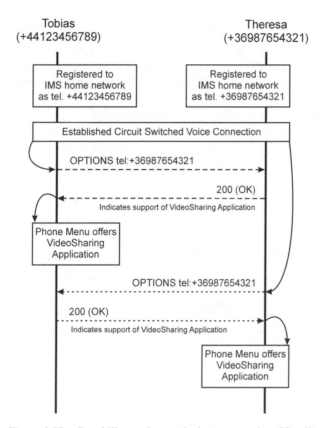

Figure 3.27 Capability exchange during an ongoing CS call.

both UEs. In order to give a consistent user experience, the menu item for "video sharing" will only be shown to Theresa and Tobias, if their UEs have verified that the application is also supported at the remote end.

In order to query the capabilities of the remote end the SIP OPTIONS request (see Section 12.4.1) is used. After establishing the CS call, both UEs send an OPTIONS request to the tel URL of the remote user. The tel URL will be derived from the MSISDN of the remote user in the CS call. In our example, Tobias's UE sends an OPTIONS request to the following address:

```
OPTIONS tel:+36987654321 SIP/2.0
```

Theresa's UE will send a 200 (OK) response to this OPTIONS request, indicating the UE's capabilities. One of these capabilities will be support for the video-sharing application. The same OPTIONS/200 (OK) exchange is done in the other direction. After these exchanges have been finished, both Theresa and Tobias will be offered the possibility of sharing a videostream with the other user in the CS call.

As said above, the example given here is very basic and only shows how capabilities between UEs can be exchanged in principle.

3.18.3   Parallel CS and IMS services

After the exchange of capabilities, one of the users can now start to share a live video. In our example Theresa wants to send the video to Tobias. The session setup between the two users will work as described in Section 11.10 (Alternative session setups). After this session has been established, Theresa can send a videostream via GPRS to Tobias. Figure 3.28 shows the connections that exist in parallel.

The SIP-based connection between the two users can either be released whilst the CS call is still ongoing or the video-sharing application will automatically release it once the CS call has ended. This is purely an application-specific behaviour.

Figure 3.28   Example for parallel connections when combining IMS and CS services.

3.19   Security services in the IMS

This section is intended to explain how security works in the IMS. It is intentionally thin in cryptography and, thus, will not discuss algorithms and key lengths in depth, nor will it perform any cryptanalysis on IMS security. There are many books specifically written for that purpose.[2]

Instead, what this chapter will do is give a high-level view of the security architecture and explain the components of that architecture, including the models and protocols used to provide the required security features. After reading this chapter the reader should be familiar with the main concepts in the IMS security architecture and understand the underlying models, especially those related to trust and identity that shape IMS security as a whole.

3.19.1   IMS security model

The IMS security architecture consists of three building blocks, as illustrated in Figure 3.29. The first building block is Network Domain Security (NDS) [3GPP TS 33.210], which provides IP security between different domains and nodes within a domain.

---

[2] See, for example, V. Niemi and K. Nyberg (2003) *UMTS Security*, John Wiley & Sons, Chichester, UK.

Figure 3.29 Security architecture of the IMS.

Layered alongside NDS is IMS access security [3GPP TS 33.203]. The access security for SIP-based services is a self-sustaining component in itself, with the exception that the security parameters for it are derived from the UMTS Authentication and Key Agreement (AKA) Protocol [3GPP TS 33.102]. AKA is also used for bootstrapping purposes – namely, keys and certificates are derived from AKA credentials and subsequently used for securing applications that run on the Hypertext Transfer (or Transport) Protocol (HTTP) [RFC2616], among other things – in what is called the Generic Authentication Architecture (GAA) [3GPP TS 33.220].

Intentionally left out of this architectural model are those security layers that potentially lie on top of the IMS access security or run below the NDS. For example, in the UMTS the radio access layer implements its own set of security features, including ciphering and message integrity. However, the IMS is designed in a way that does not depend on the existence of either access security or user-plane security.

3.19.2 Authentication and Key Agreement (AKA)

Security in the IMS is based on a long-term secret key, shared between the ISIM and the home network's Authentication Centre (AUC). The most important building block in IMS security is the ISIM module, which acts as storage for the shared secret (K) and accompanying AKA algorithms, and is usually embedded on a smartcard-based device called the Universal Integrated Circuit Card (UICC). Access to the shared secret is limited. The module takes AKA parameters as input and outputs the resulting AKA parameters and results. Thus, it never exposes the actual shared secret to the outside world.

The device on which the ISIM resides is tamper resistant, so even physical access to it is unlikely to result in exposing the secret key. To further protect the ISIM from unauthorized access, the user is usually subject to user domain security mechanisms. This, in essence, means that in order to run AKA on the ISIM, the user is prompted for a PIN code. The combination of ownership – i.e., access to a physical device (UICC/ ISIM) and knowledge of the secret PIN code – makes the security architecture of the IMS robust. An attacker is required to have possession of both "something you own" and "something you know", which is difficult, as long as there is some level of care taken by the mobile user.

AKA accomplishes mutual authentication of both the ISIM and the AUC, and establishes a pair of cipher and integrity keys. The authentication procedure is set off by the network using an authentication request that contains a random challenge (RAND) and a network authentication token (AUTN). The ISIM verifies the AUTN and in doing so verifies the authenticity of the network itself. Each end also maintains a sequence number for each round of authentication procedures. If the ISIM detects an authentication request whose sequence number is out of range, then it aborts the authentication and reports back to the network with a synchronization failure message, including with it the correct sequence number. This is another top-level concept that provides for anti-replay protection.

To respond to the network's authentication request, the ISIM applies the secret key on the RAND to produce an authentication response (RES). The RES is verified by the network in order to authenticate the ISIM. At this point the UE and the network have successfully authenticated each other and as a by-product have also generated a pair of session keys: the Cipher Key (CK) and the Integrity Key (IK). These keys can then be used for securing subsequent communications between the two entities. Table 3.13 lists some of the central AKA parameters and their meaning.

Table 3.13 Authentication and Key Agreement Parameters

AKA parameter	Length (bits)	Description
K	128	Shared secret; authentication key shared between the network and the mobile terminal
RAND	128	Random authentication challenge generated by the network
AUTN	128	(Network) authentication token
SQN	48	Sequence number tracking the sequence of the authentication procedures
AUTS	112	Synchronization token generated by the ISIM upon detecting a synchronization failure
RES	32–128[a]	Authentication response generated by the ISIM
CK	128	Cipher key generated during authentication by both the network and the ISIM
IK	128	Integrity key generated during authentication by both the network and the ISIM

[a] Note that the short key lengths are to accommodate backwards compatibility with 2G authentication tokens.

3.19.3   Network Domain Security (NDS)

3.19.3.1   Introduction

One of the main identified weaknesses of 2G systems is the lack of standardized security solutions for core networks. Even though radio access from the mobile terminal to the base station is usually protected by encryption, nodes in the rest of the system pass traffic in the clear. Sometimes these links even run over unprotected radio hops, so an attacker that has access to this medium can fairly easily eavesdrop on communications.

Having learned from these shortcomings in 2G, 3G systems have set out to protect all IP traffic in the core network. Network Domain Security (NDS)[3] accomplishes this by providing confidentiality, data integrity, authentication and anti-replay protection for the traffic, using a combination of cryptographic security mechanisms and protocol security mechanisms applied in IP security (IPsec).

3.19.3.2   Security domains

Security domains are central to the concept of NDS. A security domain is typically a network operated by a single administrative authority that maintains a uniform security policy within that domain. As a result, the level of security and the installed security services will, in general, be systematically the same within a security domain.

In many cases a security domain will correspond directly to an operator's core network. It is, however, possible to run several security domains each pertaining to a subset of the operator's entire core network. In the NDS/IP the interfaces between different security domains are denoted as Za, while interfaces between elements inside a security domain are denoted as Zb. While use of the Zb interface is, in general, optional and up to the security domain's administrator, use of the Za interface is always mandatory between different security domains. Data authentication and integrity protection is mandatory for both interfaces, while the use of encryption is optional for the Zb, as opposed to being recommended for the Za.

The IMS builds on the familiar concept of a home network and a visited network. Basically, two scenarios exist, depending on whether the IMS terminal is roaming or not. In the first scenario the UE's first point of contact to the IMS, called the P-CSCF, is located in the home network. In the second scenario the P-CSCF is located in the visited network, meaning that the UE is, in fact, roaming in such a way that its first point of contact to the IMS is not its home network. These two scenarios are illustrated in Figure 3.30.

Quite often, an IMS network corresponds to a single security domain and, therefore, traffic between the operator's IMS networks is protected using the NDS/IP. The same also applies in the above-mentioned second scenario, where traffic between the visited and the home network is also protected using the NDS/IP.

In the IMS the NDS/IP only protects traffic between network elements in the IP layer; so, further security measures are required. These will be covered in subsequent

---

[3] To specifically indicate that the protected traffic is IP based, NDS is usually denoted by the abbreviation NDS/IP.

Figure 3.30 Security domains in IMS.

chapters. Most importantly, the first hop[4] in terms of SIP traffic is not protected using the NDS/IP, but is protected using IMS access security measures [3GPP TS 33.203]. As will be explained in subsequent chapters, the above scenario in which the IMS elements are split across the home network and the visited network and, therefore, different security domains requires some special care in terms of authentication and key distribution.

---

[4] Referring here to the interface between the UE and the P-CSCF, denoted as Gm.

3.19.3.3 Security gateways

Traffic entering and leaving a security domain passes through a Security Gateway (SEG). The SEG sits at the border of a security domain and tunnels traffic toward a defined set of other security domains. This is called a hub-and-spoke model; it provides for hop-by-hop security between security domains. The SEG is responsible for enforcing security policy when passing traffic between the security domains. This policy enforcement may also include packet filtering or firewall functionality, but such functionality is the responsibility of the domain administrator.

In the IMS all traffic within the IMS core network is routed via SEGs, especially so when the traffic is inter-domain, meaning that it originates from a different security domain from the one where it is received. When protecting inter-domain IMS traffic, both confidentiality as well as data integrity and authentication are mandated in the NDS/IP.

3.19.3.4 Key management and distribution

Each SEG is responsible for setting up and maintaining IPsec Security Associations (SAs) [RFC2401] with its peer SEGs. These SAs are negotiated using the Internet Key Exchange (IKE) [RFC2409] protocol, where authentication is done using long-term keys stored in the SEGs. A total of two SAs per peer connection are maintained by the SEG: one for inbound traffic and one for outbound traffic. In addition, the SEG maintains a single Internet Security Association and Key Management Protocol (ISAKMP) SA [RFC2408], which is related to key management and used to build up the actual IPsec SAs between peer hosts. One of the key prerequisites for the ISAKMP SA is that peers are authenticated. In the NDS/IP, authentication is based on Pre-shared Secret Keys (PSKs). In 3GPP Release 6, NDS is extended so that other authentication mechanisms than the PSK can also be used between the SEGs. The NDS Authentication Framework (NDS/AF) [3GPP TS 33.310] defines the Public Key Infrastructure (PKI), trust model and mechanisms for authentication of SEGs using public key certificates and RSA signatures. The trust model for the PKI required by NDS/AF includes two modes for cross-certification:[5]

- Manual cross-certification: in this mode, the authorities decide themselves to trust another domain authority. This mode has difficulty in scaling, since each remote security domain needs a valid, locally trusted (signed) certificate. This limitation is similar to the limitations found in the baseline NDS/IP, using PSKs.
- Cross-certification using a bridge Certificate Authority (CA): in this mode, a broker entity or a clearinghouse applies 1-to-1 certification, and the different security domains trust each other via transitive trust (see Section 3.19.4.2 for a discussion of transitive trust). This limits the amount of certification required, since a single domain authority only needs cross-certification with the bridge CA.

Figure 3.29 illustrates the overall NDA/IP model.

---

[5] Cross-certification is a process for establishing trust between two domain authorities. When Domain Authority A is cross-certified with Domain Authority B, both are able to trust each others' certificates – i.e., they are able to authenticate one another.

Figure 3.31   NDS/IP and SEGs.

The security protocol used in the NDS/IP for encryption, data integrity protection and authentication is the IPsec Encapsulating Security Payload (ESP) [RFC2406] in tunnel mode. In tunnel-mode ESP the full IP datagram including the IP header is encapsulated in the ESP packet. For encryption, the 3DES [RFC1851] algorithm is mandatory, while for data integrity and authentication both MD5 [RFC1321] and SHA-1 [RFC2404] can be used. For the specifics of IPsec/IKE and ESP please refer to Chapter 22, where these protocols are discussed more extensively.

3.19.4   IMS access security for SIP-based services

3.19.4.1   Introduction

SIP is at the core of the IMS, as it is used for creating, managing and terminating various types of multimedia sessions. The key thing to accomplish in securing access to the IMS is to protect the SIP signalling in the IMS. As noted previously, in the IMS core network this is accomplished using the NDS/IP. But the first hop, meaning the interface for SIP communications between the UE and the IMS P-CSCF denoted as Gm, needs additional measures since it is outside the scope of the NDS/IP.

The security features and mechanisms for secure access to the IMS are specified in [3GPP TS 33.203]. This defines how the UE and network are authenticated as well as how they agree on the security mechanisms, algorithms and keys used.

3.19.4.2   Trust model overview

The IMS establishes a trust domain, as described in [RFC3325], that encompasses the following IMS elements:

- P/I/S-CSCF;
- BGCF;
- MGCF/MRFC;
- all ASs that are not in third-party control.

The main component of trust is identity: in order to trust an entity accessing the IMS there needs to be an established relationship with that entity (i.e., its identity is known and verified). In the IMS this identity is passed between nodes in the trust domain in the form of an asserted identity. The UE can state a preference for this identity if multiple identities exist; but, it is ultimately at the border of the trust domain (namely, in the P-CSCF) that the asserted identity is assigned. Conversely, the P-CSCF plays a central role in authenticating the UE.

The level of trust is always related to the expected behaviour of an entity. For example, Alice may know Bob and trust him to take her children to school. She expects and knows that Bob will act responsibly, drive safely and so on. But she may not trust Bob enough to give him access to her bank account.

Another important property of the IMS trust model is that it is based on transitive trust. The existence of pairwise trust between a first and a second entity as well as a second and a third entity automatically instils trust between the first and the third entity: for example, Alice knows and trusts Bob, who in turn knows Celia and trusts her to take his children to school. Now, according to transitive trust, Alice can also trust Celia to take her children to school without ever actually having met Celia in person. It is enough that Alice trusts Bob and knows that Celia is also part of the trust domain of parenthood. The fact that Alice and Bob are both parents assures Alice that Bob has applied due diligence when judging whether Celia is fit to take children to school. In essence, the trust domain of parenthood forms a network of parents, all compliant with the pre-defined behaviour of a mother or a father.

In [RFC3325] terms, both the expected behaviour of an entity in a trust domain and the assurance of compliance to the expected behaviour needs to be specified for a given trust domain T in what is called a "Spec(T)". The components that make up a Spec(T) are:

- Definition of the way in which users entering the trust domain are authenticated and definition of the used security mechanisms that secure the communications between the users and the trust domain. In the IMS this entails authentication using the AKA protocol and related specifications on Gm security in [3GPP TS 33.203].
- Definition of mechanisms used for securing the communications between nodes in a trust domain. In the IMS this bit is documented in the NDS/IP [3GPP TS 33.210].
- Definition of the procedures used in determining the set of entities that are part of the trust domain. In the IMS this set of entities is basically represented by the set of peer SEGs, of which a SEG in a security domain is aware.
- Assertion that nodes in a trust domain are both compliant with SIP and SIP-asserted identity specifications.
- Definition of privacy handling. This definition relies on SIP privacy mechanisms and the way they are used with asserted identities (Section 3.19.4.3 will discuss these issues more deeply).

3.19.4.3   User privacy handling

The concepts of trust domain and asserted identity enable passing a user's asserted identity around, potentially to entities that are not part of the trust domain. This creates obvious privacy issues, since the user may, in fact, require that her identity be kept private and internal to the trust domain.

In the IMS the user can request that her identity is not revealed to entities outside the trust domain. This is based on SIP privacy extensions [RFC3323]. A UE inserts its privacy preferences in a privacy header field, which is then inspected by the network. Possible values for this header are:

- User – indicates that user-level[6] privacy functions should be provided by the network. This value is usually set by intermediaries rather than user agents.
- Header – indicates that the UA is requiring that header privacy be applied to the message. This means that all privacy-sensitive headers be obscured and that no other sensitive header be added.
- Session – indicates that the UA is requiring that privacy-sensitive data be obscured for the session (i.e., in the SDP payload of the message).
- Critical – indicates that the requested privacy mechanisms are critical. If any of those mechanisms is unavailable, the request should fail.
- ID – indicates that the user requires her asserted identity be kept inside the trust domain. In practice, setting this value means that the P-Asserted-Identity header field must be stripped from messages that leave the trust domain.
- None – indicates that the UA explicitly requires no privacy mechanisms to be applied to the request.

3.19.4.4 Authentication and security agreement

Authentication for IMS access is based on the AKA protocol. However, the AKA protocol cannot be run directly over IP; instead, it needs a vehicle to carry protocol messages between the UE and the home network. Obviously, as the entire objective of IMS access authentication is to authenticate for SIP access, SIP is a natural choice for such a vehicle. In practice, the way in which the AKA protocol is tunnelled inside SIP is specified in [RFC3310]. This defines the message format and procedures for using AKA as a digest authentication [RFC2617] password system for the SIP registration procedure. The digest challenge originating from the network will contain the RAND and AUTN AKA parameters, encoded in the server nonce value. The challenge contains a special algorithm directive that instructs the client to use the AKA protocol for that particular challenge. The RES is used as the password when calculating digest credentials, which means that the digest framework is utilized in a special way to tunnel the AKA protocol in IMS access security.

Concurrent with authentication of the user, the UE and the IMS also need to negotiate the security mechanisms that are going to be used in securing subsequent SIP traffic in the Gm interface. The protocol used for this security agreement is again SIP, as specified in [RFC3329]. The UE and the P-CSCF exchange their respective lists of supported security mechanisms and the highest commonly supported one is selected and used. At a minimum, the selected security mechanism needs to provide data

---

[6] By "user-level privacy" we mean privacy functions that the SIP UA itself is able to provide (e.g., using an anonymous identity in the From field of a request).

integrity protection, as that is required to protect actual security mechanism negotiation. Once the security mechanism has been selected and its use started, the previously exchanged list is replayed back to the network in a secure fashion. This enables the network to verify that security mechanism selection was correct and that the security agreement was not tampered with. An example of an attack that would be possible without this feature is a "bidding-down attack", where an attacker forces peers into selecting a known weak security mechanism. The important benefit from having secure negotiation of the used security mechanism is that new mechanisms can be later added and old ones removed. The mechanisms can coexist nicely as each UE always uses the strongest mechanism it has available to it.

3.19.4.5 Confidentiality and integrity protection

In IMS access security, both confidentiality as well as data integrity and authentication are mandatory. The protocol used to provide them is IPsec ESP [RFC2406], explained in further detail in Chapter 22.

AKA session keys are used as keys for ESP SAs. The IK is used as the authentication key and the CK as the encryption key. Naturally, depending on the key lengths required by the cipher algorithms used in ESP, certain key expansion functions may be used on AKA session keys.

3.19.4.6 Key management and distribution

As described in previous chapters, the P-CSCF may also reside in the visited network. By virtue of the AKA protocol, the shared secret is only accessible in the home network, which means that, while authentication needs to take place in the home network, a certain delegation of responsibility needs to be assigned to the P-CSCF, as IPsec SAs exist between the P-CSCF and the UE. In practice, while IMS authentication takes place in the home network, the session keys that are produced in AKA authentication and used in ESP are delivered to the P-CSCF piggybacked on top of SIP registration messages.

To renew the SAs the network has to re-authenticate the UE. This means that the UE has to re-register as well, which may be either network initiated or due to the registration expiring. The net effect is the same: the AKA protocol is run and fresh keys are delivered to the P-CSCF.

3.19.5 IMS access security for HTTP-based services

3.19.5.1 Introduction

Parallel to SIP traffic, there is a need for the UE to manage data associated with certain IMS applications. The Ut interface, as explained in Section 2.3.15, hosts the protocols needed for that functionality. Securing the Ut interface involves confidentiality and data integrity protection of HTTP-based traffic [RFC2616]. As previously mentioned, user authentication and key establishment for the Ut interface are also based on AKA.

Figure 3.32   Generic bootstrapping architecture.

3.19.5.2   The Generic Bootstrapping Architecture (GBA)

As part of the GAA, the 3GPP IMS defines the Generic Bootstrapping Architecture (GBA) [3GPP TS 33.220], illustrated in Figure 3.32. The Bootstrapping Server Function (BSF) and the UE perform mutual authentication based on AKA, allowing the UE to bootstrap session keys from the 3G infrastructure. Session keys are the result of AKA and enable further applications provided by a Network Application Function (NAF). One such example is a NAF that issues subscriber certificates[7] using an application protocol secured by the bootstrapped session keys.

3.19.5.3   Authentication and key management

Authentication in the Ut interface is performed by a specialized element, called the "authentication proxy". In terms of the GBA the authentication proxy is another type of NAF. Traffic in the Ut interface goes through the authentication proxy and is secured using the bootstrapped session key.

3.19.5.4   Confidentiality and integrity protection

The Ut interface employs Transport Layer Security (TLS) for both confidentiality and integrity protection [3GPP TS 33.222]. TLS is discussed more thoroughly in Chapter 21.

---

[7] Such an entity is usually referred to as a PKI portal.

Part II

Services

4

Presence

Presence and instant messaging are changing the personal and corporate communications paradigm.

Presence will enhance messaging as well as introduce a new service (the presence service itself) that can be used in many other applications and services. Presence will be the heart of all communications and the new way for telephony to function. Presence will also be a lucrative business opportunity for both operators and service providers.

Presence is a dynamic profile of the user, which is visible to others and used to represent oneself, share information and control services. Presence can be seen as a user's status as perceived by others and others' status as perceived by the user. Status may contain information such as personal and device status, location or context, terminal capabilities, preferred contact method as well as services the user is willing to use to communicate with others, including voice, video, instant messaging as well as gaming.

Presence information is also personal: it is always linked to a particular person. It shows the person initiating the communication whether the other person is available and willing to communicate. On the other hand, presence information can be used to communicate to others when a person is able and willing to communicate as well as with whom and by what means. This will allow users to control their own communication more effectively.

Presence information sharing raises security and privacy concerns. Using Session Initiation Protocol (SIP) for presence, users can control their own specific presence information and have the final say on how it is used including who can and cannot see certain parts, if not all, of the presence information.

4.1 Who will use the presence service?

There will be many different user groups, which will use presence information for different purposes. These groups will range from corporate business users to teenagers and children. While presence is likely to be used more for rational availability management in corporate usage, young consumers are constantly seeking new ways of expressing themselves and building an identity in a fun and visually rich way. Successful presence services will be adaptable to the different needs of varied user groups and segments.

The IMS: IP Multimedia Concepts and Services Second Edition
Miikka Poikselkä, Georg Mayer, Hisham Khartabil and Aki Niemi © 2006 John Wiley & Sons, Ltd

Contacts Today   Presence Enhanced Contacts

Figure 4.1   Dynamic presence.

4.2   Presence-enhanced services

The basic presence service as we know it today works with the basic idea of "publish–receive" of presence information about humans. Operators have the ability to fetch various infrastructure-related presence attributes, such as location and terminal availability from their communication networks. This enhances the suite of presence information presented to subscribers who subscribe to the operator's presence service and gives the user a greater experience.

New presence-enabled services use presence information in dedicated application areas. This is a big opportunity for innovative companies that want to expand their business. Examples include presence-based call routing as well as network gaming services. In addition, presence creates an alternative advertising and information-sharing channel that the user is comfortable with.

With presence, team working is more efficient, bringing the ability to share information amongst the team members beyond their availability. They can share information such as a meeting location, future plans, etc.

4.3   Presence contributing to business

Presence can contribute to existing businesses and create a business of its own. There will be basic presence user services as well as new presence-enabled services. Operators and other service providers have a major role in mass adoption of presence services. The basic mobile presence service can be part of the operator's service portfolio, since other services – i.e., the new presence services – can use the basic service. The mobile domain, now reaching almost a billion subscribers worldwide, is a profitable platform for new consumer services.

Offering a basic presence service can give a competitive advantage for an operator over other operators who are not offering it: by binding their presence information to a particular operator's services, customers have high-value services that other operators may not be able to offer without that information. Presence generates new traffic for existing services like instant messaging. Presence also minimizes incomplete calls or calls being rejected due to the called party being busy.

Operators also need to carefully consider the pricing of presence allowing people to make an easy decision about adopting the presence service, without having to think for too long about the cost/benefit.

4.4 What is presence?

Presence is in essence two things: it involves making my status available to others and the statuses of others available to me. Presence information may include:

- person and terminal availability;
- communication preferences;
- terminal capabilities;
- current activity;
- location;
- currently available services.

It is envisioned that presence will facilitate all mobile communication, not only instant messaging, which has been the main driver for presence. Instant messaging has been the main interactive, almost real-time communication service in the Internet and presence is a great compliment in that you know if a friend is online before you begin a chat session with her. However, in the mobile environment, it is envisioned that presence information will not only support instant messaging, it will also be used as an indicator of the ability to engage in any session, including voice calls, video and gaming: all mobile communications will be presence-based.

Presence-specific and presence-enhanced applications and services will be available in the near future. A typical example of a presence-specific application will be a phonebook with embedded presence information, making it dynamic. Dynamic presence (Figure 4.1) will be the initial information the user sees before establishing communication. This information affects the choice of communication method and timing.

4.5 SIP for presence

The Session Initiation Protocol (SIP) has been extended for presence by the creation of an event package called "presence". Event packages are described in Section 12.13.1. When subscribing to such an event, a subscriber places the "presence" token in the Event header.

Some definitions have been created to describe the subscriber and the notifier for the purpose of presence:

- Presentity – the presence entity, a resource that provides presence information to a presence service.
- Watcher – the entity that requests information (presence), about resources (presentities).

Two SIP entities are defined for presence in [RFC3856]:

- Presence Agent (PA) – capable of storing subscriptions and generating notifications.
- Presence User Agent (PUA) – manipulates presence information for a presentity and publishes such presence information.

As mentioned earlier, the NOTIFY request body carries the state information. In this case, the state information is the presence state of a presentity. The Multipurpose Internet Mail Extension (MIME) type of such content is "application/pidf + xml" as defined in [RFC3863]. This presence Extensible Markup Language (XML) document can be extended to carry more information than defined in [RFC3863].

The presentity uploads presence information using the PUBLISH method, which is described in Section 12.13.2.

4.6   Presence service architecture in IMS

A user's presence information can be obtained from a multiplicity of entities in the Internet Protocol Multimedia Subsystem (IMS) network: it could be a PUA located in a foreign network, a PUA at the terminal or a PUA located as an entity in the network. The presence server is an example of an IMS application server. Watchers may be in the same home domain as the presentity or in a foreign domain.

Figure 4.2 represents a reference architecture to support a presence service in the IMS.

Figure 4.2   Reference architecture to support a presence service in IMS.

The entities are defined as follows:

- Presence server – manages presence information uploaded by PUAs and handles presence subscription requests.
- Watcher presence proxy – identifies the target network for a presentity and resolves its address.
- Presentity presence proxy – identifies the presence server assigned to a certain presentity.
- PUAs – assemble and provide presence information to the server.

4.7 Presentity list

It is envisioned that users (watchers) will have many presentities (friends) whose presence information is of interest to them. For congestion control and bandwidth limitations, it is uneconomical to have the user's terminal send a multiplicity of SUBSCRIBE requests, one for each presentity.

To resolve this problem, Group Management solutions were created. Section 8.3 discusses Resource lists in more detail. A presentity list is a type of resource list:

- Application Usage ID (AUID) – "resource-lists".
- Additional constraints – none.
- Naming convention – none.
- Resource interdependencies – the list is represented by a Universal Resource Identifier (URI). If the client does not populate the XML element carrying the URI value, the XML Configuration Access Protocol (XCAP) server needs to do so.
- Authorization policy – default.

The Ut interface in the IMS architecture is used to manipulate resource lists.

4.8 Setting presence authorization

Presence information can be available at different levels and different scopes to different watchers. This means that different watchers may be authorized to view different parts of the presence information of a presentity. The choice of who sees what belongs to the presentity. The presentity can set such authorization levels using an XCAP-defined solution in the form of permission statements.

[Draft-ietf-geopriv-common-policy], [Draft-ietf-simple-presence-rules], [Draft-ietf-simple-common-policy-caps] and [Draft-ietf-simple-pres-policy-caps] define the XML schema along with its semantics as well as those sections that are mandatorily defined for XCAP usage.

4.9   Publishing presence

Publishing presence information can be achieved using the SIP extension defined in
Section 12.13.2. The Event header of a PUBLISH request carries the "presence" token.
The default expiration time of a publication is 3,600 seconds. The body of a PUBLISH
request carrying presence information is of MIME type "application/pidf + xml".

4.10   Watcher information event template package

Users subscribe to receive information about the state of a particular resource using an
event package (referred to as the "main package" in this section). These subscribers can
be referred to as "watchers". A watcher information template package allows a user to
gain knowledge about watchers and the state of their subscriptions to the main
package. A watcher information template package is identified with the token "winfo".

The primary use of this template package is for presence. Users subscribe to this
template package to see who is subscribing to their presence information and the state
of that subscription.

The information carried to the watcher information subscriber contains two important items: the status of each subscription made by the watchers of the main package
and the event that caused the transition from the previous status to the current one.

The states of the watcher information package are described in [RFC3857] as follows:

- Init – no state is allocated for a subscription.
- Terminated – a policy exists that forbids a watcher from subscribing to the main
  event package.
- Active – a policy exists that authorizes a watcher to subscribe to the main package.
- Pending – no policy exists for that watcher.
- Waiting – similar to pending, but tells the template package subscriber that a user has
  attempted to subscribe to the main package and that the subscription expired before
  a policy was created.

4.11   Example signalling flows of presence service operation

4.11.1   Successful subscription to presence

Figure 4.3 shows an example flow of a watcher who has successfully subscribed to the
presence information of a presentity residing in a different network while the watcher
resides in her home network. The flow shows an initial NOTIFY request to the SUB-
SCRIBE request.

4.11.2   Successful publication of presence information

Figure 4.4 shows an example flow of a presentity that has successfully published
presence information. In this scenario the User Equipment (UE) is behaving as a

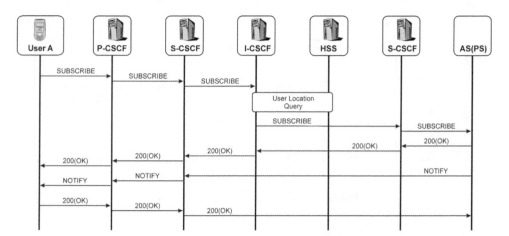

Figure 4.3 Successful subscription to presence.

Figure 4.4 Successful publication.

PUA. This typically results in the Packet-Switched (PS) domain generating notifications to watchers.

4.11.3 Subscribing to a resource list

Figure 4.5 shows an example flow of a watcher who is subscribing to the presence information of a resource list that was created earlier (possibly using XCAP). The list of presentities (resources) is referenced using a SIP URI. The flow shows the immediate NOTIFY request that is sent after receiving the SUBSCRIBE request. If the Resource List Server (RLS) does not hold any presence information about the resources in the list, then the NOTIFY request may carry an empty body.

4.11.4 Subscribing to watcher information

Figure 4.6 depicts how the message actually flows between the PUA and the PS domain when a user subscribes to receive notifications about the changes of state of the watchers. The path of the messages is the same as that of the PUBLISH request. The flow shows the immediate notification following a successful subscription request. Figure 4.7 summarizes the interactions between the presence server, watcher and presentity.

Figure 4.5   Subscription to a resource list.

Figure 4.6   Subscription to watcher information.

Figure 4.7   Presence server, watcher and presentity interaction.

5

Messaging

There are currently many forms of messaging services available. In general, messaging entails sending a message from one entity to another. Messages can take many forms, include many types of data and be delivered in various ways. It is usual to have messages carry multimedia as well as text and be delivered either in near-real time as in many instant messaging systems or into a mailbox as in email today. In this chapter we give some details about messaging in the Internet Protocol Multimedia Subsystem (IMS) context.

5.1 Overview of IMS messaging

IMS messaging takes three forms:

- immediate messaging;
- session-based messaging;
- deferred delivery messaging.

Each form of IMS messaging has its own characteristics; so, even though messaging in its simplest form can be thought of as a single service – after all, all forms of messaging are really about sending a message from A to B – the fact that these characteristics differ makes them each a service on their own. However, the way in which applications are built on top of these services may well hide the fact that these are different forms of messaging. In fact, one of the key requirements for IMS messaging is easy interworking between different messaging types.

5.2 IMS messaging architecture

Of the three IMS messaging types, immediate messaging and session-based messaging utilize the IMS architecture directly. Deferred delivery messaging utilizes the Packet-Switched (PS) domain as well, even though it is a separate infrastructure to the IMS.

The IMS: IP Multimedia Concepts and Services Second Edition
Miikka Poikselkä, Georg Mayer, Hisham Khartabil and Aki Niemi © 2006 John Wiley & Sons, Ltd

5.3   Immediate messaging

Immediate messaging, or page-mode messaging, is the familiar instant messaging paradigm adopted in the IMS framework. It uses the Session Initiation Protocol (SIP) MESSAGE method (see Section 12.13.3 for details) to send messages between peers in near-real time. Figure 5.1 illustrates a typical message flow.

Figure 5.1   Immediate messaging flow.

In immediate messaging, the User Equipment (UE) simply generates a MESSAGE request, fills in the desired content – which typically consists of text, but can also contain snippets of multimedia such as sounds and images – and populates the request-URI (Uniform Resource Identifier) with the address of the recipient. The request is then routed via the IMS infrastructure similar to the manner used for an INVITE, until the immediate message finds its way to the UE of the recipient user.

There might, of course, be a reply to this message; in fact, a full dialogue of immediate messages back and forth between the two users is likely. However, in contrast to session-based messaging, the context of this session only exists in the minds of the participating users. There is no protocol session involved: each immediate message is an independent transaction and is not related to any previous requests.

If an immediate message is received while the IMS subscriber is offline, or in an unregistered state, the MESSAGE will route to an Application Server (AS). The AS can hold the message in storage, and when the user registers, the AS can then deliver the message to its final destination.

Usually, immediate messages are addressed to another peer's public user identity. However, the user can also send a single message to a number of recipients using a list server extension in the IMS. Basically, an IMS user can create an alias, using a SIP address in the form of a Public Service Identifier (PSI), and populate this alias with a set of SIP URIs of the intended membership. Whenever a MESSAGE method is sent to the PSI corresponding to this list, the request is routed to the list server. The list server, which is an AS itself, will intercept the message and generate a new request for each of the members of the list.

5.4   Session-based messaging

Session-based messaging relates to a familiar paradigm of messaging already in use in the Internet: Internet Relay Chat (IRC) [RFC2810]. In this mode of messaging the user

Figure 5.2   Session-based messaging flow.

takes part in a session in which the main media component often consists of short textual messages. As in any other session a message session has a well-defined lifetime: a message session starts when the participants begin the session and stops when the participants close the session. After the session is set up – using SIP and SDP between the participants – media then flows directly from peer to peer. Figure 5.2 illustrates the typical message flow of a message session.

Session-based messaging can be peer to peer, in which case the experience closely mimics that of a normal voice call. An ordinary invitation to a session is received, the only difference being that the main media component is a session of messages. However, this is not an actual limitation to session-based messaging, since it is, of course, possible to combine other media sessions with message sessions. In fact, many useful and exciting applications are enabled by this functionality: for example, video calls with a text side channel might be a valuable application for hearing-impaired people.

The actual protocol for conveying the messages within a session is called the Message Session Relay Protocol (MSRP) [Draft.ietf-simple-message-sessions]. MSRP layers on top of Transmission Control Protocol (TCP), and can carry any Multipurpose Internet Mail Extensions (MIME) encapsulated data. Messages can be of arbitrary size, since one of the protocol features is the ability to support sending a complete message in small chunks that are automatically reassembled at the recipient end.

Session-based messaging forms a natural unison with conferencing as well. Using the conferencing functionality, session-based messaging can turn into a multiparty chat conference. In this mode of operation, session-based messaging can enable applications similar to modern day voice conferences. A chat conference can also be comparable with a channel in the IRC.[1] A service provider will typically offer the possibility for users to have both private chats, where the set of participants is restricted, and public chats, some of which are maintained by the service provider.

5.5 Deferred delivery messaging

Deferred delivery messaging is, in fact, the well-known Multimedia Messaging Service (MMS). It was decided that the requirements put forth in the Third Generation Partnership Project's (3GPP) IMS Messaging Stage 1 should coincide with those of the MMS. In practice, this means that MMS is used for the deferred delivery mode in IMS.

---

[1] In this respect a channel such as #Helsinki on an IRC server could be simply represented by a SIP URI: *sip:helsinki@some.chat.net*, or a chat group in a typical Internet chat service.

6

Push to talk Over Cellular

Push to talk over Cellular (PoC) provides a direct one-to-one and one-to-many voice communication service. The idea is simple. Users select the individuals or groups they wish to talk to, and then press the push to talk key to start talking. The session is connected in real time. Push to talk sessions are one-way communication: while one person speaks, the other(s) only listens. The turns to speak are requested by pressing the push to talk key and granted on a first-come-first-served basis. Push to talk speech is usually connected without the recipients answering and heard through the phone's built-in loudspeaker. Alternatively, a user can choose to receive push to talk sessions only after accepting an invitation. If more privacy is needed, they can also listen to sessions through an earphone or headset.

The push to talk service is based on multi-unicasting. Each sending client sends packet data traffic to a dedicated push to talk application server and, in the case of a group session, the server then duplicates the traffic to all the recipients (see Figure 6.1). No multicasting is performed either in the access or core network and mobility management is carried out by the radio network. This is why the push to talk solution works transparently over cellular and fixed networks. PoC session control and other signalling is based on Session Initiation Protocol (SIP) and voice traffic is carried through a Real-time Transport Protocol/Real-time Transport Control Protocol (RTP/RTCP) based streaming bearer.

PoC uses cellular access and radio resources more efficiently than circuit-switched services. Network resources are reserved one-way for the duration of talk bursts, rather than two-way for an entire session. Figure 6.2 shows an example. Compared with conventional two-way radio solutions, such as Land Mobile Radio (LMR) and Professional Mobile Radio (PMR), as well as the Family Radio Service (FRS), PoC provides better coverage as it utilizes coverage provided by GSM/WCDMA/CDMA networks. It allows users to make push to talk sessions between two people or within a group of people over nationwide networks and across regional borders (GPRS EDGE/WCDMA/CDMA2000).

6.1 PoC architecture

Open Mobile Alliance (OMA) PoC standard Release 1 architecture is based on PoC clients, PoC application server and PoC XML Document Management Server

Figure 6.1 Push to talk Over Cellular.

Figure 6.2 Voice call versus Push to talk Over Cellular.

(XDMS). XDMS can be considered as a means of application configuration setting management as it stores application-specific configuration settings. An XDMS server that stores PoC-specific data is called a "PoC XDMS". Using *reference point POC-8* (see Figure 6.3), the PoC server can fetch PoC-related documents (e.g., access lists) and using *reference point POC-5* (see Figure 6.3), the PoC server can fetch generic lists (e.g., members of *myfriends@example.com* list) from a shared XDMS. The PoC servers handle application-specific tasks such as talk burst control (the reservation of talk burst for one user at a time) and PoC session control. They also provide interfaces to the operator's provisioning and network management systems and create application-specific Charging Detail Records (CDRs). The PoC server is connected to the IMS via the IMS Service Control reference point. The IMS takes care of common functions, such as user authentication for PoC, session routing and generic charging based on SIP. PoC client is usually software in the User Equipment (UE) but it could be an application (also in the PC).

Usually, a presence service is associated with PoC, as presence adds value to PoC (e.g., users are able to learn another user's willingness and availability for PoC communication). Even though the PoC service works without presence, it is still shown here in the architecture (Figure 6.3).

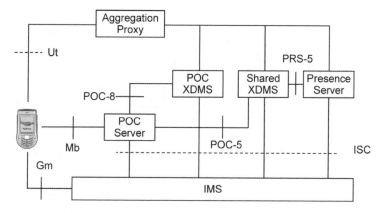

Figure 6.3   Push to talk Over Cellular architecture.

6.1.1   PoC server

A PoC server is an application server in the IMS architecture that provides the PoC service for users. It controls the PoC session setup procedure, enforces policy defined for PoC group sessions (e.g., who is allowed to join, who is allowed to invite more members, decides whether a session should be release when a particular user leaves, decides whether other users should be invited when a particular user joins), provides information about group users (e.g., informs when somebody joins in or leaves a group). Furthermore, the PoC server takes care of media distribution and adaptation, if needed. Moreover, it acts as a control point for talk bursts – i.e., the PoC server decides who has the right to send media, informs other users that permission to send media has been granted to somebody, etc. This mechanism is called "Talk Burst Control" (further described in Section 6.3.2). So, in short, a PoC server handles both control-plane and user-plane traffic associated with the PoC service, and for this purpose it uses IMS reference points ISC and Mb.

In OMA two different PoC server roles have been defined: participating PoC function and controlling PoC function. The assigment of the PoC server role takes place during a PoC session setup in such a way that there will be only one PoC server performing the controlling PoC function and two or more PoC servers performing the participating PoC function depending on the number of PoC session participants. In case of a one-to-one PoC session and an *ad hoc* PoC group session the PoC server of the inviting user performs the controlling PoC function. In case of a chat PoC group and pre-arranged group session the PoC server owning/hosting the group identity performs the controlling PoC function [OMA PoC AD].

SIP signalling from PoC clients always goes first of all to a participating PoC function, which sends SIP signalling traffic further on towards a controlling PoC function. In contrast, PoC clients may have a direct media and media signalling connection to the controlling PoC function. Figure 6.4 shows the architecture. The distribution of the different functions of a PoC server is summarized in Table 6.1.

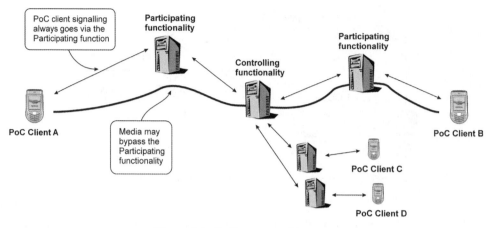

Figure 6.4 PoC server architecture.

Table 6.1 PoC server functional distribution.

Function	Role
SIP session handling	Both
Provides for privacy of the PoC addresses of participants	Both
Supports user-plane adaptation procedures	Both
Supports Talk Burst Control Protocol negotiation	Both
Provides transcoding between different codecs (optional)	Both
Provides policy enforcement for participation in group sessions (e.g. who is allowed to join)	Controlling
Provides information to the participants	Controlling
Provides the centralized media distribution	Controlling
Provides the centralized talk burst control functionality including talker identification	Controlling
Provides centralized charging reports	Controlling
Collects and provides centralized media quality information	Controlling
Provides policy enforcement for incoming PoC sessions (e.g. access control, incoming PoC session barring, availability status, etc.)	Participating
Stores the current answer mode, incoming PoC session barring and incoming instant personal barring preferences of the PoC client	Participating
Provides participant charging reports	Participating
Provides filtering of media streams in the case of simultaneous PoC sessions (optional)	Participating
Provides the talk burst control message relay function between PoC client and PoC server performing the controlling PoC function (optional). *Note.* If the participating PoC server stays on the media path then this is a mandatory function	Participating

6.1.2   PoC client

The PoC client according to the OMA standard is a functional entity on the UE that is able to register itself to IMS using a PoC feature tag and, indicating a PoC release version in SIP REGISTER request, initiates/modifies/releases PoC sessions, supports user-plane procedures (e.g., sending and receiving talk bursts to/from PoC server, supports talk burst control mechanisms, user-plane adaptation), supports the capability to set PoC service settings and receives instant personal alerts.

6.2   PoC features

6.2.1   PoC communication

PoC supports various types of communication models to meet the differing needs of group communication. The main differences between these models relate to group policy and session setup. In other words, how do users create a group and add/delete group members? How do they activate a group session and how is access control arranged? In dial-out group communication, a user invites a group of users to participate in a group session. The invited users receive an indication of the incoming session and they may join in using either an automatic or manual answer. The invited group may be a *pre-arranged PoC group* or a list of users selected from the calling user's phone book as a temporary arrangement (so-called *ad hoc PoC group*). In the latter case, in particular, the ability to see other users' availability, or presence statuses, before making the dial-out session brings clear additional value for the user. A dial-out session suits unplanned situations or cases where the participants must be handpicked.

There are some special rules for pre-arranged PoC groups (defined in OMA). First, a PoC session between group members is established when any individual member of the same pre-arranged PoC group invites other members of the group to join in. Second, the communication starts after the first pre-arranged group member accepts the invitation and the controlling PoC server grants media permission to the initiator of the pre-arranged group. Third, participation in a pre-arranged PoC group session is only allowed for pre-defined members (i.e., members of the pre-arranged group) [OMA POC RD]. Similarly, there are some rules for *ad hoc* PoC groups as well. An *ad hoc* PoC group is created when a PoC user invites one or more users to a PoC session (note that a one-to-one PoC session is an *ad hoc* session having two participants). Only those users that have been invited to the *ad hoc* PoC session are allowed – i.e., a user must have received a request to join (such as a SIP INVITE or SIP REFER from the controlling PoC server) the *ad hoc* group from one of the current members of the *ad hoc* PoC group. A local policy at the controlling PoC server may only allow the *ad hoc* PoC group initiator to add more users (for charging reasons) [OMA POC RD].

In join-in group communication, *chat group*, the participants themselves explicitly join a PoC group session for communication. In this way users have full control over which groups they participate in. They will never receive any traffic unless they have joined the group. Join-in operation is well suited for communication during any routine or pre-planned activities. Participation in a chat PoC group is analogous to real-life activities such as watching TV, going to a movie or participating in a meeting. Chat

Figure 6.5 Different PoC communication models.

PoC sessions can go on for hours, with actual communication comprising only a small portion of the total session time. So, being in a chat PoC session should not prevent a user receiving other sessions in the meantime. The user should also be able to participate in several chat sessions simultaneously (e.g., "my family", "basketball friends", "beer buddies"). This requires simultaneous session support.

A chat group can be an unrestricted group without any access control or a restricted group with a list of members. Unrestricted groups are open to anyone who knows the group identification (SIP URI of the group). The group identification can be found, for instance, on an operator portal or chat room. Unrestricted PoC chat groups are suitable as open discussion forums on general and specific topics (e.g., fishing, cars, football). Restricted groups are groups where access is limited to pre-defined users. For restricted groups where access control methods are used, see Section 6.2.4. To join a restricted group, users need to know the group identification (SIP URI of the group) and they need to have the right to join the group session. Restricted PoC chat groups are best suited to the needs of business users where continuous communication within secure groups is needed to support daily work. Figure 6.5 summarizes the different PoC communication models.

6.2.2   Simultaneous PoC sessions

Compared with the traditional telephony service, PoC offers the capability to participate in more than one PoC session at the same time without placing any of the sessions on hold. This capability is called a "simultaneous PoC session functionality". For instance, a user, Alice, could have the following functionality on her PoC device: it automatically joins in with those groups that Alice has pre-configured when the device is turned on. Let's assume that Alice has three groups that she actively follows: "my family", "work colleagues" and "basketball team". After joining the groups Alice is able to send and receive media from any of them. This functionality has a clear benefit over the single-session model, as Alice does not need to guess which group is active in a certain time period. Moreover, this model also allows users to hang on in chat groups and still be able to receive one-to-one PoC sessions with other users.

When Alice wants to speak she just chooses the correct group and presses the PoC button. Receiving media from the network is a bit more challenging in that it requires support from Alice's PoC server. The PoC server needs to filter incoming PoC traffic if there is incoming media in more than one PoC session in which Alice is participating at the same time. The participating PoC function filters the traffic so that Alice hears a single conversation. The following rules govern traffic filtering (in order of preference):

- The user can lock herself into a single group. Only traffic from that group will be delivered to the user (analogous to what happens in a telephony service).
- The user can set one of the sessions as a primary PoC session. Traffic in the primary session is delivered to the user (if the user is not currently speaking in the secondary session), although there is ongoing traffic in a secondary PoC session.
- Among secondary PoC sessions, the traffic of ongoing conversation is delivered as long as the conversation remains active. After a silent period (length is defined by the operator), the PoC server selects active media from another session.

Our example user, Alice, would select "my family" as a primary PoC group and the rest of the groups would then be automatically classified as secondary groups. Now she will hear media from her family members whenever they have something to say. To set up the "my family" group as a primary session, her device needs to include value "1" in a specific attribute, a = poc_sess_priority = "1/0" in the SDP payload of the session request (SIP INVITE, UPDATE or RE-INVITE). To switch the priority value, "0" can be used or the device can indicate priority "1" to other groups and, then, the PoC Server is assumed to swap the priorities.

When Alice wants to concentrate on single-group communication, her device needs to include value "1" in a specific attribute, a = poc_lock = "1/0" in the SDP payload of the session request (SIP INVITE, UPDATE or RE-INVITE). To unlock the session, her device needs to assign locking to other sessions or send another session request and indicate value "0" [OMA POC Control Plane].

6.2.3 PoC session establishment models

In this section different session establishment models are explained. The reader is recommended to read the section on PoC communication (see p. 137) before exploring this section.

Two different session models exist: on-demand and pre-established session. The main difference between session models is media parameter negotiation. In a pre-established session model, a user establishes a session towards her participating PoC function and negotiates all media parameters prior to making requests for PoC sessions to other PoC users. In an on-demand model, the "normal" SIP method is used – i.e., media parameters are negotiated when a user makes a request for a PoC session. The pre-established session model allows a PoC client to invite other PoC clients or receive PoC sessions without negotiating the media parameters again. This will bring additional savings in session setup time. Figure 6.6 shows a simplified example of pre-established session usage.

Figure 6.6   Pre-established PoC session setup.

In the upper part of Figure 6.6, PoC clients establish pre-established sessions. In the lower part of the figure, Tobias wants to contact a user named Tuomo. Tobias's PoC client issues a SIP REFER request that contains Tuomo's identity as a target user. Tobias's PoC server will play the role of controlling PoC function, in addition to participating as the PoC function, and sends a SIP INVITE request via IMS towards the terminating network. Tuomo's PoC server (participating) gets the SIP INVITE request and accepts the session immediately, as Tuomo is using a pre-established session and has set his answer mode to auto-answer. Tuomo's PoC client receives a talk burst control message from his PoC server (participating). The talk burst message indicates that a call is coming from Tobias. Tuomo's PoC client acknowledges this message. When the controlling PoC function gets a 200 OK, it will send a talk burst control message to Tobias indicating that permission to speak has been granted. Similarly, the controlling PoC function sends a talk burst control message to Tuomo's PoC client indicating that permission to send media has been granted to Tobias.

Figure 6.7 shows a PoC session setup in which an on-demand PoC session setup model is used. When Tobias wants to a have a PoC session with Tuomo his device generates a SIP INVITE request including the device's media capabilities and media transport information. Tobias's PoC server will play the role of controlling PoC function, in addition to participating as the PoC function, and sends a SIP INVITE request via IMS towards the terminating network. Tuomo's PoC Server (participating) gets the SIP INVITE request and becomes aware of Tuomo's answer mode setting. In this case Tuomo had earlier indicated auto-answer mode and, therefore, his PoC server returns a 183 Session Progress SIP response back to the controlling PoC server. This response contains the information that Tuomo is using auto-answer mode. At the same

Figure 6.7 On-demand PoC session setup using an unconfirmed mode in the terminating network.

time, Tuomo's PoC server sends a SIP INVITE request to Tuomo's device. After receiving the INVITE request Tuomo's device automatically responds with a 200 OK SIP message. When Tuomo's PoC server gets the 200 OK response it sends a 200 OK response towards the originating network. At the time of receiving the 183 Session Progress response the controlling PoC function sends a 200 OK response to the initiator (Tobias). The controlling PoC function may grant permission to speak (talk burst control message in Figure 6.7) to the originating user (Tobias) after receiving the 183 Session Progress response if it is willing to buffer media streams. Buffering is needed as the controlling PoC function is not allowed to forward media to recipient participating PoC function(s) before they have responded with a 200 OK SIP final response. If the controlling PoC function is unwilling to buffer media it grants permission to speak to the originating user when the first 200 OK SIP response is received.

In addition to the two examples described above, there is a number of different combinations as the originating user may use either the on-demand or pre-established model and similarly the terminating user may use either on-demand or pre-established models. In addition, the calling user is able to request the called user's PoC device to answer the incoming session automatically – i.e., the inviting PoC user's speech is immediately audible at the called PoC user's device without any action by the called PoC user. This feature is called as "the manual answer override feature". Authorization of the manual answer override takes place in the terminating user's network – i.e., users must set proper authorization rules if they want to allow somebody to use the manual answer override. Details on how users can request a manual answer override were still under discussion at the time of writing.

Moreover, the terminating user may user either a manual or automatic answer mode. Table 6.2 shows possible PoC session combinations.

6.2.4   Incoming PoC session treatment

At the beginning of the chapter it was stated that push to talk speech is usually connected without the need for the recipient to answer and is heard through the PoC device's built-in loudspeaker. This occurs when a user has set her terminal to an auto-answer mode, the caller is included in the user's access control list and the caller has a

Table 6.2 Summary of different PoC session setup combinations.

Originating end	Terminating end
Pre-established session	Pre-established session (auto-answer)
Pre-established session	Pre-established session (manual-answer)
Pre-established session	On-demand session (auto-answer)
Pre-established session	On-demand session (manual-answer)
On-demand session	On-demand session (auto-answer)
On-demand session	On-demand session (manual-answer)
On-demand session	Pre-established session (auto-answer)
On-demand session	Pre-established session (manual-answer)

specific value in the access control list. In this section we explain how this can be done and what other mechanisms are in place to control incoming PoC sessions.

Let's start with answering modes. Two different answer modes have been defined: auto-answer and manual answer mode. When auto-answer mode is turned on, the PoC device will accept the incoming PoC sessions without waiting for any specific actions from the user – i.e., incoming media streams can be played immediately. Manual answer mode is an ordinary mode in which a user needs to accept an incoming PoC session request before the PoC device confirms the acceptance of an incoming PoC session to the PoC server. The answer mode used is also signalled back to the PoC server performing the controlling PoC and, further, to the originating user function using Unconfirmed[1] or Confirmed[2] indications in the SIP response. A Confirmed indication is given when the terminating PoC device is using manual answer mode and has accepted a new PoC session by sending a SIP 200 OK response. Otherwise, an Unconfirmed indication is generated by the terminating participating PoC server function. Section 6.4 describes how a PoC client informs the PoC server about the desired answer mode.

Using auto-answer mode when you are available would be a nice and useful feature when you can be sure that the calling user is going to behave correctly – e.g., the caller would not say something improper. However, users cannot be sure who might call them and, therefore, they may not feel comfortable to use auto-answer mode for all possible users. On the other hand, using manual answer mode all the time is not convenient either. Moreover, a user may want to automatically decline PoC sessions from certain users or PoC groups.

---

[1] An Unconfirmed indication is an indication returned by the PoC server to confirm that it is able to receive media and believes the PoC client is able to accept media; the PoC server sends the Unconfirmed indication prior to determining whether all egress elements are ready or even able to receive media.

[2] A Confirmed indication is a signalling message returned by the PoC server to confirm that the PoC server, all the other network elements intermediary to the PoC server and a terminating PoC client are able and willing to receive media [OMA PoC Control Plane].

In order to overcome these types of issues, the access control mechanism was developed. Access control means that a user can:

- allow or block, selectively, incoming PoC sessions from other PoC users and PoC groups; and
- define, selectively, those users whose PoC sessions are to be accepted automatically.

Access control is executed at the PoC server performing the participating role for the called PoC user. In order to execute access control a PoC user needs to create an access control list document and send it to the PoC XDMS. The way to do this is further described in Section 8.6. In addition, a PoC user is able to bar *all* incoming PoC sessions by activating Incoming PoC Session Barring (this procedure is covered in Section 6.2.7).

The result of answer mode and access control list usage at the PoC server and PoC client is perhaps best illustrated by a simple example. Tobias has just bought a new PoC-enabled device, as most of his friends are using them. Tobias's device is turned on and it is registered to the IMS. First, Tobias opens a window and sets up an access control list for his buddies. He decides to include Theresa, Tuomo and Matias as his closest friends and, therefore, sets their identities in the "accept" category in his access control list. At the same time, he inserts John's identity in the "reject" category as he does not want to receive any sessions from him. After completing the access control list, Tobias accepts the list and his device uploads the document to the PoC XDMS. Second, Tobias chooses to set his answer mode to auto-answer mode. This information is then sent to Tobias's PoC server by his device. Now that all the necessary settings are done, Tobias is ready to make use of his settings.

After a short while (maybe after sending an instant personal alert) Tuomo tries to reach Tobias. A one-to-one PoC session request is sent from Tuomo's device towards Tobias's device. When it reaches Tobias's PoC server, the PoC server will fetch the access control list from the PoC XDMS and discover that Tuomo is included in the "accept" category in the access control list and that Tobias has indicated a willingness to accept incoming PoC sessions automatically; therefore, the PoC server decides to let this PoC session go through. When the PoC server sends the incoming PoC session request to Tobias's PoC device it includes an auto-answer indication in the request. Based on the auto-answer indication, Tobias's phone accepts the PoC session, activates the loudspeaker and waits for the incoming PoC speech stream.

Later on the same day, one of Tobias's friends, Marja, who has subscribed to Tobias's presence information, realizes that Tobias's presence information shows PoC availability. After realizing this, Marja attemps to reach Tobias. When the PoC session request reaches Tobias's PoC server, this time the PoC server is unable to find Marja in the access control list. Nevertheless, Tuomo has indicated a willingness to accept incoming PoC sessions automatically, so the PoC server decides to accept the incoming PoC session but, compared with the previous case, it will add a manual answer indication in the request. Based on the manual answer indication Tobias's phone will not accept the PoC session immediately; instead it alerts Tobias and displays that the caller is Marja.

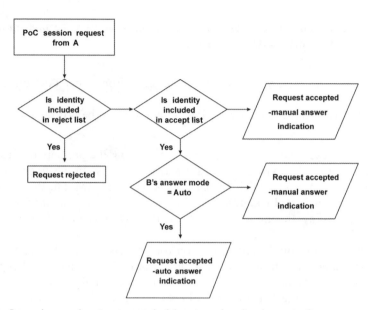

Figure 6.8 Incoming session treatment decision tree showing impact of access control list and user's answer mode.

When John tries to reach Tobias, the PoC server will realize that John is included in the reject category and it will automatically reject the incoming PoC session from John. This time the PoC session request does not reach Tobias's device at all.

Figure 6.8 shows incoming session treatment using access control lists and answer modes.[3]

6.2.5   Instant personal alerts

Sometimes a caller user is not able to reach a recipient (e.g., the user has Incoming Session Barring activated), then a caller may wish to leave an explicit message to request the called party to call back. This message is called an "instant personal alert". The instant personal alert request can also be used instead of a push to talk voice transmission when a less noticeable method of alerting is desired.

Whenever users want to send an instant personal alert request to another user, users select the target (e.g., from the phonebook) and send the instant personal alert request. For this purpose a SIP MESSAGE method is used. From a recipient point of view there is a need to differentiate between an ordinary SIP instant message and PoC instant personal alert, as both use the SIP message method. Differentiating PoC instant personal alerts makes it possible to create automatic operations at the PoC client. For this purpose the originating PoC client is mandated to include an Accept-

---

[3] If a manual answer override is requested and the calling user is allowed to use this feature, then an automatic answer mode is applied. If the request is not allowed the answer mode selection follows Figure 6.8.

Contact header with the PoC feature tag, "+g.poc.talkburst" along with "require" and "explicit" parameters:

```
MESSAGE sip:bob@example.com SIP/2.0
From: <sip:tobias@home1.fr>;tag=31415
To: <sip:bob@example.com>
Accept-Contact: *;+g.poc.talkburst;require;explicit
User-Agent: PoC-client/OMA1.0
```

If a user does not want to receive instant personal alerts, then she can activate Incoming Instant Personal Alert Barring. When this service is activated, the participating PoC server serving the user is mandated to block the delivery and to return a SIP 480 Temporarily Unavailable error response to the sender [OMA POC Control Plane].

6.2.6   Group advertisement

When a PoC user, Tobias, wants to plan his summer vacation to Finland he decides to create a chat PoC group with his sister Theresa and his Finnish friends Tuomo and Matias. After creating a group, *summervacation@poc.example.com*, he decides to advertise the created group to Theresa, Tuomo and Matias. To do this, Tobias sends a group advertisement that contains all the necessary information about the created group: name of the group (URI and display name), type of group, advertisement text. When the other group members receive a group advertisement, they may decide to store the contact information for further communication. Using the received information they could immediately decide to dial into the group.

A group advertisement could be sent to one or more users or it could be sent to all group members at once using a SIP MESSAGE that has a PoC-specific content in the form of a Multipurpose Internet Mail Extension (MIME) vnd.poc.advertisement + xml body. Additionally, a sender must include an Accept-Contact header with the PoC feature-tag "+g.poc.groupad" along with the "require" and "explicit" parameters. Using a feature tag and the above-mentioned parameter guarantees that the advertisement is only delivered to PoC-enabled UEs that are able to understand the group advertisement message.

Our example group advertisement, as sent by Tobias's PoC device, looks like:

```
MESSAGE sip:summervacation@poc.home1.fr SIP/2.0
From: <sip:tobias@home1.fr>;tag=31415
To: <sip: summervacation@poc.home1.fr >
Accept-Contact: *;+g.poc.groupad;require;explicit
User-Agent: PoC-client/OMA1.0
Content-Type: application/vnd.poc.group-advertisement+xml
Content-Length: (482)
```

```
<?xml version="1.0" encoding="UTF-8"?>
<group-advertisement
            xmlns="urn:oma:params:xml:ns:poc:group-advertisement"
            xmlns:xsi="http://www.w3.org/2001/XMLSchema-instance"
            xsi:schemaLocation=
            "urn:oma:params:xml:ns:poc:group-advertisement">
   <note>Tobias summer vacation planning</note>
   <group type="dialed-in">
     <display-name>Great trip to Finland</display-name>
     <uri>sip:summervacation@poc.example.com</uri>
   </group>
</group-advertisement>
```

Group advertisement XML documents are based on XML 1.0 and use UTF-8 encoding. The namespace URI for elements defined by an OMA PoC control-plane document is a URN, using the namespace identifier "oma". This URN is: urn:oma: params:xml:ns:poc:groupadvertisement. The key PoC elements in the group advertisement XML document are:

- < note > element – advertisement text added by a sender;
- < group > element – contains information about type of group (dialled-in, dialled-out, other), display name of the PoC group, and URI of the PoC group.

When "dialled-in" is used in the type attribute it indicates a chat PoC group. Similarly, "dialled-out" means a pre-arranged PoC group. "Other" is added for future expandability [OMA POC Control Plane].

Support of this feature is optional for both PoC clients and PoC servers in OMA PoC specifications. Controlling PoC servers may support delivery of Group Advertisement messages to all PoC group members and may apply different authorization rules that stipulate who is allowed to send group advertisement information to all group members. The most obvious authorization rules are:

- only the group owner is allowed to send Group Advertisement information to all group members;
- all group members are allowed to send Group Advertisement information to all group members.

If the PoC client does not want to receive group advertisement messages then it shall not include a PoC feature-tag "+g.poc.groupad" in the SIP REGISTER request.

6.2.7 Barring features

A PoC user can selectively block PoC incoming sessions using an access control list, as explained in Section 6.2.4. In addition, a PoC user is able to instruct a PoC server to reject all new incoming PoC sessions. This feature is called Incoming Session Barring (ISB). When barring is activated, the participating PoC server will reject all incoming session requests with a SIP 480 Temporarily Unavailable response and, at the same time, the participating PoC server will keep ongoing PoC session(s) intact and deliver incoming Instant Personal Alerts to the user. If the user wants to block Instant Personal Alerts as well, then she needs to activate Instant Personal Alert Barring. Section 6.4 shows how barring can be activated and de-activated.

6.2.8 Participant information

A PoC user can request information about PoC session participants and their status in the PoC session. The PoC client uses the conference-state event package to learn about changes in PoC participants: in other words, a user can learn, through notifications, who has joined or left a conference. This event package also allows participants to learn the status of a user's participation in a conference (on-hold, alerting).

Users can subscribe to a conference state by sending a SIP SUBSCRIBE request to the conference URI that identifies the controlling PoC server. The controlling PoC server acts as a notifier for this event package.

The name of this event package is "conference". This token appears in the Event header of the SUBSCRIBE request. The body of a notification carries the conference-state information document in the MIME-type "application/conference-info+xml" as defined in [Draft-ietf-sipping-conference-package]. The PoC Release 1.0 specification uses only a limited subset of offered functions as follows:

- PoC group identity;
- PoC user identities are now part of the PoC session;
- PoC user status (connected, disconnected, on-hold, alerting).

6.3 User plane

The user plane of PoC includes three components, as illustrated in Figure 6.9: the media flow (consisting of talk bursts), talk burst control and quality feedback.

The participating PoC server relays talk bursts, talk burst control messages and quality feedback measurements back and forth between the controlling PoC server and the PoC client, except in the following scenarios:

- the PoC client has a pre-established session with the PoC server;
- the PoC client and PoC server support simultaneous sessions;
- the PoC server needs media transport logging to support charging funtionality;
- the PoC server is used for transcoding services or other media adaptations;
- the PoC server needs to support lawful interception;
- the PoC server acts as a gateway between two different talk burst control protocols.

Figure 6.9 User-plane protocol entities.

The PoC server – in performing both the participating and controlling function – acts as an RTP translator. A translator, as opposed to an RTP mixer, only passes media packets from one IP flow to another, without re-packetizing or alteration to the media format.

6.3.1 Talk bursts

A talk burst is simply a single burst of media flowing from a participant to the controlling PoC server. The controlling PoC server distributes the talk burst to all of the participants of the session. The participants receive the media flow, and render it via their sound systems and play it out from their loudspeakers. The PoC session is a shared floor, which means that only one single talk burst at a time can successfully be distributed and played out at the receiving end. In addition, since in a group PoC session there may be a number of participants to a single floor, there needs to be an arbitrator function – a floor moderator that controls who is allowed to send media at any given moment. Requests to hold the floor may also be queued and granted in succession. Alternatively, a request is either granted or denied, and the user will usually discover this via some visual or sound indication – just like when using a regular walkie-talkie device.

6.3.2 Talk burst control

For floor control, the PoC specifications define the Talk Burst Control Protocol (TBCP) as part of the user-plane specifications. TBCP is used to request, grant, deny and release the PoC session floor. The floor moderator, or the TBCP server, is always located in the controlling PoC function. The participating PoC server usually only relays the messages back and forth to the PoC client, but may also perform additional functions, as we describe later in this chapter. TBCP includes the following messages:

1. TBCP Talk Burst Request (TB_Request) – this message is sent by the clients to request permission to send a talk burst. Such a message may be generated for instance when a user presses the send key in order to speak to the PoC session.
2. TBCP Talk Burst Granted (TB_Granted) – this message is sent by the PoC server to the PoC client to notify that the client has been granted the floor. This means that the client is allowed to send a talk burst that will be heard by the other participants in the PoC session.
3. TBCP Talk Burst Deny (TB_Deny) – this message is sent by the PoC server to the PoC client to announce that the client has been denied the floor. This means that the client is not allowed to send a talk burst and one would not be heard by the other participants at this time.
4. TBCP Talk Burst Release (TB_Release) – this message is sent by the PoC client to the PoC server to announce that the client has completed sending its talk burst.

5. TBCP Talk Burst Taken (TB_Taken) – this message is sent by the PoC server to all of the other PoC session participants in order to inform them that another PoC client has gained possession of the floor and is in the process of sending a talk burst. This message can also be sent reliably by the PoC server, requesting that an immediate acknowledgement be sent by the recipient of the message.
6. TBCP Talk Burst Revoke (TB_Revoke) – this message is used by the PoC server to revoke a grant to speak. For example, the server may interrupt a talk burst that is overly long.
7. TBCP Talk Burst Idle (TB_Idle) – this message is sent by the PoC server to all of the PoC session participants in order to inform them that the floor is open and that the server is willing to accept TBCP Talk Burst Request messages.
8. TBCP Talk Burst Acknowledgement (TB_Ack) – this message is sent by the PoC client to the PoC server upon request of an acknowledgement in the corresponding request.
9. TBCP Talk Burst Queue Status Response (TB_Queued) – this message is sent by the PoC server to the PoC client informing it that the talk burst request has been queued. The message may also indicate the current position of the PoC user in the queue.
10. TBCP Talk Burst Queue Status Request (TB_Position) – this message is sent by the PoC client to the PoC server to request its current position in the talk burst or floor queue.
11. TBCP Connect (Connect) – in a pre-established session, this message is sent by the PoC server to the PoC client to indicate that a new session has been received and that a corresponding floor has thus been created.
12. TBCP Disconnect (Disconnect) – in a pre-established session, this message is sent by the PoC server to the PoC client to indicate that a particular session has ended and the corresponding floor destroyed.

The TBCP shares protocol formatting with the RTCP – in fact, the protocol utilizes the RTCP APP packet, as illustrated in Figure 6.10, for transmitting control messages. Since RTCP operates over an unreliable transport – the User Datagram Protocol (UDP) – the protocol uses a request–reply scheme, with application level retransmissions based on specific timers. In addition, the message can be sent reliably – i.e., immediate acknowledgement can be requested by the sender.

Figure 6.10   RTP Control Protocol APP packet format.

Table 6.3 Mapping of Subtype bit patterns to TBCP messages.

Bit pattern	TBCP message
00000	TB_Request
00001	TB_Granted
00010	TB_Taken (TB_Ack requested)
00011	TB_Deny
00100	TB_Release
00101	TB_Idle
00110	TB_Revoke
00111	TB_Ack
01000	TB_Position
01001	TB_Queued
01010	Reserved for future use
01011	TB_Disconnect
01100	Reserved for future use
01101	Reserved for future use
01110	Reserved for future use
01111	TB_Connect
10010	TB_Taken (No TB_Ack requested)

The Version (V) and Padding (P) fields are identical to other RTCP packets. Specifically, the version field is set to 2 (the current RTP version), and the padding field is a boolean, indicating whether or not any padding was added at the end of the RTCP packet.

The Name field identifies the application; in TBCP, the Name field is set to "POC1", indicating that the APP packet is part of PoC Release 1. The Subtype field identifies the TBCP message type. In Table 6.3 the Subtype bit pattern is mapped to the corresponding TBCP message. The SSRC field identifies the source of the TBCP request. In case the sender is the PoC server, it contains the controlling PoC server's identity, except for TB_Connect and TB_Disconnect messages, which are sent by the participating PoC server.

6.3.3 Quality feedback

The PoC server and client can optionally create, send and process RTCP quality feedback reports. The receivers of RTP can generate reception quality feedback that can take one of two forms: Sender Reports (SRs) and Receiver Reports (RRs). The difference between an RR and an SR is that the SR also includes information about the data sent; both reports include statistics about received media packets, including total number of RTP packets, total number of octets of data received and interarrival jitter, approximating the relative transit time of the media packets.

In the PoC user plane, a PoC client is instructed to generate an SR every time it has completely sent a talk burst – i.e., after having sent the TB_Release message to the PoC server. The PoC client is instructed to send an RR when one of two things happen:

3. The PoC client receives an SR.
4. The PoC client receives an indication that a talk burst has ended (i.e., the TB_Idle message).
5. The PoC client's RTP media timer fires.

As PoC servers are acting as RTP translators in PoC sessions, they generally only relay RTCP reports between PoC session participants. This means that SRs are fanned out from a single sender to all others, and similarly, RRs are suppressed from participants that didn't actually receive any media. This can happen if the PoC client is participating in multiple simultaneous sessions, in which case only the primary session is heard if there happens to be conflicting incoming talk bursts.

6.4   PoC service settings

In Section 6.2 (PoC features) different features were described but it was not shown how answer mode, barring settings or support for simultaneous sessions are activated or de-activated. In OMA PoC Release 1 it was decided that SIP PUBLISH should be used for these service settings. At the time of writing, the required Internet Engineering Task Force (IETF) solution is not ready, therefore this is subject to further change.

The main principle is that after each successful initial registration the PoC client sends SIP PUBLISH that has a PoC-specific content (XML document) in the form of MIME "application/poc-settings+xml" body. Additionally, a sender must include an Accept-Contact header with the PoC feature tag "+g.poc.talkburst" along with the "require" and "explicit" parameters. Using a feature tag and the above-mentioned parameters guarantees that the request is correctly delivered to the user's PoC server. The XML document itself contains four PoC-specific elements as follows:

```
<isb-settings>; active=true/false
<am-settings>; manual/auto
<ipab-settings>; active=true/false
<sss-settings> active=true/false

PUBLISH sip:tobias@home1.fr SIP/2.0
From: <sip:tobias@home1.fr>;tag=31415
To: <sip: tobias@home1.fr>
Accept-Contact: *;+g.poc.talkburst;require;explicit
User-Agent: PoC-client/OMA1.0
Event: poc-settings
Expires: 7200
Content-Type: application/poc-settings+xml
Content-Length: (...)
```

```xml
<?xml version="1.0" encoding="UTF-8"?>
<poc-settings xmlns="urn:oma:params:xml:ns:poc:poc-settings">
  <entity id="do39s8zksn2d98x">
    <isb-settings>
      <incoming-session-barring active="false">
    </isb-settings>
    <am-settings>
      <answer-mode>automatic</answer-mode>
    </am-settings>
    <ipab-settings>
      <incoming-personal-alert-barring active="false"/>
    </ipab-settings>
    <sss-settings>
      <simultaneous-sessions-support active="true"/>
    </sss-settings>
  </entity>
</poc-settings>
```

7

Conferencing

A conference is a conversation between multiple participants. There are many different types of conferences, including loosely coupled conferences, fully distributed multiparty conferences and tightly coupled conferences. In this chapter only the latter is described since it is the only one that is of interest to the Internet Protocol Multimedia Subsystem (IMS).

Conferencing is not just limited to audio; the popularity of video and text conferencing, better known as chatting, has been growing rapidly over the past few years. This popularity is due to conferencing's ability to simulate a face-to-face meeting in so many realistic ways: for example, by enabling file and whiteboard sharing and conveying emotions using video, all in real time.

7.1 Conferencing architecture

In tightly coupled conferences, there is always a central point of control where each conference participant has a connection. This central point provides a variety of conference services including media mixing, transcoding and participant list notifications.

The central point is referred to as a "focus" – a Session Initiation Protocol User Agent (SIP UA) – that is addressed by a conference uniform resource identifier (a SIP URI). The conference Universal Resource Identifier (URI) identifies a conference and the focus sets up a signalling dialog between each participant.

A conference policy is a set of rules associated with a certain conference. These rules include directives on the lifespan of the conference, who can and who cannot join the conference (membership policy), definitions of roles available in the conference and the responsibilities associated with those roles, and policies on who is allowed to request which roles. The conference policy also includes the media policy: the mixing characteristics of a conference.

There are many ways to create a conference: one method uses SIP, which creates *ad hoc* conferences. *Ad hoc* conferences are ones that are created without scheduling and are short-lived. Scheduled conferences could be created using a conference control protocol. Such a protocol would allow users to create and manipulate a conference and its policy, as well as actively manage a running conference's membership, or roster. However, such a protocol does not yet exist at the time of writing, although the Internet

The IMS: IP Multimedia Concepts and Services Second Edition
Miikka Poikselkä, Georg Mayer, Hisham Khartabil and Aki Niemi © 2006 John Wiley & Sons, Ltd

Engineering Task Force (IETF) is working on the problem in the Centralized Conferencing (XCON) Working Group and several protocol proposals are available.

A conference factory URI (a conference factory is a conference server that creates a new conference instance and assigns it a URI when it receives an INVITE request addressed to it using the conference factory URI) can be allocated and globally published in order to automatically create an *ad hoc* conference, using SIP call control means (INVITE request): a conference factory URI is globally routable.

7.2 SIP event package for conference state

Conference state notification is yet another event package (see Section 12.13.1) using the SIP events framework. The conference-state event package is used to learn about changes in the conference participants: in other words, a user can learn, through notifications, who has joined or left a conference. This event package also allows participants to learn the status of a user's participation in a conference and the sidebars in the conference.

Users can subscribe to a conference state by sending a SIP SUBSCRIBE request to the conference URI that identifies the focus. The focus acts as a notifier for this event package.

The name of this event package is "conference". This token appears in the Event header of the SUBSCRIBE request. The body of a notification carries the conference-state information document in the Multipurpose Internet Mail Extension (MIME) type "application/conference-info+xml" as defined in [Draft-ietf-sipping-conference-package].

Two pieces of information are provided about user status: the user's current level of participation in a conference (labelled activity-status) and the method by which the participant entered or left the conference (labelled history-status). The activity-status carries one of the following statuses: connected, disconnected or on-hold. The history-status carries one of the following statuses: dialled-in, dialled-out, departed, booted or failed.

Conference status information also carries information about the media streams that each participant is connected to and about sidebars in a conference, by indicating the URIs of those sidebars and the participants in each sidebar.

7.3 Example signalling flows of conferencing service operation

7.3.1 Creating a conference with a conference factory URI

When creating a conference with a conference factory URI the conference participant generates an initial INVITE request in accordance with Section 8.10 and sets the request URI of the INVITE request to the conference factory URI.

The conference server creates a focus for the newly created conference, assigns it a conference URI and returns the conference URI in the Contact header of the 200 OK response. The URI contains an "isfocus" parameter indicating that this is a focus URI.

Figure 7.1 Conference creation using a conference factory URI.

On receiving a 200 OK response to the INVITE request with the "isfocus" parameter indicated in the Contact header, the conference participant stores the content of the received Contact header as the conference URI. This conference URI can be used to refer other users to the conference as well as to subscribe to the conference-state event package. Figure 7.1 shows an example call flow of creating a conference using a conference factory URI.

7.3.2 Referring a user to a conference using the REFER request

When generating a REFER request that is destined to a user in order to invite that user to a specific conference, the Refer-To header of the REFER request is set to the conference URI, including the "isfocus" parameter. The referrer treats the incoming NOTIFY requests that are related to the previously sent REFER request in accordance with [RFC3515]. Figure 7.2 shows an example flow.

Another way of inviting a user to a conference is by sending the REFER request to the focus itself and putting the URI of the participant being invited into the Refer-To header. This causes the focus to generate an INVITE request inviting the user to the conference.

7.3.3 Subscribing to a conference state

Figure 7.3 shows an example flow of a user subscribing to a conference state. The flow shows the immediate notification carrying the conference state at that moment. More notifications can then follow depending on the change of state of the conference.

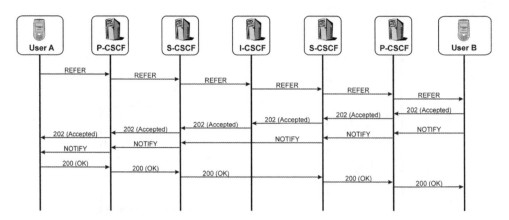

Figure 7.2   Referring a user to a conference.

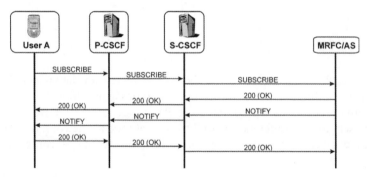

Figure 7.3   Subscribing to conference state.

8

Group management

With the growing need of users to employ a multiplicity of terminals – i.e., mobile phone, PDA, PC – and to have their services available on each of these devices, the situation has arisen in which users expect their service data to be available on all devices, thus obviating the need to build such data more than once.

One solution for such a problem is to use web pages. The problem with web pages is that users have yet to master web browsing on the small screen. Another problem is that solving such a need using the web does not allow such data to be integrated with existing applications running on a mobile phone or any other device.

Let's take a closer look at the issues by taking a real-life example: a user wants to create what is called a "buddy list" on her PC and mobile phone. Without a group management solution, this user has to create her buddy list twice; once on the PC and once on the mobile. Now, the user walks into an Internet cafe and wants to use her web messenger. She cannot do so because her buddy list is stored locally on her PC and mobile phone. Had she a web interface that allowed her to build a buddy list that could be stored in the network, her problem would be solved. But, what if her buddy list on her mobile phone used the address book embedded in the device? Different means are needed to enable the device to store such data in the network. The same means can then be used to build a single buddy list. So, instead of the user building multiple duplicated buddy lists, she can build one on her mobile phone address book, for example, and a protocol can be used to upload such a list to the network. Now, when the user logs in from any device, like the PC, that device can automatically contact the network and retrieve the buddy list that was built using the mobile phone, without the need for any user interaction.

Another advantage with such an architecture is that the user can create, modify and delete such a list and synchronization will automatically take place since the means for uploading and modifying such a buddy list would also have a built-in functionality to notify other devices of changes to the buddy list.

Users have in their possession service data that can be reused by other services. An example of this is the buddy list just mentioned. A user can use the same list – i.e., her buddy list – in a presence application. She can also use the same list to create a conference call where her buddy list represents the list of participants.

Another creative use of group management is the creation of an Access Control List (ACL). This is a list of users that the user creates as an authorization check by network entities before a communication attempt is relayed. For example, a user, Alice, can

The IMS: IP Multimedia Concepts and Services Second Edition
Miikka Poikselkä, Georg Mayer, Hisham Khartabil and Aki Niemi © 2006 John Wiley & Sons, Ltd

create an ACL that allows Bob and John to call her or initiate a Push to talk over Cellular (PoC) session to her while barring Sarah from taking part. The network will then automatically reject any communication attempts made by Sarah towards Alice.

8.1 Group management's contribution to business

Group management can contribute to existing businesses and create a business of its own. Operators and other service providers have a major role to play in mass adoption of group management services in combination with other services like presence and instant messaging. A group management service can be part of the operator's service portfolio. The mobile domain, now reaching almost a billion subscribers worldwide, is a profitable platform for new consumer services. A group management service allowing users to store buddy lists and any other user data related to the presence service or any other service encourages customer loyalty, as shown by instant messaging in the fixed Internet. Users do not want to move to a new subscriber and waste time building their group management data all over again, just to save a couple of dollars.

Offering a basic group management service can give a competitive advantage for an operator over other operators who are not offering it: by binding their group management information to a particular operator's services, customers have available to them high-value services that other operators may not be able to offer without that information. Group management generates new traffic for existing services, like conferencing.

8.2 What is group management?

Group management (also known as "data manipulation") is a service that allows users to store service-specific data in the service provider network. These data can be created, modified and deleted at will by the user. Data could be anything that a user needs to complete a service. Examples of such data are buddy lists (presentity lists) and presence authorization lists.

Various services – such as presence, PoC, IM, etc. – need support for access to and manipulation of certain data that are needed by these services. Some examples of such data include:

- *Conference Access List*: the list of participants who can take part in a conference. In PoC, this is referred to as PoC group.
- *Resource List*: a list of users who are potential notifiers, so that this list can be used to collectively subscribe to the status of each resource in that list. An example of such a list is the presentity list.
- *Subscription Authorization Policy*: An example of an access control policy for presence, which specifies whether a particular watcher is authorized to subscribe to a certain set of events.

The services specify the items that make up the documents representing the information in the examples above, including their semantics and usage.

Data are stored in the network where they can be located, accessed and manipulated (created, changed, deleted) by authorized users. This enables data to be shared and accessed by the services that need them.

Open Mobile Alliance (OMA) has adopted the term XML Document Management (XDM) to be synonymous with the term "group management". The XDM service specifies documents that can be shared by multiple services (called "enablers" by OMA). One such case is a particular type of list, the Universal Resource Identifier (URI) list (also known as a "resource list"), which is a convenient way for a user to group together a number of end-users (e.g., "friends" or "family) or other resources, where such a list is expected to be reused for a number of different services.

The XML Configuration Access Protocol (XCAP), as defined by Internet Engineering Task Force (IETF) and briefly described in Chapter 25 (XCAP chapter), has been selected by OMA and Third Generation Partnership Project (3GPP) as the protocol for transporting, accessing, reading and manipulating the XML documents that contain the data.

8.3 Resource list

The Session Initiation Protocol (SIP) specific event notification mechanism (Section 12.13.1) allows a user (the subscriber) to request notification of changes to a resource state.

In many cases, a subscriber's list of resources for a particular event about which she needs state change information is a lengthy one. Without some aggregating mechanism, this will require the subscriber to generate a SIP SUBSCRIBE request for each resource. This will also result in SIP NOTIFY requests arriving at the subscriber's terminal more rapidly. For congestion control and bandwidth limitations, it is uneconomical to have the user's terminal send a multiplicity of SUBSCRIBE requests, one for each resource. Figure 8.1 demonstrates the problem.

To resolve this problem, [Draft-ietf-simple-event-list] describes an event notification extension that allows users to subscribe to a list of resources with a single SUBSCRIBE request. The list is identified by a URI and contains zero or more URIs pointing to atomic resources or to other lists. In a presence system these resources are presentities. The entity that processes the list's SUBSCRIBE request is referred to as the Resource List Server (RLS). The RLS can generate individual subscriptions for each resource in the list. These subscriptions may or may not be SIP SUBSCRIBE requests. Figure 8.2 demonstrates the solution.

A client sending SUBSCRIBE requests to a list includes a Supported header with an option tag of value "eventlist". If this option tag is not included and the URI in the request-URI represents a list, the RLS returns a "421 Extension Required" error response.

If the subscription is accepted, then the RLS generates NOTIFY requests carrying state information about the list. The NOTIFY request contains a Require header with an option tag of value "eventlist". It also contains a body of MIME type "multipart/related", which internally carries a MIME type of "application/rlmi+xml" that holds resource list meta-information.

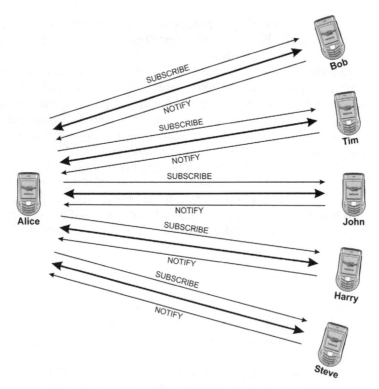

Figure 8.1 Presence subscription example flow, no RLS.

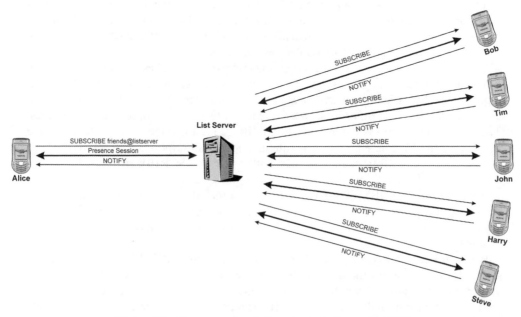

Figure 8.2 Presence subscription example flow, with RLS.

8.4   XCAP usage for resource lists

Section 8.3 (above) discusses resource lists and how to gain state information about them. This section discusses how these lists are created and maintained.

The list creation solution takes advantage of XCAP (see Chapter 25) as an application usage. [Draft-ietf-simple-xcap-list-usage] defines the XML schema along with its semantics. It also defines the following:

- Application Usage ID (AUID) – "resource-lists".
- Additional constraints – none.
- Naming convention – none.
- Resource interdependencies – the list is represented by a URI. If the client does not populate the XML element carrying the URI value, the XCAP server needs to do so.
- Authorization policy – default.

The Ut interface in the Internet Protocol Multimedia Subsystem (IMS) architecture is used to manipulate resource lists.

As an example, Alice is using two terminals, one at home and one in the office. She creates her buddy list from her home terminal and adds Bob and Sarah as her buddies. Figure 8.3 illustrates the example. An XCAP request is created carrying the resource (buddy) list to be stored on the server.

The XCAP request may look like the following:

```
PUT http://xcap.example.com/services/resource-lists/users/
                      sip:alice@example.com/friends.xml HTTP/1.1
Content-Type:application/resource-lists+xml
Host: xcap.example.com

<?xml version="1.0" encoding="UTF-8"?>
  <resource-lists xmlns="urn:ietf:params:xml:ns:resource-lists">
    <list name="friends">
      <entry uri="sip:bob@example.com">
      <display-name>Bob</display-name>
      </entry>
      <entry uri="sip:sarah@example.com">
      <display-name>Bob</display-name>
      </entry>
</list>
    </resource-lists>
```

Alice then goes to the office and decides that John should also be added as her buddy. She does so by using her work terminal and modifying the list created earlier using her home terminal. An XCAP request is created modifying the existing list of the presence service list server.

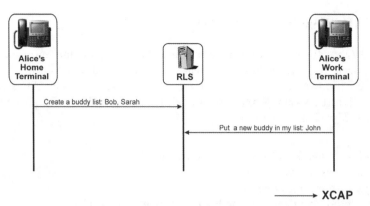

Figure 8.3 Example resource list flow.

8.5 PoC XML Document Management (XDM) specification

XML Document Management (XDM) is the term adopted by OMA for group management. XCAP, as defined by IETF and in Chapter 25 (XCAP chapter) of this book, has been chosen by OMA as the common XDM protocol.

The following sections provide a brief description of the XCAP application usages for PoC-specific documents. PoC XDM application usages include PoC groups and PoC user access policy.

8.6 PoC XDM application usages

8.6.1 PoC group

A PoC group is the list of PoC participants who can take part in a PoC session as well as additional PoC-specific properties such as auto-answering of incoming PoC call requests.

[OMA-TS-PoC_XDM-V1_0-20050428-C] defines the XML schema along with its semantics. It defines the AUID as "org.openmobilealliance.poc-groups".

PoC group application usage contains the following:

- a URI representing the PoC group identity;
- a display name for the group;
- one or more lists of group members (URIs);
- an indication as to whether the group members will be invited to the PoC session by the PoC server or not;
- a maximum number of participants for a particular PoC session from that group;
- an authorization policy associated with the group.

Each list of group members needs to contain entries identifying a single user by a SIP or tel URI, or a reference to an external list URI.

The structure of the authorization policy (also referred to as a "ruleset") must conform to the Common Policy, described in more detail in Chapter 26. The conditions for such an authorization policy include the identity of the participant, the identity of a participant derived from an external list, other identities that are not present in any rule and a condition that checks whether the user is a list member.

The actions that can be taken if a condition returns as "true" are: allow the user to subscribe to the PoC session state, allow the user to dynamically invite participants to join the PoC session, allow the user to join, bar the user from joining, allow the user to start the session, allow the user to be anonymous and treat the user as a key participant.

As an example, Alice wants to create an early PoC session with Bob, Sarah and John. This group is her friends group. She would like her friends to join on their own – i.e., they will not be invited by the server. The maximum number of people joining the session cannot exceed 4. This allows Alice to add more friends to the list but still limit the number of participants as she does not want too many people to be involved in discussions.

Alice wants to authorize those people who constitute only a small part of the members list (currently Bob, Sarah and John) to be able to join the conference and subscribe to the session state. No anonymity is allowed. Below is what the XCAP request would look like. The XML document is also shown as the payload:

```
PUT http://xcap.example.com/services/org.openmobilealliance.poc-
    groups/users/sip:alice@example.com/friendsgroup.xml HTTP/1.1
...
Content-Type: application/list-service+xml

<?xml version="1.0" encoding="UTF-8"?>
  <group xmlns="urn:oma:params:xml:ns:list-service"
                  xmlns:rl="urn:ietf:params:xml:ns:resource-lists"
                  xmlns:cr="urn:oma:params:xml:ns:common-policy">
    <list-service uri="sip:myconference@example.com">
      <list>
        <rl:entry uri="sip:bob@example.com"/>
        <rl:entry uri="sip:sarah@example.com"/>
        <rl:entry uri="sip:john@example.com"/>
      </list>
      <display-name xml:lang="en-us">Friends</display-name>
      <invite-members>false</invite-members>
      <max-participant-count>4</max-participant-count>
      <cr:ruleset>
        <cr:rule id="1a">
          <cr:conditions>
            <cr:is-list-member/>
          </cr:conditions>
          <cr:actions>
            <allow-conference-state>true</allow-conference-state>
            <join-handling>allow</join-handling>
```

```
            <allow-anonymity>false</allow-anonymity>
          </cr:actions>
        </cr:rule>
      </cr:ruleset>
    </list-service>
  </group>
```

8.6.2 PoC user access policy

The PoC user access policy is a set of rules that a user creates to control who can and
who cannot initiate a PoC session to him/her.

[OMA-TS-PoC_XDM-V1_0-20050428-C] defines the XML schema along with its
semantics. It defines the AUID as "org.openmobilealliance.poc-rules".

The structure of the access policy (also referred to as a "ruleset") must conform to
the Common Policy (described in more detail in Chapter 26). The conditions for such
an authorization policy include the identity of the participant from whom the user is
accepting or rejecting a PoC session, an identity of a participant derived from an
external list and other identities that are not present in any rule.

The only action defined for a condition that has been satisfied is what to do with the
invitation. One of three values may appear: pass, allow and accept. "Pass" instructs the
PoC server to process the PoC session invitation using manual answer procedures (see
Section 6.2.4). "Accept" instructs the PoC server to accept the invitation according to
the user's answer mode setting, while "reject" instructs the PoC server to reject the
invitation.

To give an example, Alice is creating a PoC user access policy that only allows her
friends to invite her to a PoC session. She does this by utilizing the list that she created
earlier (see example in Section 8.4). Below is what the XCAP request would look like.
The XML document is also shown as the payload:

```
PUT http://xcap.example.com/services/org.openmobilealliance.poc-
              rules/ users/sip:alice@example.com/pocrules HTTP/1.1
Content-Type: application/auth-policy+xml

<?xml version="1.0" encoding="UTF-8"?>
  <ruleset xmlns="urn:ietf:params:xml:ns:common-policy"
                   xmlns:poc="urn:oma:params:xml:ns:poc-rules">
  <rule id="f3g44r1">
    <conditions>
      <poc:external-list>
        <poc:anc>/services/resource-lists/users/sip:alice@example.
                                                   com/friends.xml
        </poc:anc>
      </poc:external-list>
    </conditions>
    <actions>
      <poc:allow-invite>accept</poc:allow-invite>
```

```
        </actions>
        </rule>
 <rule id="f3g44r1">
      <conditions>
        <other-identity/>
      </conditions>
      <actions>
        <poc:allow-invite>reject</poc:allow-invite>
      </actions>
    </rule>
  </ruleset>
```

Part III

Detailed Procedures

9

Introduction to detailed procedures

This part gives a detailed example of Session Initiation Protocol (SIP) and Session Description Protocol (SDP) related procedures in the Internet Protocol Multimedia Subsystem (IMS). Signalling flows, protocol messages and their elements are described and explained based on IMS registration and a subsequent IMS session between two users.

The reader will see how IMS signalling works and how previously described concepts and architecture are realized at the protocol level. Nevertheless, this part does not handle error or abnormal procedures in detail.

To give a better understanding of the procedures applied, the part is split into several chapters that concentrate on different concepts, such as routing, authentication or media negotiation. Each chapter will describe those parts of individual SIP and SDP messages that are necessary for their understanding. An overview section is included in each chapter to give an introduction to the basic operation. At the end of each chapter the related standards and specifications are listed, to allow the interested reader to obtain more detail by reading the base specifications.

The final chapters in this part describe further scenarios and mechanisms for establishing SIP sessions within the IMS. These alternative session establishment mechanisms are needed because of specific application requirements or interworking scenarios.

9.1 The example scenario

This section gives a detailed example of a normal IMS session between two users and all the required prerequisites. It is based on the assumption that both users are attached to the General Packet Radio Service (GPRS), which is used as the example access technology throughout the section.

Tobias, who is a student in France and currently visiting Finland, is calling his sister Theresa, who is working in Hungary and currently on a business trip to Austria (see Table 9.1 and Figure 9.1).

Tobias's home operator is located in France. As he is roaming in Finland, the Finnish operator provides the Proxy Call Session Control Function (P-CSCF), as the

The IMS: IP Multimedia Concepts and Services Second Edition
Miikka Poikselkä, Georg Mayer, Hisham Khartabil and Aki Niemi © 2006 John Wiley & Sons, Ltd

Table 9.1 Location of CSCFs and GPRS access for the example scenario.

User	Home operator S-CSCF location	P-CSCF location	GPRS access
Tobias	France	Finland	Finland
Theresa	Hungary	Hungary	Austria

Figure 9.1 The example scenario.

home operator and the Finnish operator have signed an IMS roaming agreement. Consequently, the Gateway GPRS Support Node (GGSN) that Tobias is using is also located in Finland.

Theresa's home operator in Budapest has no IMS roaming agreement with the operator in Austria. Therefore, her terminal gets attached to the P-CSCF in her Hungarian home network, where the GGSN is also located. Theresa's access to the IMS is based on the GPRS-level roaming agreement between the operators of her home network and the visited network.

It is assumed that Theresa has already registered her SIP URI (uniform resource identifier), *sip:theresa@home2.hu*, as Tobias is just switching on his mobile phone. He wants to call his sister to show her one of the beautiful wooden buildings in Oulu and, therefore, points his camera, which is connected to the phone, toward the building. In parallel with this, his phone will also send a second video stream, showing his face to Theresa. The built-in camera of his phone records this second stream. However, Tobias first has to register his public user identity, *sip:tobias@home1.fr*, before he can call his sister.

9.2 Base standards

The following specifications define the basic procedures and architecture as used in the following chapters:

3GPP TS 23.228   IP Multimedia Subsystem (IMS).

3GPP TS 24.229   IP Multimedia Call Control Protocol based on SIP and SDP.

RFC3261   SIP: Session Initiation Protocol

3GPP TS 24.228   Signaling flows for IP multimedia call control based on SIP and SDP. This standard provides example call flows for procedures within the IMS.

10

An example IMS registration

10.1 Overview

Session Initiation Protocol (SIP) registration is performed in order to bind the Internet Protocol (IP) address that is currently used by the user and the user's public user identity, which is a SIP URI (uniform resource identifier). If Tobias wants to call Theresa, he will send a SIP INVITE request to her address – i.e., *sip:theresa@home2.hu*; he does not need to be aware of which terminal Theresa is using. The INVITE then gets routed to Theresa's registrar, which is located in *home2.hu*. This registrar became aware of Theresa's current terminal address during her registration and will replace the address *sip:theresa@home2.hu* with the registered contact, which is an IP address. Afterwards, the request can be routed to Theresa's terminal.

Therefore, even for non-IMS cases, Theresa needs to be registered at a SIP registrar so that her current terminal address can be discovered. The IP Multimedia Subsystem (IMS) couples more functionality with SIP registration procedures, which makes it necessary for Tobias to register as well, before he can call his sister.

The following procedures are performed during Tobias's IMS registration (see Figure 10.1):

- The dedicated signalling Packet Data Protocol (PDP) context is established between Tobias's User Equipment (UE) and the Gateway GPRS Support Node (GGSN) in the case of General Packet Radio Service (GPRS) – Section 10.2.
- The UE discovers the address of the Proxy Call Session Control Function (P-CSCF), which it uses as a SIP outbound proxy during registration and for all other SIP signalling while it is registered – Section 10.3.
- The UE sends a REGISTER message to Tobias's home network to perform SIP registration for Tobias's public user identity – Section 10.5.2.
- The Interrogating-CSCF (I-CSCF) selects the Serving-CSCF (S-CSCF) that serves the user while it is registered – Section 10.5.5.
- The S-CSCF downloads the authentication data of the user from the Home Subscriber Server (HSS) – Section 10.6.4.
- The UE and the P-CSCF agree on a security mechanism – Section 10.8.
- The UE and the network (S-CSCF) authenticate each other – Section 10.6.
- IP Security (IPsec) associations between the UE and the P-CSCF are established – Section 10.7.

The IMS: IP Multimedia Concepts and Services Second Edition
Miikka Poikselkä, Georg Mayer, Hisham Khartabil and Aki Niemi © 2006 John Wiley & Sons, Ltd

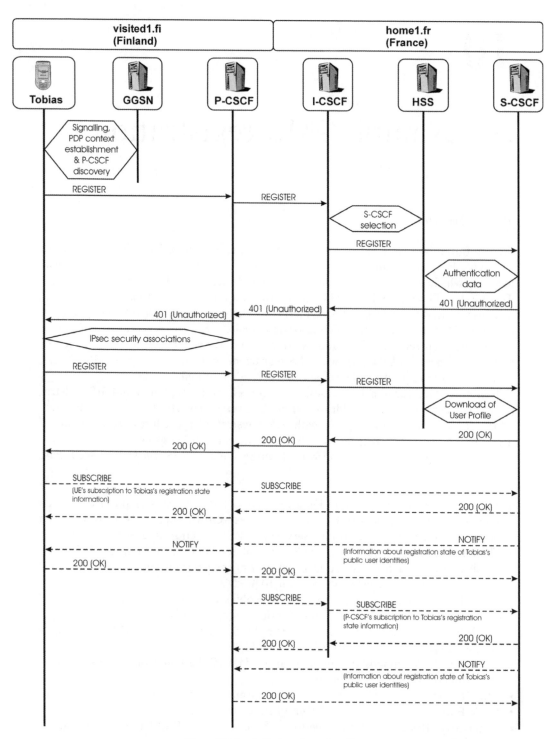

Figure 10.1   Initial registration flow.

- SIP compression starts between the UE and the P-CSCF – Section 10.9.
- The UE learns the route to the S-CSCF – Section 10.5.8.
- The S-CSCF learns the route to the UE – Section 10.5.9.
- The S-CSCF downloads the user profile of the user from the HSS – Section 10.5.6.
- The S-CSCF registers the default public user identity of the user – Section 10.5.6.
- The S-CSCF may, based on the user profile, implicitly register further public user identities of the user – Section 10.12.
- The UE becomes aware of all public user identities that are assigned to Tobias and his current registration state – Section 10.12.
- The P-CSCF becomes aware of all public user identities that are assigned to Tobias and his current registration state – Section 10.12.

As a consequence of all these required basic actions, Tobias would not have been able to send the INVITE to his sister had he not registered earlier.

10.2   Signalling PDP context establishment

Before Tobias's UE can start the IMS registration procedure, it needs to establish an IP connection with the network. In the case of GPRS such an IP connection is provided by either a dedicated signalling PDP context or a general purpose PDP context. The concepts and procedures for PDP context establishment and usage are described in Section 3.9 and Chapter 17.

In this example it is assumed that Tobias's UE establishes a dedicated signalling PDP context with the GGSN in Finland. After the UE has established the signalling PDP context, it will be able to send SIP signalling over the air interface.

10.3   P-CSCF discovery

The P-CSCF is the single entry point for all SIP messages that are sent from Tobias's UE to the IMS. Therefore, the P-CSCF address needs to be known by the UE before the first SIP message is sent. As this address is not pre-configured in our example, it needs to be discovered by the UE.

In the case of the GPRS the UE can request the addresses of a P-CSCF during the establishment of the general or signalling PDP context. The GGSN will then return the IPv6 prefix of an P-CSCF in response to the activate PDP context request.

Alternatively, the UE can choose to use Dynamic Host Configuration Protocol for IPv6 (DHCPv6) in order to discover the P-CSCF. If the P-CSCF address is returned from DHCP as a Fully Qualified Domain Name (FQDN) rather than an IP address, then the P-CSCF address will be resolved via the Domain Name System (DNS) as the address of any other SIP server. The related procedures are described in Chapter 12.

10.4   Transport protocols

The IMS puts no further restrictions on the transport protocol for SIP used between the UE and the P-CSCF. In this example it is assumed that the User Datagram Protocol (UDP) is the default transport protocol. UDP will be used for the transport of SIP messages that are sent between Tobias's UE and the P-CSCF as long as these messages do not exceed 1,300 bytes. When they exceed this limit, the Transmission Control Protocol (TCP) must be used. Due to the fact that SIP also allows a lot of content in the SIP message body (e.g., pictures can be attached to the body of a MESSAGE request), it is likely that both UDP and TCP will be used in parallel while a user is registered.

10.5   SIP registration and registration routing aspects

10.5.1   Overview

This section concentrates on the SIP routing aspects of Tobias's registration (see Table 10.1 and Figure 10.2).

Tobias's UE will first of all construct a REGISTER request, which it sends to the home domain of Tobias's operator. The relevant information is obtained from the IP Multimedia Services Identity Module (ISIM) application on Tobias's Universal Subscriber Identity Module (USIM). The request will traverse the P-CSCF and the I-CSCF, which – if not previously assigned – will select an S-CSCF for Tobias.

The S-CSCF will create, based on the information given in the REGISTER request, the binding between Tobias's public user identity and the IP address of Tobias's UE. This makes it possible for requests from other users to be routed from the S-CSCF to Tobias's UE. The S-CSCF will update the registration information in the HSS, download Tobias's user profile and will, based on the received initial filter criteria from the HSS, inform any Application Servers (ASs) that are interested in Tobias's registration state.

During the registration procedures the UE will learn the direct route to the S-CSCF from the Service-Route header. After that, the I-CSCF will no longer need to be contacted when Tobias's UE sends out an initial request.

The S-CSCF will become aware of the address of Tobias's P-CSCF from the Path header. This is necessary as all initial requests that are destined for Tobias (e.g., an INVITE request) need to traverse the P-CSCF before they can be sent to the UE.

10.5.2   Constructing the REGISTER request

After establishing the signalling PDP context and discovering the P-CSCF address, Tobias's UE can finally start to construct the initial REGISTER request:

```
REGISTER sip:home1.fr SIP/2.0
Via: SIP/2.0/UDP [5555::1:2:3:4];branch=0uetb
Route: sip:[5555::a:f:f:e];lr
```

```
Max-Forwards: 70
From: <sip:tobias@home1.fr>;tag=pohja
To: <sip:tobias@home1.fr>
Contact: <sip:[5555::1:2:3:4]>;expires=600000
Call-ID: apb03a0s09dkjdfglkj49111
CSeq: 25 REGISTER
Content-Length: 0
```

How the used public and private user identities as well as the registrar address are obtained from the ISIM is described in Section 10.12.2. The above message is not a complete IMS REGISTER request: there are some headers and parameters missing

Table 10.1 Routing-related headers.

Header	Function	Set up
Via	Routing of requests	By every traversed SIP entity, which puts its address to the Via header during the routing of the request
Route	Routing of requests	Initial requests: by the request-originating UE, which puts the P-CSCF (outbound proxy) address and entries of the Service-Route header
		Initial requests: by CSCFs, which find the next hop from the public user identity in the request URI (by querying DNS and HSS) or the received Path header
		Subsequent requests: by the request-originating UE, which put entries to the Route header as collected in the Record-Route header during initial request routing
Record-Route	Records the Route header entries for subsequent requests within a dialog	By CSCFs, which put their addresses into the Record-Route header if they need to receive subsequent requests within a dialog
Service-Route	Indicates the Route header entries for initial requests from the UE to the user's S-CSCF (originating case)	By the S-CSCF, which sends this header back to the UE in the 200 (OK) response for the REGISTER request
Path	Collects the Route header entries for initial requests from the S-CSCF to the user's P-CSCF (terminating case)	By the P-CSCF, which adds itself to the Path header in the REGISTER request and sends it to the S-CSCF

Figure 10.2   Routing during registration.

from it. It only includes the information required to explain the procedures in this section, as is the case with all the following messages.

The final destination of the request is the registrar, which is identified in the request URI as *sip:home1.fr* which is the domain name of the home network of Tobias as read from the ISIM.

In the To header we find the public user identity *sip:tobias@home1.fr*, as read from the ISIM, which is going to be registered. SIP registration takes place to tell the registrar that the public user identity *sip:tobias@home1.fr* will be reachable under the IP address that is indicated in the Contact header. This IP address includes the IPv6 prefix, which the UE got assigned during establishment of the dedicated signalling PDP context (see Section 10.2).

Also within the Contact header, the UE indicates that this binding of the IP address to the SIP URI is intended to last 600,000 seconds (nearly a week). In IMS the UE is forced to register for such a long time. Nevertheless, the network can adjust this time:

- During registration procedures by setting the expires value in the Contact header of the 200 (OK) response to the REGISTER request to a smaller value.
- After the user has registered, by making use of registration-state event notifications (e.g., Section 10.13.2 for network-initiated re-authentication).

The UE puts its IP address into the Via header of the request. This ensures that all responses to this request will be routed back to the UE. A branch parameter that uniquely identifies the transaction is also put in the Via header. Every entity on the route will add its own Via header.

The P-CSCF, which was resolved in the previous step, is put in the Route header. The P-CSCF is the next hop to receive the REGISTER message, as it is the topmost – and only – entry of the Route header. The ;lr parameter indicates that the P-CSCF is a loose router (see Section 12.12.2).

The From header identifies the user who is performing the registration. We find in the From header the same public user identity as in the To header, as Tobias is performing a so-called first-party registration (i.e., he is registering himself).

Note that the From header includes a tag, while the To header does not. The recipient of the request (i.e., the registrar) will set the To tag when sending the response to the UE.

A Call-ID header is included, which, together with the value of the CSeq header, identifies the REGISTER transaction.

Finally, there is the indication that the REGISTER request does not carry any content, as the Content-Length header is set to 0.

The example shown on the previous page gives the header names in their long form. In order to avoid unnecessary signalling over the air interface, Tobias's UE may as well use the compact form, which would make the REGISTER request look like:

```
REGISTER sip:home1.fr SIP/2.0
v: SIP/2.0/UDP [5555::1:2:3:4];branch=0uetb
Route: sip:[5555::a:b:c:d];lr
Max-Forwards: 70
f: <sip:tobias@home1.fr>;tag=pohja
t: <sip:tobias@home1.fr>
m: <sip:[5555::1:2:3:4]>;expires=600000
i: apb03a0s09dkjdfglkj49111
CSeq: 25 REGISTER
l: 0
```

To make reading of SIP messages more convenient, only the long form of the header names will be used in this example.

10.5.3   From the UE to the P-CSCF

Now Tobias's UE can send out the REGISTER request to the next hop, which is the topmost entry of the Route header (i.e., the P-CSCF). It sends the request via the UDP protocol, as its length does not exceed the strict limit of 1,300 bytes. As no port is indicated in the Route header, the request gets sent to the default SIP port (default SIP port 5060).

10.5.4   From the P-CSCF to the I-CSCF

When receiving the initial REGISTER request the P-CSCF becomes aware for the first time that Tobias's UE is using it as a SIP outbound proxy. As Tobias is not authenticated at this moment, it can only act as a SIP outbound proxy and, therefore, tries to route the REGISTER request to the next hop.

The P-CSCF removes its own entry from the Route header. After doing so the Route header will be empty. The only routing-related information left now is the registrar address in the request URI, which points to Tobias's home network. In order to discover the address of a SIP proxy in Tobias's home network the P-CSCF needs to resolve the domain name (as given in the request URI) via the DNS. By using DNS NAPTR, SRV and AAAA queries, the P-CSCF will resolve the address of an I-CSCF in Tobias's home network (see Chapter 12).

Nevertheless, the P-CSCF will not put the address of the I-CSCF in the Route header, as it cannot be sure whether the I-CSCF will act as a loose router or not. Therefore, the P-CSCF will put the address of the I-CSCF as the destination address into the UDP packet that transports SIP requests.

Before sending the REGISTER message the P-CSCF also adds itself to the Via header, in order to receive the response to the request. It also adds a branch parameter to the Via header:

```
REGISTER sip:home1.fr SIP/2.0
Via: SIP/2.0/UDP sip:pcscf1.visited1.fi;branch=0pctb
Via: SIP/2.0/UDP [5555::a:b:c:d];branch=0uetb
Max-Forwards: 69
From: <sip:tobias@home1.fr>;tag=pohja
To: <sip:tobias@home1.fr>
Contact: <sip:[5555::1:2:3:4]>;expires=600000
Call-ID: apb03a0s09dkjdfglkj49111
CSeq: 25 REGISTER
Content-Length: 0
```

10.5.5   From the I-CSCF to the S-CSCF

The I-CSCF is the entry point to Tobias's home network and will receive every REGISTER request that is originated by Tobias's UE. It will query the HSS for the S-CSCF that is assigned to serve the user who is registering.

If no S-CSCF has been selected up to now, it is the task of the I-CSCF to select one. These procedures are described in Section 3.10.

After putting its own entry in the topmost Via header, the I-CSCF sends the REGISTER request to the S-CSCF address that it either got from the HSS or that it selected:

```
REGISTER sip:home1.fr SIP/2.0
Via: SIP/2.0/UDP sip:icscf1.home1.fr;branch=0ictb
Via: SIP/2.0/UDP sip:pcscf1.visited1.fi;branch=0pctb
```

```
Via: SIP/2.0/UDP [5555::a:b:c:d];branch=0uetb
Route: sip:scscf1.home1.fr;lr
Max-Forwards: 68
From: <sip:tobias@home1.fr>;tag=pohja
To: <sip:tobias@home1.fr>
Contact: <sip:[5555::1:2:3:4]>;expires=600000
Call-ID: apb03a0s09dkjdfglkj49111
CSeq: 25 REGISTER
Content-Length: 0
```

10.5.6   Registration at the S-CSCF

After receiving the initial REGISTER request, the S-CSCF will request Tobias to authenticate himself, as described in Section 10.6. This will result in another REGISTER request from Tobias. This second REGISTER request will include the same registration-related information and will also be routed exactly in the same way as the initial REGISTER request. Nevertheless, for the second REGISTER a new Call-ID will be created. Consequently, new CSeq numbers, branch parameters and a new From tag will be included in it. The second REGISTER received by the S-CSCF will look like:

```
REGISTER sip:home1.fr SIP/2.0
Via: SIP/2.0/UDP sip:icscf1.home1.fr;branch=3ictb
Via: SIP/2.0/UDP sip:pcscf1.visited1.fi;branch=2pctb
Via: SIP/2.0/UDP [5555::a:b:c:d];branch=1uetb
Route: sip:scscf1.home1.fr;lr
Max-Forwards: 68
From: <sip:tobias@home1.fr>;tag=ulkomaa
To: <sip:tobias@home1.fr>
Contact: <sip:[5555::1:2:3:4]>;expires=600000
Call-ID: apb03a0s09dkjdfglkj49222
CSeq: 47 REGISTER
Content-Length: 0
```

Assuming that the authentication procedures are successful, the S-CSCF will then register Tobias. This means the S-CSCF will create a binding for the public user identity that was indicated in the To header of the REGISTER request (*sip:tobias@home1.fr*) and the contact address (*sip:[5555::a:b:c:d]*). This binding will exist for exactly 600,000 seconds, which is the value that the UE entered into the "expires" parameter of the Contact header, unless the S-CSCF decides to reduce this time due to local policy.

The S-CSCF will also update the information in the HSS to indicate that Tobias has now been registered. The HSS will download Tobias's user profile to the S-CSCF via the Cx interface (see Section 3.13).

10.5.7   The 200 (OK) response

Afterwards, the S-CSCF will send back a 200 (OK) response to the UE, to indicate that
the registration procedure has succeeded:

```
SIP/2.0 200 OK
Via: SIP/2.0/UDP icscf1.home1.fr;branch=3ictb
Via: SIP/2.0/UDP pcscf1.visited1.fi;branch=2pctb
Via: SIP/2.0/UDP [5555::1:2:3:4]:1357;branch=1uetb
From: <sip:tobias@home1.fr>;tag=ulkomaa
To: <sip:tobias@home1.fr>;tag=kotimaa
Contact: <sip:[5555::a:b:c:d]>;expires=600000
Call-ID: apb03a0s09dkjdfglkj49222
CSeq: 47 REGISTER
Content-Length: 0
```

The S-CSCF has added a tag to the To header.

The response is routed back to the UE over all the CSCFs that received the
REGISTER request; it manages to do this because CSCFs put their own address in
the topmost Via header list when they receive REGISTER requests. Now, when receiving the 200 (OK) response, they just remove their own entry from the Via list and send
the request forward to the address indicated in the topmost Via header.

The UE, when receiving this response, will know that the registration was successful.

10.5.8   The Service-Route header

We have seen that neither the UE nor the P-CSCF were aware of the address of the
S-CSCF during the registration procedures; consequently, the I-CSCF had to be
contacted to discover the S-CSCF address from the HSS.

In order to avoid the I-CSCF as an extra hop for every initial message sent from the
UE, the S-CSCF will return its address in the Service-Route header in the 200 (OK)
response to the REGISTER request:

```
SIP/2.0 200 OK
Service-Route: sip:orig@scscf1.home1.fr;lr
```

The UE, when receiving the 200 (OK) response, will store the entries in the Service-
Route header. Whenever the UE sends out any initial request other than a REGISTER
message, it will:

- include the addresses that were received in the Service-Route header within a Route
  header of the initial request; and
- include the P-CSCF address as the topmost Route entry in the initial request.

Examples of how initial requests are routed are given in Section 10.12.5 for a SUB-
SCRIBE request and in Section 11.3.2 for an INVITE request.

The S-CSCF in this example puts a user part ("orig") in its Service-Route entry as it needs to distinguish between two types of requests:

- requests originated from the served user (i.e., Tobias); and
- requests destined for Tobias's UE.

Whenever the S-CSCF receives an initial request (e.g., an INVITE request) it needs to determine whether this request is originated from or destined to the served user. The user part entry in the Route header makes it easy for the S-CSCF to discover whether a received request was originated from the served user, as Tobias's UE will include the S-CSCF's Service-Route entry as a Route entry within all requests that it originates.

10.5.9   The Path header

The S-CSCF will receive all initial requests that are destined to Tobias, as it acts as his registrar. Normal SIP procedures allow the registrar to send requests directly to the UE. In the case of IMS this is not possible, because the P-CSCF needs to be contacted first; this is because the P-CSCF has established IPsec SAs with the UE that guarantee that all messages will be sent and received integrity-protected (see Section 10.7). Furthermore, the P-CSCF has an important role in media authorization (see Section 11.7.2) as it is the only network element in the IMS that has a direct connection to the GGSN.

Therefore, the S-CSCF needs to ensure that every request that is sent to the UE first traverses the P-CSCF. To make this possible, the P-CSCF includes its own address in every REGISTER request within a Path header:

```
REGISTER sip:home1.fr SIP/2.0
Path: sip:pcscf1.visited1.fi;lr
```

After successful registration of the user, the S-CSCF saves this P-CSCF address. Whenever a request for Tobias is received, the S-CSCF will include a Route header with the address that was received in the Path header. An example of routing an initial INVITE request toward the served user is given in Section 11.3.3.5.

10.5.10   Third-party registration to application servers

After successful registration the S-CSCF will check the downloaded filter criteria of the user (see Section 3.13). We assume that there is a presence server that provides its services to Tobias; this presence server needs to know that Tobias has now been registered and is, therefore, available. To inform the presence server about this, filter criteria have been set which trigger all the REGISTER requests that originate from Tobias's public user identity (Table 10.2).

Due to these filter criteria, the S-CSCF will generate a third-party REGISTER request (Figure 10.3) and send it to the presence server whenever Tobias performs a successful registration:

```
REGISTER sip:presence.home1.fr SIP/2.0
Via: SIP/2.0/UDP scscf1.home1.fr;branch=99sctb
Max-Forwards: 70
From: <sip:scscf1.home1.fr>;tag=6fa
To: <sip:tobias@home1.fr>
Contact: <sip:scscf1.home1.fr>;expires=600000
Call-ID: las22kdoa45siewrf
CSeq: 87 REGISTER
Content-Length: 0
```

This REGISTER request is destined to the presence server at *presence.home1.fr*, as indicated in the request URI. As no Route header is included, the request will be sent directly to that address.

The To header includes the public user identity of Tobias, as this is the URI that was registered.

The S-CSCF indicates its own address in the From header, as it is registering Tobias's public user identity on behalf of Tobias (i.e., as a third party). Furthermore, the

Table 10.2 Filter criteria in Tobias's S-CSCF.

Element of filter criteria	Filter criteria
SPT: session case	Originating
SPT: public user identity	*sip:tobias@home1.fr*
SPT: SIP method	REGISTER
Application server	*sip:presence.hom1.fr;lr*

Figure 10.3 Third-party registration by S-CSCF.

S-CSCF indicates its own address within the Contact header. This ensures that the presence server never routes directly to Tobias's UE, but will always contact the S-CSCF first.

The presence server will send back a 200 (OK) response for this REGISTER request to the S-CSCF, but will not start acting as a registrar for Tobias. It will take the REGISTER request as an indication that Tobias has been successfully registered at the S-CSCF that is Tobias's registrar. If the presence server needs more information about Tobias's registration state (e.g., all other public user identities that have been implicitly registered for Tobias), it can subscribe to the registration-state information of Tobias in the same way as the UE and the P-CSCF do (see Section 10.12).

10.5.11   *Related standards*

Specifications relevant to Section 10.5 are:

RFC3327   Session Initiation Protocol (SIP) Extension Header Field for Registering Non-Adjacent Contacts.

RFC3608   Session Initiation Protocol (SIP) Extension Header Field for Service Route Discovery during Registration.

10.6   Authentication

10.6.1   *Overview*

As shown in Section 3.8, the IMS is based on several security relations. Two of them – authentication between user and network, and the SAs between the UE and the P-CSCF – have an influence on SIP signalling (Figure 10.4). Authentication and SA establishment procedures in the IMS are directly coupled to SIP registration procedures.

IMS authentication is based on a shared secret and a sequence number (SQN), which is only available in the HSS and the ISIM application on the Universal Integrated Circuit Card (UICC) card in Tobias's UE. As the HSS never directly communicates with the UE, the S-CSCF performs the authentication procedures and all security-related parameters that are needed by the S-CSCF. The so-called Authentication Vector (AV) is downloaded by the S-CSCF from the HSS during registration.

In order to authenticate, Tobias sends his private user identity (in our example this is *tobias_private@home1.fr*) in the initial REGISTER request. This private user identity is stored within the ISIM application and is only used for authentication and registration procedures.

When receiving this REGISTER request, the S-CSCF downloads the AV from the HSS. The AV does not include the shared secret and the SQN itself, but does include (among other parameters):

- a random challenge (RAND);
- the expected result (XRES);

- the network authentication token (AUTN);
- the Integrity Key (IK); and
- the Ciphering Key (CK).

These parameters enable the S-CSCF to perform authentication without knowing the shared secret or the SQN.

In order to authenticate, the S-CSCF rejects the initial REGISTER request from the user with a 401 (Unauthorized) response, which includes (among other parameters) the RAND, the AUTN, the IK and the CK.

The P-CSCF, when receiving the 401 (Unauthorized) response, removes the IK and the CK from the response before sending it to the UE. The IK is the base for the SAs that get established between the P-CSCF and the UE immediately afterwards (see Section 10.7).

After receiving the response, the UE hands the received parameters over to the ISIM application, which:

- verifies the AUTN based on the shared secret and the SQN – when AUTN verification is successful the network is authenticated (i.e., the UE can be sure that the authentication data were received from the home operator's network);
- calculates the result (RES) based on the shared secret and the received RAND;
- calculates the IK, which is then shared between the P-CSCF and the UE and will serve as the base for the SAs.

Afterwards, the UE sends the authentication challenge response (RES) in the second REGISTER request back to the S-CSCF, which compares it with the XRES that was

Figure 10.4 Authentication information flows during IMS registration.

received in the AV from the HSS. If the verification is successful, the S-CSCF will treat the user as authenticated and will perform the SIP registration procedures (see Section 10.5.6).

Whenever the UE sends out another REGISTER request (i.e., due to either re- or de-registration), it will always include the same authentication parameters as included in the second REGISTER request, until the S-CSCF re-authenticates the UE.

10.6.2 HTTP digest and 3GPP AKA

The Hypertext Transfer Protocol (HTTP) digest is specified in [RFC2617], and how it is used with SIP is described in [RFC3261]. The IMS on the contrary is part of the Third Generation Partnership Project/Universal Mobile Telecommunications System (3GPP/UMTS) architecture, which uses the 3GPP Authentication and Key Agreement (AKA) mechanism for authentication.

In order to achieve 3GPP AKA-based authentication within the IMS, [RFC3310] defines how 3GPP AKA parameters (as described above) can be mapped to HTTP digest authentication. Therefore, the signalling elements (SIP headers and parameters) used to transport 3GPP AKA information are identical to those used for the HTTP digest. Nevertheless, their meanings (i.e., their interpretation at the UE, the P-CSCF and the S-CSCF) are different.

In order to distinguish the 3GPP AKA authentication mechanism from other HTTP digest mechanisms (e.g., MD5), it was given a new algorithm value: "AKAv1-MD5".

10.6.3 Authentication information in the initial REGISTER request

Within the initial REGISTER request Tobias's UE utilizes the HTTP Digest Authorization header to transport Tobias's private user identity. In order to fulfil HTTP digest requirements, the UE includes the following fields in the Authorization header:

- The authentication scheme – set to the value "Digest", as the 3GPP AKA is mapped to the HTTP digest mechanism.
- The username field – set to Tobias's private user identity, which will be used by the S-CSCF and the HSS to identify the user and to find the corresponding AV.
- The realm and URI fields – set to the home domain of Tobias.
- The response and nonce fields – which are left empty. These fields are mandated by the HTTP digest, but not used in the initial REGISTER request.

The REGISTER now looks like:

```
REGISTER sip:home1.fr SIP/2.0
Authorization: Digest username="tobias_private@home1.fr",
               realm="home1.fr",
               nonce="",
               uri="sip:home1.fr",
               response=""
```

As the UE and the P-CSCF did not establish any kind of mutual security mechanism at the SIP signalling level, the P-CSCF cannot guarantee that the REGISTER request really does originate from Tobias: for example, a malicious user could have constructed the request and sent it to the P-CSCF, without the P-CSCF knowing. Therefore, the P-CSCF adds the integrity-protected field with the value "no" to the Authorization header, before sending the request toward Tobias's home network:

```
REGISTER sip:home1.fr SIP/2.0
Authorization: Digest username="tobias_private@home1.fr",
               realm="home1.fr",
               nonce="",
               uri="sip:home1.fr",
               response="",
               integrity-protected="no"
```

10.6.4   S-CSCF challenges the UE

The S-CSCF, after receiving the REGISTER request, identifies the user by the private user identity found in the username field and downloads the AV from the HSS. Based on the data in the AV, it returns the WWW-Authenticate header in the 401 (Unauthorized) response and populates its fields as follows:

- in the nonce field it has the RAND and AUTN parameters, both 32 bytes long and Base 64-encoded (the nonce field may include additional server-specific data);
- in the algorithm field it has the value "AKAv1-MD5", which identifies the 3GPP AKA mechanism; and
- in the ik and ck extension fields it has the integrity and ciphering keys. Note that these two fields are not part of the original definition of the WWW-Authenticate header, which is defined in [RFC3261]. These fields are defined in [3GPP TS 24.229].

The WWW-Authenticate fields look like:

```
SIP/2.0 401 Unauthorized
WWW-Authenticate: Digest realm="home1.fr",
                  nonce=A34Cm+Fva37UYWpGNB34JP, algorithm=AKAv1-MD5,
                  ik="0123456789abcdeedcba9876543210",
                  ck="9876543210abcdeedcba0123456789"
```

After receiving the 401 (Unauthorized) response, the P-CSCF must remove and store the ik and ck fields from the WWW-Authenticate header, before sending the response toward the UE:

```
SIP/2.0 401 Unauthorized
WWW-Authenticate: Digest realm="home1.fr",
                  nonce=A34Cm+Fva37UYWpGNB34JP, algorithm=AKAv1-MD5
```

10.6.5 UE's response to the challenge

From the received AUTN parameter the ISIM application in Tobias's UE now discovers that it was really Tobias's home operator network that sent the 401 (Unauthorized) response. It can also derive from the AUTN that the SQN (sequence number) is still in sync between the HSS and the ISIM.

The received parameters as well as the shared secret allow the ISIM to generate the values for the response and hand them over to the UE. The UE adds the Authorization header to the second REGISTER request, including (among others) the following fields:

- The username field – which includes Tobias's private user identity.
- The nonce field – which is returned with the same value as it was received in the WWW-Authenticate header of the 401 (Unauthorized) response.
- The response field – which includes the authentication challenge RES that was derived by the ISIM from the received RAND and the shared secret.

The ISIM will also calculate the IK, which is also known by the P-CSCF. Based on this key (and other information – see Section 10.7) the UE and the P-CSCF establish IPsec SAs, over which the UE sends the second REGISTER request:

```
REGISTER sip:home1.fr SIP/2.0
Authorization: Digest username="user1_private@home1.fr",
               realm="home1.fr",
               nonce=A34Cm+Fva37UYWpGNB34JP, algorithm=AKAv1-MD5,
               uri="sip:home1.fr",
               response="6629fae49393a05397450978507c4ef1"
```

10.6.6 Integrity protection and successful authentication

The P-CSCF is now in a position to discover whether the received REGISTER request was modified on its way from the UE to the P-CSCF, as it can now check its integrity. If this check is successful, the P-CSCF adds the "integrity-protected" field with the value "yes" to the Authorization header and sends the REGISTER request toward Tobias's home network:

```
REGISTER sip:home1.fr SIP/2.0
Authorization: Digest username="user1_private@home1.fr",
               realm="home1.fr",
               nonce=A34Cm+Fva37UYWpGNB34JP, algorithm=AKAv1-MD5,
               uri="sip:home1.fr",
               response="6629fae49393a05397450978507c4ef1",
               integrity-protected="yes"
```

The S-CSCF now compares the received RES and the XRES that was included in the AV. If these two parameters are identical, then the S-CSCF has successfully authenticated the user. Only after that, will it proceed with normal SIP registration procedures.

10.6.7   Related standards

Specifications relevant to Section 10.6 are:

3GPP TS 33.102	Security architecture.
3GPP TS 33.203	Access security for IP-based services.
RFC2401	Security Architecture for the Internet Protocol.
RFC2403	The Use of HMAC-MD5-96 within ESP and AH.
RFC2404	The Use of HMAC-SHA-1-96 within ESP and AH.
RFC2617	HTTP Authentication: Basic and Digest Access Authentication.
RFC3310	Hypertext Transfer Protocol (HTTP) Digest Authentication Using Authentication and Key Agreement (AKA).

10.7   Access security – IPsec SAs

10.7.1   Overview

Section 3.8.4 describes how access security works in principle. Security via the Gm interface is achieved by means of IPsec SAs, which require specific handling at the SIP signalling level. This section describes how the UE and P-CSCF negotiate the security mechanism, how IPsec-related parameters are exchanged and how SAs are established and handled.

As the establishment of IPsec SAs is based on authentication of the user, new SAs are established during every re-authentication process. Consequently, new pairs of IPsec SAs have to be established between the UE and the P-CSCF.

10.7.2   Establishing an SA during initial registration

The initial REGISTER request as well as the 401 (Unauthorized) response are sent between the UE and the P-CSCF without any kind of protection. These two messages transport information that allows the UE and the P-CSCF to negotiate the security mechanism and to agree on the parameters and ports that will be used for the SAs.

During the registration process two pairs of IPsec SAs are established between the UE and the P-CSCF. Unless otherwise stated, such a set of two pairs of SAs is referred to as a "set of SAs", while a single or specific IPsec SA from these four is referred to as an "SA".

The four IPsec SAs are not static connections (e.g., TCP connections). They can be regarded as logical associations between the UE and the P-CSCF that allow the secure exchange of SIP messages.

Figure 10.5   SA establishment during initial registration.

A set of SAs facilitates four ports:

- the protected client port at the UE (uc1);
- the protected server port at the UE (us1);
- the protected client port at the P-CSCF (pc1); and
- the protected server port at the P-CSCF (ps1).

These ports are negotiated between the UE and the P-CSCF during initial registration (Figure 10.5) by using the Security-Client, Security-Server and Security-Verify headers of the SIP Security Mechanism Agreement (see Section 10.8).

The set of SAs needs to be established with a shared key. Unfortunately, the P-CSCF knows nothing about the security parameters that are shared between Tobias's ISIM application and the HSS in the home network. Therefore, the S-CSCF sends the IK and the CK to the P-CSCF within the WWW-Authenticate header in the 401 (Unauthorized) response. The P-CSCF must remove these two keys from the header and store them locally before sending the 401 (Unauthorized) response toward the UE. The IK is then used by the P-CSCF as the shared key for the set of SAs. The UE at the other end of the Gm interface calculates the IK from the received challenge in the 401 (Unauthorized) response and also uses it as the shared key (see Section 10.6.6).

By means of the IK, the P-CSCF and the UE can then establish the set of SAs between the four ports that were exchanged beforehand in the initial REGISTER request and its response:

- between uc1 and ps1 for sending SIP requests from the UE to the P-CSCF;
- between us1 and pc1 for sending SIP responses from the P-CSCF to the UE;
- between us1 and pc1 for sending SIP requests from the P-CSCF to the UE; and
- between uc1 and ps1 for sending SIP responses from the UE to the P-CSCF.

After their establishment the set of SAs gets assigned a temporary lifetime. Although the UE will send all subsequent requests and responses via this temporary set of SAs, the set of SAs cannot be taken into use until the authentication procedure between the UE and the S-CSCF has been finished. This is done in order to ensure that the security mechanism between the UE and the P-CSCF is based on successful authentication of the user.

When sending the 200 (OK) response to the UE, the P-CSCF will update the lifetime of the set of SAs by giving it the lifetime of the registration (as indicated in the expires value of the Contact header) plus 30 seconds. The UE will do the same after receiving the 200 (OK) response.

In the case of initial registration (as described here), both ends (i.e., P-CSCF and UE) will immediately afterwards take this set of SAs into use. This means that the P-CSCF will send all SIP messages that are directed toward the UE via the established set of SAs. The UE will in the same way send all SIP messages via the established set of SAs.

10.7.3   Handling of multiple sets of SAs in case of re-authentication

We have now seen how the first set of SAs is established during initial registration. As the establishment of a set of SAs is based on the authentication data that are sent from the S-CSCF in the 401 (Unauthorized) response, every re-authentication will generate a new set of SAs between the UE and the P-CSCF. Re-authentication procedures are described in Section 10.13. After successful re-authentication the UE and the P-CSCF will maintain two sets of SAs (Figure 10.6):

- the set of SAs that was already established and taken into use before the re-registration took place, which is now called the "old set of SAs"; and
- a new set of SAs that was established based on re-authentication, which is now called the "new set of SAs".

Figure 10.6   Two sets of SAs during re-authentication.

The major complication in this situation is that the P-CSCF cannot be sure whether the 200 (OK) response for the second REGISTER request has been received by Tobias's UE, as SIP defines no acknowledgement mechanism for received responses for any request other than an INVITE. If the UE has not received the 200 (OK) response for the second REGISTER, then it will not take the new set of SAs into use. Therefore, it has to wait until the UE sends a new request on the new set of SAs before it can take them into use. This means that, as long as the P-CSCF does not receive a request from the UE on the new set of SAs, it will:

- send incoming requests to the UE over the old set of SAs (i.e., from its protected client port pc1 to the UE's protected server port us1); and
- keep both sets of SAs active until one or both of them either expires or a new request from the UE is received.

In our example we assume that the UE has received the 200 (OK) for the second REGISTER request and, therefore, is aware that the authentication procedure was successful and the new set of SAs can be used. Unfortunately, the P-CSCF does not know this and will send incoming requests to the UE over the old set of SAs; therefore, the UE also needs to maintain both sets of SAs.

When the UE needs to send out a new request, it will send it by means of the new set of SAs, which will confirm to the P-CSCF that the new set of SAs can be taken into full use (Figure 10.7). Furthermore, at this moment the old set of SAs will not be immediately dropped, as the UE might have received or sent a request over it, which the remote

Figure 10.7 Taking a new set of SAs into use and dropping an old set of SAs.

end has not yet responded to. Therefore, the old set of SAs is kept for another $64 * T_1$ seconds (usually 128 seconds in an IMS environment), before it is dropped.

Note also that the UE cannot take the new set of SAs into use by sending a response – e.g., a 200 (OK) response – for a request – e.g., a MESSAGE request – that was received over the old set of SAs. The UE is forced either by the Via header of the P-CSCF or due to a TCP connection to send the response to the same port and over the same set of SAs as the request was received.

Whenever a set of temporary SAs is established the UE will drop all other SAs, other than the one over which it sent the last REGISTER request. Consequently, the UE never needs to handle more than two sets of SAs at the same time.

10.7.4   SA lifetime

During an ongoing authentication procedure the lifetime of a temporary set of SAs is restricted to 4 minutes. This guarantees that the authentication procedure can be finished. After successful authentication the lifetime of the new set of SAs is set to:

- Either the expiration time of the concluded registration plus 30 seconds. The expiration time of the registration is indicated in the expires parameter that is returned in the Contact header of the 200 (OK) response to the REGISTER.
- Or, if another set of SAs does already exist, to the lifetime of that already-existing set of SAs as long as its lifetime is longer than the expiration time of the just-concluded registration plus 30 seconds.

Whenever a re-registration takes place and is successful the P-CSCF and the UE have to update the lifetime of all existing SAs with the expiration time of the concluded re-registration plus 30 seconds, if that value is bigger than the already-assigned lifetime of the SAs.

Consequently, the SAs between the UE and the P-CSCF will be kept 30 seconds longer than Tobias is registered to the IMS network.

When the P-CSCF becomes aware that Tobias is no longer registered (e.g., by receiving a NOTIFY with Tobias's registration-state information which indicates network-initiated de-registration – see Section 10.14.3), the P-CSCF will drop all SAs toward the UE after $64 * T_1$ seconds.

10.7.5   Port setting and routing

Special attention has to be paid when it comes to the usage of SA ports, as they heavily influence the routing between the P-CSCF and the UE. As shown in Figure 10.6, Tobias's UE:

- will send all requests from its protected client port (2468);
- expects all responses to be received on its protected server port (1357);
- expects all requests to be received at its protected server port (1357);
- will send all responses to received requests from its protected client port (2468).

The P-CSCF, on the other hand:

- will send all requests toward the UE from its protected client port (8642);
- expects to receive all responses from the UE at its protected server port (7531);
- expects to receive all requests from the UE at its protected server port (7531); and
- will send all responses toward the UE from its protected client port (8642).

To ensure that all requests are sent via IPsec SAs:

- The UE will set its protected server port as part of its address:
  - in the Contact header of every request (including all REGISTER requests);
  - in the Via header of every request, besides the initial REGISTER.
- The UE will set the protected server port of the P-CSCF as part of the outbound proxy (i.e., P-CSCF) address in the Route header of every initial request that it sends.
- The P-CSCF will set its protected server port as part of its address:
  - in the Record-Route header of every initial request that is sent toward the UE;
  - in the Record-Route header of every response that carries the P-CSCF's Record-Route entry toward the UE (for detailed setting of port numbers in the Record-Route header see Section 11.3).

10.7.5.1 Port setting during registration

For example, Tobias's UE initially registers with the following information:

```
REGISTER sip:home1.fr SIP/2.0
Via: sip:[5555:1:2:3:4];branch=0uetb
Route: <sip:[5555::a:b:c:d];lr>
Security-Client: digest, IPsec-3gpp; alg=hmac-sha-1-96
               ;spi-c=23456789 ;spi-s=12345678
               ;port-c=2468; port-s=1357
Contact: sip:[5555::1:2:3:4]:1357
```

This means that the UE:

- Is going to establish IPsec SA with:
  - port 2468 as the protected client port (port-c parameter of the Security-Client header);
  - port 1357 as the protected server port (port-s parameter of the Security-Client header).
- Expects all incoming requests to be routed to its protected server port (port value in the Contact header).
- Will send this initial REGISTER request to the unprotected port 5060 of the P-CSCF, as no port value is given in the Route header.
- Will await all responses to this initial REGISTER request on unprotected port 5060, as no port value is given in the Via header.

The 401 (Unauthorized) response that is received afterwards by the UE will look like this:

```
SIP/2.0 401 Unauthorized
Via: sip:[5555:1:2:3:4];branch=0uetb
Security-Server: tls ;q=0.2, IPsec-3gpp; q=0.1
                 ;alg=hmac-sha-1-96
                 ;spi-c=98765432 ;spi-s=87654321
                 ;port-c=8642 ;port-s=7531
```

This means that the P-CSCF is going to establish an IPsec SA with:

- port 8642 as the protected client port (port-c parameter of the Security-Server header); and
- port 7531 as the protected server port (port-s parameter of the Security-Server header).

After this exchange the UE and the P-CSCF will set up the temporary set of SAs and the UE will then send the second REGISTER request already protected, which then will look like:

```
REGISTER sip:home1.fr SIP/2.0
Via: sip:[5555:1:2:3:4]:1357;branch=1uetb
Route: <sip:[5555::a:b:c:d]:7531;lr>
Contact: sip:[5555::1:2:3:4]:1357
```

Note that the Security-Client and Security-Verify headers are also included in this request (see Section 10.8), but, as they no longer have any influence on SA establishment and routing, they are not shown here. This means that the UE:

- expects all incoming initial requests to be routed to its protected server port (port value in the Contact header);
- sends this REGISTER request already over the temporary IPsec SA (i.e., to the protected server port of the P-CSCF – port value in the Route header); and
- expects all responses to this REGISTER request to be sent via the temporary IPsec SA (i.e., on its protected server port 1357 – port value in the Via header).

10.7.5.2 Port setting during re-authentication

As said before, every re-authentication will result in a new pair of IPsec SAs. When exchanging the security parameter indexes and protected port numbers for the new set of SAs according to the SIP Security Mechanism Agreement, the P-CSCF and the UE only change their protected client ports:

- the UE receives requests and responses for both sets of SAs via its protected server port (us1);

- the P-CSCF receives requests and responses for both sets of SAs via its protected server port (ps1);
- the UE uses a new protected client port (uc2) for sending requests and responses toward the P-CSCF over the new set of SAs; and
- the P-CSCF also uses a new protected client port (pc2) for sending requests and responses toward the UE over the new set of SAs.

This is due to the fact that two sets of SAs must not use the same port parameters. Furthermore, if the protected server ports change, this would cause major problems and would mean that:

- the UE would need to perform re-registration, as its registered contact includes the protected server port;
- the UE would need to send re-INVITE on all established sessions, as its contact information that was sent to the remote end includes the protected server port;
- the P-CSCF would receive from the UE all subsequent requests to every already-established dialog (including all subscriptions of the UE) on the P-CSCF's old, protected server port, as there is no possibility in SIP to change the route information for an already-established dialog.

This list is not complete, but it shows that changing the protected server port would cause a lot of problems for SIP routing. Therefore, it is essential that this value is not changed as long as the user stays registered.

10.7.5.3 Port settings for SIP requests other than REGISTER

The setting of the protected ports in non-REGISTER requests is described in more detail in Section 11.3.

10.7.5.4 Usage of ports with UDP and TCP

The previous sections showed how requests and responses are routed via one or more sets of SAs. In the chosen example, only UDP was used as a transport protocol. For TCP, however, there is a slight difference in these procedures.

When a request is sent out via UDP (Figure 10.8) the Via header indicates the IP address and port number to which all related responses should be routed. When TCP is used to send the request (Figure 10.9) the information in the Via header is overridden and the response is routed back to the same address and port that the request was received from. This draws attention to the nature of TCP as a connection-oriented

Figure 10.8   Request and response routing between UE and P-CSCF over UDP.

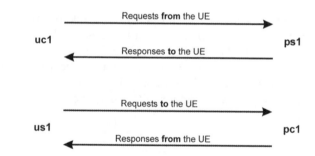

Figure 10.9   Request and response routing between UE and P-CSCF over TCP.

transport protocol. By applying this rule it is ensured that no additional TCP connection needs to be opened to send the response to a request that was received via TCP. This causes the routing of SIP messages between the P-CSCF and the UE to behave differently. The UE will set its protected server port (us1) in the Via header of every request that it sends out, regardless of whether UDP or TCP is used. All requests will originate from the UE's protected client port (uc1).

In the case of UDP the responses to such a request will be sent to the UE's protected server port (us1), as indicated in the Via header.

In the case of TCP the responses to such a request will be sent to the UE's protected client port (uc1), as the request originated from there. The same is true in the other direction (i.e., for requests sent from the P-CSCF toward the UE and their responses).

10.7.6   Related standards

Specifications relevant to Section 10.7 are:

3GPP TS 33.102	Security architecture.
3GPP TS 33.203	Access security for IP-based services.
3GPP TS 33.210	Network Domain Security (NDS); IP network layer security.
RFC2401	Security Architecture for the Internet Protocol.
RFC2403	The Use of HMAC-MD5-96 within ESP and AH.
RFC2404	The Use of HMAC-SHA-1-96 within ESP and AH.
RFC2451	The ESP CBC-Mode Cipher Algorithms.

10.8   SIP Security Mechanism Agreement

10.8.1   Why the SIP Security Mechanism Agreement is needed

The IMS in 3GPP Releases 5 and 6 makes use of IPsec as the security mechanism between the P-CSCF and the UE. IPsec is only one of several possible security mechanisms. IMS was designed to allow alternative security mechanisms over the Gm interface as well. Allowing such an openness usually creates backward compatibility problems because, for example, a Release 6-compliant UE would not be able to understand any alternative security mechanism, while it could be attached to a P-CSCF of a higher release that would already support alternatives to IPsec.

Therefore, the SIP Security Mechanism Agreement (Sip-Sec-Agree) was introduced to allow the UE and the P-CSCF to negotiate a common security mechanism for use between them. For current releases the only security mechanism is IPsec; however, it might be that some entities already support alternative mechanisms on a proprietary basis.

10.8.2   Overview

To make the example not too simple and boring, we assume that the UE supports IPsec and the HTTP digest, and that the P-CSCF supports IPsec and Transport Layer Security (TLS), with a preference toward TLS. It is not necessary for the reader of this chapter to have any knowledge of any of these mechanisms.

As we have seen, the initial REGISTER request is sent without any protection from the UE to the P-CSCF. To guarantee that a common security mechanism can be established, Tobias's UE advertises the mechanisms it supports in this initial REGISTER request within the Security-Client header, which includes a list of supported mechanisms.

The P-CSCF sends back a Security-Server header in the 401 (Unauthorized) response. The header includes a list of supported mechanisms from the P-CSCF end. Furthermore, the P-CSCF adds a preference (q-value) to each of the mechanisms.

Based on this information, both ends now know which common mechanisms are supported by the UE and the P-CSCF. If there is more than one common mechanism, the mechanism which was given the highest preference by the P-CSCF will be selected and applied. To guarantee that this mechanism can be established immediately, the P-CSCF will send further information in the 401 (Unauthorized) response to enable the UE to set up the mechanism: for example, in a non-IMS environment it could send a Proxy-Authenticate header when HTTP digest is the chosen mechanism.

As we saw in Section 10.7, the UE and the P-CSCF will then establish the security mechanism, which is in our case based on IPsec SAs. Afterwards, all messages between the two entities will be sent protected over these SAs.

Nevertheless, the initial REGISTER request and its response are still not protected. There is the slight chance that a malicious user has tampered with the messages or that an error has occurred over the vulnerable air interface.

As shown in earlier chapters, the second REGISTER request from the UE repeats all the information necessary for authentication and registration, both of which are performed with the S-CSCF. In order to guarantee that Sip-Sec-Agree-related information has not been changed as well, the UE:

- repeats the Security-Client header that it sent in the initial REGISTER in the second REGISTER request as well; and
- copies the content of the Security-Server header that was received in the 401 (Unauthorized) response from the P-CSCF into a Security-Verify header and sends it along with the second REGISTER request as well.

As long as the established Security-Association is used, the UE will always repeat the same Security-Verify header in every request that it sends to the P-CSCF.

During the exchange between the Security-Client (from the UE) and the Security-Server (from the P-CSCF) headers, the two ends also agree on some parameters for IPsec SAs: that is, they indicate to each other the protected client and server ports (port-c and port-s) as well as the security parameter indexes (SPIs: spi-c and spi-s).

10.8.3   Sip-Sec-Agree-related headers in the initial REGISTER request

In order to activate the agreement on the security mechanism, the UE includes the following information in the initial REGISTER request:

```
REGISTER sip:home1.fr SIP/2.0
Require: sec-agree
Proxy-Require: sec-agree
Security-Client: digest, IPsec-3gpp ;alg=hmac-sha-1-96
                ;spi-c=23456789 ;spi-s=12345678
                ;port-c=2468 ;port-s=1357
```

The Proxy-Require header includes the option tag "sec-agree"; this indicates that the next hop proxy, in this case the P-CSCF, must support the procedures for the Sip-Sec-Agree in order to process the request further. If the next hop proxy does not support Sip-Sec-Agree procedures, it would – based on handling of the Proxy-Require header, which is defined in [RFC3261] – send back a 420 (Bad Extension) response, including an Unsupported header with the option tag "sec-agree". As the P-CSCF in our example is fully IMS Releases 5 and 6-compliant, it of course supports SIP SA procedures and will not send this response to the UE.

Also, a Require header is included, indicating the sec-agree option tag. This is mandated to be included by [RFC3329], which defines the Sip-Sec-Agree. The Require header is used in the same way as the Proxy-Require header, but by the remote UE (not the proxy). It is there just in case a request (in this case the REGISTER request) is sent directly from the sending UE to the final destination (the S-CSCF), which would not look at the Proxy-Require header at all; this would mean that no negotiation of the security mechanism would take place. The Require header forces the receiving end to perform the sec-agree procedures.

As the P-CSCF is able to perform Sip-Sec-Agree procedures, it removes the sec-agree option tags from the Require and Proxy-Require headers before sending the request toward Tobias's home operator network.

Tobias's UE sends the list of supported security mechanisms to the P-CSCF within the Security-Client header. The P-CSCF will discover, based on the information given in this header, that Tobias's UE supports two security mechanisms: one is the HTTP digest ("digest") and the other is IPsec as used by 3GPP ("IPsec-3gpp"). These two mechanisms are separated by commas in the header. The list of parameters (separated by ";") for the latter includes:

- the algorithm (alg parameter) – used for IPsec encryption and protection (in this case it is the HMAC SHA 1-96 algorithm, which is defined in [RFC2404]);
- the protected client port (port-c) and the protected server port (port-s) – used from the UE's end for IPsec SAs; and
- the SPI – used for the IPsec SA that relates to the protected client port (spi-c) as well as the SPI used for the IPsec SA that relates to the protected server port (spi-s).

The P-CSCF will also remove the Security-Client header before sending the REGISTER request further.

Note that the IPsec-3gpp security mechanism is only relevant for the IMS. The example given here uses digest and TLS as possible additional security mechanisms in Sip-Sec-Agree-related headers. This is only done to explain the procedures behind the negotiation process.

10.8.4   The Security-Server header in the 401 (Unauthorized) response

When receiving a 401 (Unauthorized) response from the S-CSCF for a REGISTER request, the P-CSCF includes a list of supported security mechanisms in a Security-Server header in the response:

```
SIP/2.0 401 Unauthorized
Security-Server: tls ;q=0.2, IPsec-3gpp; q=0.1 ;alg=hmac-sha-1-96
                ;spi-c=98765432 ;spi-s=87654321
                ;port-c=8642 ;port-s=7531
```

In this example the P-CSCF supports two security mechanisms: 3GPP-specific usage of IPsec and TLS. It even gives a higher preference to TLS: should the UE also support TLS, this would be chosen to protect the messages between the UE and the P-CSCF.

Furthermore, the P-CSCF sends IPsec-related information about SPIs and protected client and server ports in the same way as the UE.

At the point of sending out the 401 (Unauthorized) response to the UE, the P-CSCF is already aware that the IPsec will be used as the security mechanism, as it knows that this is the only mechanism that is supported by both the UE and itself.

10.8.5   Sip-Sec-Agree headers in the second REGISTER

After receiving the 401 (Unauthorized) response the UE is able to set up IPsec SAs. When this has been done, it can use these SAs to send the second REGISTER request over it. In this REGISTER request it now includes the following related information:

```
REGISTER sip:home1.fr SIP/2.0
Require: sec-agree
Proxy-Require: sec-agree
Security-Verify: tls ;q=0.2, IPsec-3gpp ;q=0.1
                ;alg=hmac-sha-1-96
                ;spi-c=98765432 ;spi-s=87654321
                ;port-c=8642 ;port-s=7531
```

```
Security-Client: digest, IPsec-3gpp ;alg=hmac-sha-1-96
                 ;spi-c=23456789 ;spi-s=12345678
                 ;port-c=2468 ;port-s=1357
```

Once again the Require and Proxy-Require headers with the sec-agree option tag are there. They serve the same purpose as in the initial REGISTER (see Section 10.8.3) and will be repeated in every REGISTER request that is sent from the UE. The P-CSCF will always remove them before sending the request on, in the same way as it did for the initial REGISTER request. If either the Proxy-Require or the Require header (or both) are found empty after the sec-agree option tag has been removed, the P-CSCF will also remove this or these empty headers.

The Security-Verify header includes a copy of the received Security-Server header. The Security-Client header is simply re-sent as in the initial REGISTER request.

The P-CSCF will compare the two Security-Client headers that were received in the initial and this second REGISTER request and see whether they match. It will also compare the content of the Security-Server header that it sent in the 401 (Unauthorized) response and with the content of the Security-Verify header that it received in this second REGISTER request.

Before sending the REGISTER request any further, the P-CSCF will remove the Security-Client and Security-Server headers from it.

10.8.6   *Sip-Sec-Agree and re-registration*

The S-CSCF can decide to re-authenticate the UE during any re-registration procedure, and by doing so it will force the UE and the P-CSCF to establish a new set of IPsec SAs, as these IPsec SAs are based on the IK, which changes during each re-authentication procedure (see Section 10.13.2). Establishing a new set of IPsec SAs also means that a new set of spi's and new protected client and server ports are negotiated.

When sending the new REGISTER request for re-registration the UE cannot be sure whether the S-CSCF will request re-authentication. Therefore, it will add in every new REGISTER request a new Security-Client header with new values for the spi's and the protected client and server ports:

```
REGISTER sip:home1.fr SIP/2.0
Require: sec-agree
Proxy-Require: sec-agree
Security-Verify: tls ;q=0.2, IPsec-3gpp ;q=0.1 ;alg=hmac-sha-1-96
                 ;spi-c=98765432 ;spi-s=87654321
                 ;port-c=8642 ;port-s=7531
Security-Client: digest, IPsec-3gpp ;alg=hmac-sha-1-96
                 ;spi-c=23456790 ;spi-s=12345679
                 ;port-c=2470 ;port-s=1357
```

Note that the values for the spi's and the protected client port number have changed in the Security-Client header, in order to allow the set-up of a new set of SAs, should the

S-CSCF re-authenticate the UE. The protected server port of the UE has not changed and will be kept throughout the user's registration (see Section 10.7.3).

The content of the Security-Verify header is sent unchanged, as it is a copy of the latest received Security-Server header.

Both the P-CSCF and the UE will know, at the moment of receiving the response to this REGISTER request from the S-CSCF, whether new IPsec SAs have to be established: that is, whether a 401 (Unauthorized) response is received or whether a 200 (OK) response is received.

When a 401 (Unauthorized) response is received from the S-CSCF, the P-CSCF will add a new Security-Server header to the response, providing new values for its protected ports and new spi's:

```
SIP/2.0 401 Unauthorized
Security-Server: tls ;q=0.2, IPsec-3gpp ;q=0.1 ;alg=hmac-sha-1-96
               ;spi-c=98765434 ;spi-s=87654322
               ;port-c=8644 ;port-s=7531
```

Also, the P-CSCF will not change the value of its protected server port (7531). Consequently, the UE and the P-CSCF will now establish a new set of temporary SAs (see Section 10.7.3). The REGISTER request, which includes the response to the re-authentication challenge (see Section 10.6), will be sent over this new, temporary set of SAs and will include the following headers:

```
REGISTER sip:home1.fr SIP/2.0
Require: sec-agree
Proxy-Require: sec-agree
Security-Verify: tls ;q=0.2, IPsec-3gpp ;q=0.1 ;alg=hmac-sha-1-96
               ;spi-c=98765434 ;spi-s=87654322
               ;port-c=8644 ;port-s=7531
Security-Client: digest, IPsec-3gpp ;alg=hmac-sha-1-96
               ;spi-c=23456790 ;spi-s=12345679
               ;port-c=2470 ;port-s=1359
```

Once again, as during the initial registration procedure (Figure 10.10), the second REGISTER request repeats the Security-Client header that was sent in the latest REGISTER request (with the new values) and copies in the Security-Verify header the values of the Security-Server header that was received in the last 401 (Unauthorized) response. This second REGISTER request within the re-registration procedure no longer carries any information related to any previously established set of SAs.

10.8.7   Related standards

Specifications relevant to Section 10.8 are:

3GPP TS 33.203   Access security for IP-based services.
RFC2246   The TLS Protocol Version 1.0.

Figure 10.10 Sip-Sec-Agree during initial registration.

RFC2617   HTTP Authentication: Basic and Digest Access Authentication.
RFC3329   Security Mechanism Agreement for the Session Initiation Protocol
                 (SIP).

10.9 Compression negotiation

10.9.1 Overview

The ability to compress SIP messages over the air interface is essential for IMS. How Signalling Compression (SigComp) works is described in Section 3.17. This section shows how the UE and the P-CSCF indicate that they support SigComp and are both willing to use it.

The P-CSCF and an IMS UE must support SIP SigComp, but they are not mandated to use it. Therefore, they need a mechanism to express whether they are willing to apply SigComp.

[RFC3486] defines a new URI parameter "comp", which can be set to "comp = SigComp" by either the UE or a SIP proxy (in the case of IMS this applies only to the P-CSCF) in order to express its willingness to send or receive certain SIP messages compressed.

Tobias's UE will express its willingness to use SigComp with the P-CSCF that is already in the initial REGISTER request. The P-CSCF will give a similar indication in the 401 (Unauthorized) response. As these two SIP messages are sent without any protection, the P-CSCF and the UE will not create states (compartments) for SigComp at this point in time; this is to ensure that a malicious user–who wants, say, to start a Denial Of Service (DOS) attack against the P-CSCF – cannot overload the P-CSCF by forcing it to reserve memory for a huge number of unnecessary SigComp compartments.

State creation will only be done after an IPsec SA (see Section 10.7) between the UE and the P-CSCF has been established.

10.9.2   Indicating willingness to use SigComp

The "comp" parameter can be set:

- by the UE in the Contact header of the REGISTER request – this means that the UE is willing to receive every initial request that is destined for it compressed, as initial requests that are destined to the UE are routed based on the registered contact address;
- by the UE in the Contact header of any other initial request or the first response to an initial request – this means that the UE is willing to receive every subsequent request within this dialog compressed, as subsequent requests are routed based on the address in the Contact header of the initial request (from the originating end) or the first response to an initial request (from the terminating end);
- by the UE in the Via header of any request – this means that the UE is willing to receive all responses to this request compressed, as responses are routed based on the Via header in the related request;
- by the P-CSCF in its own entry to the Record-Route header that is sent toward the UE – this means that the P-CSCF is willing to receive subsequent requests within this dialog compressed, as subsequent requests are routed toward SIP proxies based on the entries in the Route header (which is generated from the Record-Route header); and
- by the P-CSCF in the Via header of any request – this means that the P-CSCF is willing to receive all responses to this request compressed, as responses are routed based on the Via header in the related request.

10.9.3   comp = SigComp parameter during registration

The initial REGISTER request by the UE will include the following compression-related information:

```
REGISTER sip:home1.fr SIP/2.0
Via: SIP/2.0/UDP sip:[5555::1:2:3:4];comp=SigComp ;branch=0uetb
Route: sip:[5555::a:b:c:d];lr
Contact: <sip:[5555::1:2:3:4]:1357;comp=SigComp>;expires=600000
```

The comp = SigComp parameter is included in the Via header and indicates that the UE is willing to receive all responses to this request compressed. Consequently, the P-CSCF may send the 401 (Unauthorized) response already compressed, but it will not create a state (i.e., a compartment) because of this.

The comp = SigComp parameter can also be found in the Contact header. This parameter will be included in every initial request that is received by the UE, as the

S-CSCF will replace the request URI (which points to *sip:tobias@home1.fr*) of every initial request with the registered contact address (i.e., *sip:[55551234]1357; comp = SigComp*).

The 401 (Unauthorized) response from the P-CSCF does not include any further information on the P-CSCF's ability to perform SigComp. The P-CSCF address that was discovered before the initial registration (see Section 10.3) cannot be discovered with the comp = SigComp parameter. As SIP messages should only be sent compressed when the comp = SigComp parameter is set in the address of the next hop, the UE would therefore not send any initial request to the P-CSCF compressed.

Subsequent requests (such as ACK, PRACK, UPDATE or BYE) could be sent compressed, as the routing from the UE to the P-CSCF would be based on the Record-Route entry of the P-CSCF (see Section 11.3.2), in which the P-CSCF can include the comp = SigComp parameter. The same is true for responses sent from the UE to the P-CSCF, as they are routed based on the Via header entry of the P-CSCF, which is also set by the P-CSCF itself.

Although it is a requisite for the comp parameter to indicate whether compression is used, 3GPP TS 24.229 does not make a clear requirement on compression of the initial message. One possibility would be that the UE just sends every initial request compressed, as the P-CSCF must support Sigcomp in any event.

Therefore, the UE adds the comp = SigComp parameter to the P-CSCF address that was discovered previously. Therefore, it can send out the second REGISTER request already compressed:

```
REGISTER sip:home1.fr SIP/2.0
Via: SIP/2.0/UDP sip:[5555:1:2:3:4];comp=SigComp;branch=1uetb
Route: sip:[5555:a:b:c:d]:7531;lr;comp=SigComp
Contact: <sip:[5555:1:2:3:4]:1357;comp=SigComp>;expires=600000
```

This REGISTER request is routed on the basis of the topmost Route header, which includes the P-CSCF address and the comp = SigComp parameter. As the parameter is already there, the UE can send the request compressed.

The 200 (OK) response to this REGISTER request will be sent from the P-CSCF to the UE on the basis of the Via header, and, as the UE also includes the comp = SigComp parameter, the P-CSCF will send it compressed.

10.9.4 *comp = SigComp parameter in other requests*

The handling of the comp = SigComp parameter in requests other than REGISTER is described in Section 11.4.

10.9.5 *Related standards*

The comp parameter is defined in [RFC3486]: Compressing the Session Initiation Protocol (SIP).

10.10   Access and location information

10.10.1   P-Access-Network-Info

The P-Access-Network-Info header is a 3GPP-specific header and indicates to the IMS network over which access technology the UE is attached to IMS. In our example the access technology is GPRS. It also includes the Cell Global ID (CGI), which indicates the location of the user.

Tobias's UE will include the P-Access-Network-Info header in every request (besides ACK and CANCEL requests) and every response (besides responses to the CANCEL request) that it sends out, but only if that request or response is sent integrity-protected (i.e., via an SA, see Section 10.7).

The first time this header is sent out is, therefore, within the second REGISTER request, which is sent after the 401 (Unauthorized) response has been received by the UE. The header looks like:

```
REGISTER sip:home1.fr SIP/2.0
P-Access-Network-Info: 3GPP-UTRAN-TDD
                     ;utran-cell-id-3gpp=234151D0FCE11
```

Tobias's S-CSCF will remove the P-Access-Network-Info header from every request or response that it sends toward another entity. The only exception from this rule are ASs that are in the same trust domain as the S-CSCF (see Section 10.5.10).

When the P-Access-Network-Info header is sent in an INVITE request that is sent for an emergency call, the P-CSCF and S-CSCF can determine from the Cell-ID which emergency centre is closest to the user and should be contacted. When writing this chapter the details for IMS emergency calls were still under discussion in 3GPP standardization groups. In the future there may be more applications that use the information contained in this header.

10.10.2   P-Visited-Network-ID

The P-Visited-Network-ID header indicates to Tobias's home network the identification of the network within which Tobias is currently roaming. The header is included by the P-CSCF to which Tobias's UE is attached. The information in this header will be used by the S-CSCF to check the roaming agreement with that visited network.

In this example it is assumed (see Section 9.1) that Tobias is roaming in Finland and is attached to the fictitious Finish operator Musta Kissa. As the P-CSCF is also provided by this operator, it will include a P-Visited-Network-ID header in every REGISTER request that it sends toward Tobias's home network. Within this header will be a string, from which the S-CSCF will recognize the visited network:

```
REGISTER sip:home1.fr SIP/2.0
P-Visited-Network-ID: "Kaunis Musta Kissa"
```

10.10.3   Related standards

3GPP-specific SIP headers are defined in [RFC3455]: Private Header (P-Header) Extensions to the Session Initiation Protocol (SIP) for the 3rd-Generation Partnership Project (3GPP).

10.11   Charging-related information during registration

The charging concept and the relevant entities in the network are described in Section 3.10. The current section only explains the handling and content of SIP headers that are related to charging during registration. Charging of IMS sessions is described in Section 11.7.7.

When the P-CSCF receives the initial REGISTER request, it creates an IMS Charging ID (ICID), which is valid for all IMS-related signalling as long as the user stays registered. The ICID value is transported from the P-CSCF to the S-CSCF in the P-Charging-Vector header:

```
REGISTER sip:home1.fr SIP/2.0
P-Charging-Vector:icid-value="AyretyU0dm+6O2IrT5tAFrbHLso=023551024"
```

The S-CSCF, when receiving this header, will store the ICID and will perform the charging procedures as described in Section 11.7.7. The P-Charging-Vector header is defined in [RFC3455]. Extensions to this header and its usage within the IMS are described in [3GPP TS 24.229].

10.12   User identities

10.12.1   Overview

Tobias needs to register within his home network in order to be able to originate a call toward his sister. In the example so far he has used the SIP URI *sip:tobias@ home1.fr* for registration. This is the user identity Tobias uses when he uses IMS services that are not work-related. Nevertheless, Tobias has a whole set of user identities that are registered at his operator in France, which are shown in Table 10.3.

During the initial registration procedure, Tobias can explicitly register only one of these URIs, which in the our example is *sip:tobias@home1.fr*. Nevertheless, the IMS

Table 10.3   Tobias's public user identities.

Registration set	SIP URI	tel URL
1	*sip:tobias@home1.fr*	*+44-123-456-789*
2	*sip:tobi@home1.fr*	*+44-123-456-111*
3	*sip:gameMaster@home1.fr*	

allows implicit and explicit registration of further public user identities:

- Some of the above-listed identities might automatically (implicitly) be registered by the network during the initial registration phase.
- Others might stay unregistered until Tobias explicitly requests them to be registered.

When receiving the 200 (OK) response for the second REGISTER request, both Tobias's terminal and the P-CSCF discover Tobias's default public user identity, which is received as the first URI in the P-Associated-URI header.

To find out more about the registration status of the other public user identities that are assigned to Tobias, the UE automatically subscribes to the registration-state event information that is provided by the S-CSCF in the home network. It is mandatory that the UE performs this subscription immediately after the initial registration has succeeded, because:

- the UE needs to get the registration status of the associated URIs;
- the subscription enables the network (S-CSCF) to force the UE to perform re-authentication (see Section 10.13.2);
- the subscription enables the network (S-CSCF) to de-register the user (see Section 10.14.3).

In parallel, the P-CSCF makes a subscription to the user's registration-state information as well, mainly to be informed about network-initiated de-registration (see Section 10.14.3).

10.12.2   Public and private user identities for registration

The identities that go into the first REGISTER request are read from the ISIM, one of the applications contained on the UICC within the UE. Data read from the ISIM include:

- the private user identity of the user;
- the public user identity of the user, which is used for registration; and
- the address of the SIP registrar of the user.

The private user identity is only used for authentication, which is described in Section 10.6. The public user identity is the SIP URI that Tobias is going to initially register. There may be more public user identities available for Tobias, some of them may even be stored on the ISIM; however, at the beginning only one is explicitly registered.

If the UE is not equipped with an ISIM, it will derive the identities and the address of the registrar from the USIM application that also resides in the UICC. The USIM includes all user-related data that are needed for Circuit-Switched (CS) and Packet-Switched (PS) domain registration and authentication. This is described in more detail in Section 10.12.3.

Armed with these parameters the UE can fill in the following fields of the initial REGISTER request:

```
REGISTER sip:home1.fr SIP/2.0
From: <sip:tobias@home1.fr>;tag=pohja
To: <sip:tobias@home1.fr>
Authorization: Digest username="tobias_private@home1.fr",
                realm="home1.fr",
                nonce="",
                uri="sip:home1.fr",
                response=""
```

The public user identity, as read from the ISIM, is put into the To and From headers. The value of the username field of the Authorization header takes the value of the private user identity, and the address of the registrar is put into the request URI of the request as well as in the realm and uri fields of the Authorization header.

10.12.3 Identity derivation without ISIM

When Tobias registers, his UE takes the SIP URI *sip:tobias@home1.fr* from the ISIM application that is running on the UICC that he got from his operator and put into his UE. The ISIM always holds at least one valid public user identity.

However, IMS services can also be provided to users who own UICC cards on which no ISIM application and, therefore, no valid public user identity are present. Therefore, the UE needs to create a temporary public user identity from the data available from the USIM application (see Section 3.5) and use this temporary identity for registration.

As the temporary public user identity is constructed from security-related data on the USIM, it must not be exposed to any entity outside the IMS. Therefore, it is treated as a "barred identity": that is, it is strongly recommended that the network reject any usage of this identity outside the user's registration.

In this case the private user identity will be derived from USIM data as well. It will take the format of an International Mobile Subscriber Identity (IMSI) as the user part, followed by a host part, which includes the Mobile Country Code (MCC) and the Mobile Network Code (MNC), both of which are included in the IMSI: for example, the private user identity of Tobias could look something like: *22330999999999@ims.mnc33. mcc222.3gppnetwork.org*

The domain name of Tobias's home network would be derived from the USIM as well and would appear just like the domain part of the user identity: that is, *ims.mnc33.mcc222.3gppnetwork.org*

10.12.4 Default public user identity/P-Associated-URI header

If Tobias had used a temporary public user identity for his initial registration, he would now have the problem that he would be registered but could not perform any other action (e.g., call his sister or subscribe to a service), as he is registered with an identity that he must not use further (barred identity). His terminal needs to know an identity that has been implicitly registered.

Whenever a user has successfully been authenticated and registered, the S-CSCF, therefore, sends in the 200 (OK) response for the REGISTER request the P-Associated-

URI header, which lists all the SIP URIs and tel URLs (i.e., public user identities), which are associated but not necessarily registered for the user. Only the first URI listed in this header is always a valid, registered public user identity and can be used by the UE and the P-CSCF for further actions.

The P-Associated-URI in the 200 (OK) response to Tobias's REGISTER request looks like:

```
SIP/2.0 200 OK
P-Associated-URI: <sip:tobias@home1.fr>, <sip:tobi@home1.fr>,
                  <sip:gameMaster@home1.fr>,
                  <sip:+44-123-456-789@home1.fr;user=phone>,
                  <sip:+44-123-456-111@home1.fr;user=phone>
```

From this information Tobias knows that at least the public user identity *sip:tobias @home1.fr* is registered. He also becomes aware that there are two more SIP URIs and two more tel URIs that he can use, but he does not know whether they are currently registered or not.

As the P-Associated-URI is only defined to transport SIP URIs, it includes the tel URLs that are associated with Tobias (*tel:+44-123-456-789* and *tel:+44-123-456-111*) in the format of SIP URIs.

10.12.5 UE's subscription to registration-state information

After the initial registration and authentication has succeeded, Tobias's terminal sends out a SUBSCRIBE request with the following information:

```
SUBSCRIBE sip:tobias@home1.fr SIP/2.0
Via: SIP/2.0/UDP [5555::1:2:3:4]:1357;comp=sigcomp;branch=4uetb
Route: <sip:[5555::a:b:c:d]:7531;lr>
Route: <sip:orig@scscf1.home1.fr;lr>
From: "Tobias" <sip:tobias@home1.fr>;tag=sipuli
To: "Tobias" <sip:tobias@home1.fr>
P-Preferred-Identity: "Tobias" <sip:tobias@home1.fr>
Event: reg
Expires: 600000
Accept: application/reginfo+xml
Contact: <sip:[5555::1:2:3:4]:1357>
Content-Length: 0
```

Again, not all the information that is included in the SUBSCRIBE request is shown here – the above headers are only those that are necessary to understand the nature of the registration-state event subscription and the routing of the request.

The subscription is intended for an event named "reg", which is the registration-state event package; it is identified in the Event header of the request.

The request URI identifies the user whose registration-state information is requested and, therefore, has to be set to a registered public user identity of Tobias and must be indicated in the To header.

In order to identify itself, Tobias's UE sets the To and P-Preferred-Identity headers to a SIP URI that it knows is currently registered. This is:

- Either the default public user identity that was received in the P-Associated-URI header (see Section 10.12.4).
- Or the public user identity that was explicitly registered during initial registration as long as that was not a temporary public user identity (see Section 10.12.3). If no temporary public user identity was used, it is possible that this explicitly registered public user identity is identical to the default public user identity.

The relationship between the P-Preferred-Identity header, the P-Asserted-Identity header and the To header is described in Section 11.2. The To header does not include a tag, as SUBSCRIBE is an initial request and, therefore, the tag will be assigned by the remote end (i.e., in this case the S-CSCF).

The Expires header is set to the same value as the expiry time of the initial registration (i.e., 600,000 seconds which is about 7 days).

The Accept header indicates that only information of the type "reginfo+xml" can be processed by the UE for this subscription, which is the Extensible Markup Language (XML) format for registration-state information.

The Contact header is set to the same contact information as used during registration: that is, the IP-Address of the UE which was assigned by the access network (see Section 10.2) and the protected server port that is used by the IPsec SA (see Section 10.7).

Finally, the Route headers are worth looking at: they include the route set that was received in the Service-Route header within the 200 (OK) response for the REGISTER request (see Section 10.5.8) and on top of it the address of the P-CSCF, which acts as an outbound proxy. This forces the SUBSCRIBE request to be routed first to the P-CSCF and then onward directly to the S-CSCF that was assigned during registration.

The P-CSCF, when receiving this SUBSCRIBE request from the UE, will check whether the information set in the P-Preferred-Identity header is a valid public user identity of Tobias. If this is the case, then it replaces the P-Preferred-Identity header with the P-Asserted-Identity header:

```
SUBSCRIBE sip:tobias@home1.fr SIP/2.0
P-Asserted-Identity: "Tobias" <sip:tobias@home1.fr>
```

The S-CSCF, when receiving this SUBSCRIBE request, will check whether the user identified by the P-Asserted-Identity header is registered at the S-CSCF. Afterwards, it checks whether it can provide the requested registration-state information of Tobias to the subscribing user (Figure 10.11). As Tobias is subscribing to his own

Figure 10.11 Tobias's subscription to his registration-state information.

registration-state information in this case, this is allowed. Therefore, the S-CSCF will immediately:

- return a 200 (OK) response for the SUBSCRIBE request, indicating that the subscription was successful;
- generate an XML document of type reginfo, including the current registration-state information for the URIs that are associated with Tobias; and
- send the generated XML document in a NOTIFY message toward the subscriber (in this case Tobias's UE).

As the 200 (OK) response and the NOTIFY request are sent approximately at the same time, the NOTIFY request may be received at the terminal before the 200 (OK) response. In this exceptional case, the UE must be able to create the related subscription dialog based on the NOTIFY request: that is, it must not discard the information received in the NOTIFY request just because it did not receive a prior 200 (OK) response to the SUBSCRIBE request.

10.12.6 *P-CSCF's subscription to registration-state information*

The P-CSCF also needs to subscribe to Tobias's registration-state information and, therefore, creates a SUBSCRIBE request, which looks similar to the one that the terminal generates:

```
SUBSCRIBE sip:tobias@home1.fr SIP/2.0
Via: SIP/2.0/UDP pcscf1.visited1.fi
From: <sip:pcscf1.visited1.fi>;tag=retiisi
To: "Tobias" <sip:tobias@home1.fr>
P-Asserted-Identity: <sip:pcscf1.visited1.fi>
Event: reg
Expires: 600000
Accept: application/reginfo+xml
Contact: <sip:pcscf1.visited1.fi>
Content-Length: 0
```

The main difference here is that it is the P-CSCF that subscribes to the registration-state information of Tobias (Figure 10.12); therefore, it has to identify itself in the From header and the P-Asserted-Identity header. As the P-CSCF is a trusted entity (see Section 3.6.4.2) it immediately puts a P-Asserted-Identity header into the request.

As the P-CSCF did not save any routing information during the initial registration phase for its own routing purposes, it has no knowledge about the S-CSCF that was assigned for the user and, therefore, cannot include any Route headers. Consequently, it will route the request on the basis of the host part of the request URI, which is "home1.fr" and can be resolved via DNS to one or more I-CSCF addresses of Tobias's home network. The I-CSCF then queries the HSS for the address of the S-CSCF that is assigned for the URI *sip:tobias@home1.fr* and sends the request to the S-CSCF.

Note that with this SUBSCRIBE request a new dialog is created, this time between the P-CSCF and the S-CSCF. This dialog has no relation to the UE's subscription to the very same registration-state information; therefore, the S-CSCF will generate separate NOTIFY requests, including the registration-state information of Tobias, for the UE's and for the P-CSCF's subscription.

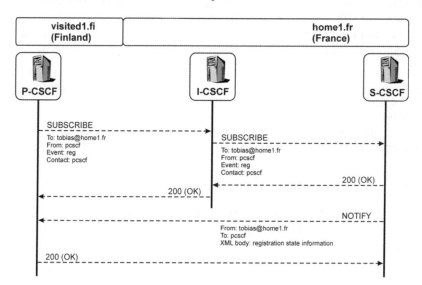

Figure 10.12 P-CSCF subscription to Tobias's registration-state information.

10.12.7   Elements of registration-state information

The S-CSCF generates a NOTIFY with Tobias's registration-state information immediately after a new subscription has been received and whenever the registration-
state information changes (e.g., when a new public user identity becomes registered).

In this section we only look at the NOTIFY request and registration-state information that is received by Tobias's terminal immediately after the subscription. This
information is identical to the information received by the P-CSCF at more or less
the same time.

The NOTIFY request as received by Tobias's UE includes – among others – the
following headers:

```
NOTIFY sip:[5555::1:2:3:4]:1357;comp=sigcomp SIP/2.0
Via: SIP/2.0/UDP scscf1.home1.fr;branch=nosctb
Via: SIP/2.0/UDP pcscf1.visited1.fi:7531;branch=nopctb
From: "Tobias" <sip:tobias@home1.fr>;tag=peruna
To: "Tobias" <sip:tobias@home1.fr>;tag=sipuli
Subscription-State: active;expires=599999
Event: reg
Content-Type: application/reginfo+xml
Contact: <sip:scscf1.home1.fr>
Content-Length: (...)
```

The things to note about this NOTIFY request are that:

• The To and From headers changed as this request was sent from the Notifier
  (S-CSCF) to the Subscriber (Tobias's UE). Although both headers have nearly
  identical content, their tags are different. The S-CSCF also has added a "To" tag
  ("peruna"), which now appears in the From header.
• A Subscription-State header has been added, which indicates that the subscription is
  active and will expire after 599,999 seconds.

10.12.8   Registration-state information in the body of the NOTIFY request

The registration-state information for the URIs associated with Tobias is included in
the body of the NOTIFY request and shown in detail in Section 10.12.9. Registration-
state information is a hierarchical list that consists of:

• The root element "reginfo", which includes registration-state information that is
  associated with one user.
• One or more "registration" sub-elements of the "reginfo" root element. A "registration" sub-element includes information about exactly one URI (i.e., one public user
  identity).
• Zero or more "contact" sub-elements of every "registration" sub-element.
  A "contact" sub-element includes information about an address that has been registered (or de-registered) for the URI in the "registration" sub-element.

Each registration sub-element can include the following attributes:

- The Address Of Record (AOR) attribute, which is followed by the URI for the public user identity.
- The ID attribute, which uniquely identifies the registration sub-element from among all the others.
- The state attribute of the registration sub-element, which indicates whether the indicated URI is in one of the following states:
  - "active" (i.e., registered);
  - "terminated" (i.e., de-registered);
  - "init" (i.e., in the process of being registered, such as when an initial REGISTER request has been received, but authentication procedures have not yet been finished).

Each contact sub-element includes the registered contact address and can include the following attributes:

- The ID attribute, which uniquely identifies the contact sub-element from among all the others.
- The state attribute of the contact sub-element, which indicates whether the indicated contact – in relation to the URI of the registration sub-element – is in one of the following states:
  - "active" (i.e., the URI is registered with this contact information);
  - "terminated" (i.e., the binding between the URI and this contact information has just been removed).
- The event attribute of the contact sub-element, which indicates the event that caused the latest change in the contact state attribute. The events can be:
  - registered – this event switches the contact address from the "init" state to the "active" state and indicates that the AOR has been explicitly registered (i.e., a valid REGISTER request has been received for this AOR and the related contact information is bound to it);
  - created – this event has the same meaning as the registered event, but indicates that the AOR has been implicitly registered (i.e., the binding was created automatically, such as when there is a received REGISTER request for another AOR);
  - refreshed – this event occurs when re-registration for an AOR takes place and may also occur implicitly (i.e., when re-registration for an associated AOR is performed);
  - shortened – this event occurs when the network shortens the expiry time of an AOR (e.g., to trigger network-initiated re-authentication, see Section 10.13.2);
  - deactivated – this event occurs when the binding is removed by the network (e.g., due to a network-initiated de-registration), allowing the user to perform a new initial registration attempt afterwards;
  - probation – with this event the network can de-register the user and request her to send a new initial registration after a certain time (dependent on the retry-after value);

○ unregistered – this event occurs when the user has explicitly unregistered the contact; and

○ rejected – this event occurs when the network does not allow the user to register the specific contact.

● Additional attributes, such as:

○ the expires attribute – which indicates the remaining expiration time of the registration for the specific contact address (it must be set for the shortened event, but is optionally set for other events);

○ the retry-after attribute – which is only set for the probation event and indicates how long the UE should wait before it can try again to register.

10.12.9   Example registration-state information

Tobias's registration-state information is included in the body of the NOTIFY requests that the S-CSCF sends out to the UE and the P-CSCF. It includes first of all an XML document heading:

```
<?xml version="1.0"?>
<reginfo xmlns="urn:ietf:params:xml:ns:reginfo" version="0" state="full">
```

The heading indicates the XML version in use (1.0). The registration information always starts with the root element, named "reginfo", which includes a number of attributes:

● The xmlns attribute points to the Uniform Resource Name (URN) that defines the XML document and the XML namespace.

● The version attribute always starts with the value "0" and is incremented by one every time a new (updated) version of the registration-state information is sent to the same recipient.

● The state attribute indicates that the following registration-state information is a full list of all the AORs that relate to Tobias. The first version ("0") of a reginfo document always needs to be sent as a complete list ("full") – subsequent information (starting from "1") can be sent as "partial" and will only include information that has changed since the last notification.

All the public user identities that relate to Tobias and their registration states are now listed in the document:

```
<registration aor="sip:tobias@home1.fr" id="a1" state="active">
<contact id="15" state="active" event="registered">
sip:[5555::1:2:3:4]
</contact>
</registration>
```

The first AOR or URI is *sip:tobias@home1.fr*, which we already know from the above example. It is currently registered (state = "active"). The content of this registration sub-element is one contact sub-element, which shows the binding that was created by the S-CSCF between *sip:tobias@home1.fr* and the contact information *sip[55551234]*. The event attribute is set to "registered", which indicates that this AOR was explicitly registered with this contact: this can be verified, because the registration procedures described in this chapter showed the AOR in the To header and the IP address in the Contact header:

```
<registration aor="tel:+44-123-456-789" id="a2" state="active">
  <contact id="16" state="active" event="created">
    sip:[5555::1:2:3:4]
  </contact>
</registration>
```

The next AOR is a tel URL that was implicitly registered (event = "created") with the same IP address as the first AOR. This implicit registration was made by the S-CSCF, based on the user profile of Tobias. In this case the telephone number is directly related to the SIP URI *sip:tobias@home1.fr*:

```
<registration aor="sip:tobi@home1.fr" id="b1" state="terminated">
</registration>
<registration aor="tel:+44-123-456-111" id="b2" state="terminated">
</registration>
```

These two AORs are currently not registered (state = "terminated") and, therefore, the registration sub-elements do not include any information at all. Finally, Tobias is also the game master of an online role-playing game. He takes his job in this game very seriously and is, therefore, always registered from a gaming console that has the address *sip[5555101102103104]*. The contact of the gaming console was explicitly registered (event="registered"). Nevertheless, Tobias also wants to stay informed about the ongoing status of the game when he is online with his IMS UE; therefore, this AOR was also implicitly registered (event = "created") by the S-CSCF when the REGISTER request for *sip:tobias@home1.fr* was received:

```
<registration aor="sip:gameMaster@home1.fr" id="c1" state="active">
  <contact id="45" state="active" event="registered">
    sip:[5555::101:102:103:104]
  </contact>

  <contact id="19" state="active" event="created">
    sip:[5555::1:2:3:4]
  </contact>
</registration>
</reginfo>
```

The last line of the registration-state information shows the tag that ends the XML document.

10.12.10   Multiple terminals and registration-state information

One or more public user identities can be registered from different terminals (i.e., different UEs). In our example it could be that Tobias also owns a simple paging device that uses his public user identity *sip:tobias@home1.fr*. This device would also need to perform registration procedures before being able to use IMS services. The registration of this device could take place over a different P-CSCF, but would end up in the same S-CSCF as the first registration.

After this paging device has registered, Tobias's UE and his P-CSCF would receive another NOTIFY message indicating that additional contact information for the public user identity is now available: that is, information that relates to the first AOR in the body of the NOTIFY would include the following information:

```
<?xml version="1.0"?>
<reginfo xmlns="urn:ietf:params:xml:ns:reginfo" version="1" state="partial">
  <registration aor="sip:tobias@home1.fr" id="a1" state="active">
    <contact id="15" state="active" event="registered">
      sip:[5555::1:2:3:4]
    </contact>
    <contact id="20" state="active" event="registered">
      sip:[5555::171:171:172:173]
    </contact>
  </registration>
</reginfo>
```

The second lot of contact information in the registration information relates to the paging device. Note that this is now the second lot of registration-state information that the UE receives. To make sure that no registration-state information was lost, the "version" parameter is set to "1" (the first lot of information had version="0"). While the first NOTIFY included complete registration-state information, the UE will receive only information about changed registration elements, in this case for the AOR *sip:tobias@home1.fr*. Consequently, the state parameter is set to "partial" (in the first lot of information it was set to "full").

10.12.11   Related standards

Specifications relevant to Section 10.12 are:

3GPP TS 23.003	Numbering, addressing and identification.
RFC3265	Session Initiation Protocol (SIP)-specific Event Notification.
RFC3325	Private Extensions to the Session Initiation Protocol (SIP) for Asserted Identity within Trusted Networks.
RFC3455	Private Header (P-Header) Extensions to the Session Initiation Protocol (SIP) for the 3rd-Generation Partnership Project (3GPP).
RFC3680	A Session Initiation Protocol (SIP) Event Package for Registrations.

Figure 10.13   User-initiated re-registration (without re-authentication).

10.13   Re-registration and re-authentication

10.13.1   User-initiated re-registration

Tobias's UE can at any time perform re-registration by sending a new REGISTER request (see Section 10.4) to the network (Figure 10.13). This happens when, say, the registration needs to be refreshed due to expiry of the registration time getting close. As the re-registration is handled in the same way as an initial SIP registration procedure, this is not further described here.

10.13.2   Network-initiated re-authentication

The IMS UE registers its contact information for a time of 600,000 seconds, which means that the binding of the registered public user identities and the physical IP address is kept for around 7 days in the S-CSCF. As user authentication procedures are directly coupled to registration procedures, this would mean that the S-CSCF has no means of re-authenticating the user within this time period. Certain conditions may, nevertheless, make it necessary for the S-CSCF to re-authenticate the UE.

To achieve this, the S-CSCF can reduce the expiry time of the user's registration. Let us assume that Tobias has already been registered for 3 hours and his home operator wants to perform a random re-authentication. The S-CSCF assigned to Tobias will reduce the expiry time of Tobias's registration to 600 seconds (exactly 10 minutes). Up to that moment Tobias's UE is unaware of the reduced registration time and would, therefore, not perform a re-registration, which is needed for re-authentication. To inform the UE about this, the S-CSCF makes use of the UE's subscription to the registration-state event package.

The S-CSCF generates a NOTIFY request for the registration-state event package, in which it indicates that it shortened the registration time and sends this NOTIFY request to Tobias's UE. On receiving this request the UE will immediately update the registration expiry time information.

Furthermore, all other subscribers to Tobias's registration-state information (e.g., the P-CSCF and the subscribed ASs) will receive a NOTIFY request from the S-CSCF with the updated state information.

After half the indicated time has elapsed (i.e., 300 seconds), the UE will send out a REGISTER request. From then on, the normal registration procedures as described in Section 10.4 will take place, during which the S-CSCF can authenticate the user again (see Section 10.6).

10.13.3   Network-initiated re-authentication notification

The NOTIFY message (Figure 10.14) that is sent from the S-CSCF to the UE will include the following information:

```
<?xml version="1.0"?>
<reginfo xmlns="urn:ietf:params:xml:ns:reginfo" version="2" state="partial">
  <registration aor="sip:tobias@home1.fr" id="a1" state="active">
    <contact id="15" state="active" event="shortened" expires="600">
      sip:[5555::1:2:3:4]
    </contact>
    <contact id="20" state="active" event="registered">
      sip:[5555::171:171:172:173]
    </contact>
  </registration>
  <registration aor="tel:+44-123-456-789" id="a2" state="active">
    <contact id="16" state="active" event="shortened" expires="600">
      sip:[5555::1:2:3:4]
    </contact>
  </registration>
  <registration aor="sip:gameMaster@home1.fr" id="c1" state="active">
    <contact id="45" state="active" event="registered">
      sip:[5555::101:102:103:104]
    </contact>
    <contact id="19" state="active" event="shortened" expires="600">
      sip:[5555::1:2:3:4]
    </contact>
  </registration>
</reginfo>
```

All registration and related contact states are still set to "active", but the latest event that occurred for contact is indicated as "shortened". The expires value shows that there are 10 minutes left for the UE to re-register. In this document only partial registration-state information is delivered (state="partial" in the document header), as the rest of the registration-state information has not been changed.

Figure 10.14   Network-initiated re-authentication.

10.13.4   Related standards

Specifications relevant to Section 10.13 are:

RFC3265   Session Initiation Protocol (SIP)-specific Event Notification.
RFC3680   A Session Initiation Protocol (SIP) Event Package for Registrations.

10.14   De-registration

10.14.1   Overview

All things come to an end at some point, and this is also true for the registration of a user to the IMS. Tobias might want to be undisturbed after he called his sister and switches off his mobile phone. When doing so, his phone sends another REGISTER request to the S-CSCF, including all the information we have already seen, but indicating that this time it is for de-registration (Figure 10.15). The S-CSCF will then clear all the information it has stored for Tobias, update the data in the HSS and send a 200 (OK) response to Tobias's UE.

Sometimes the network sees the need to de-register the user (Figure 10.16): maybe the S-CSCF needs to be shut down or maybe Tobias is using a pre-paid card and has ran out of money. In these cases the S-CSCF would simply send another NOTIFY message with registration-state information to Tobias's UE, this time indicating that he has been de-registered.

Figure 10.15   User-initiated de-registration.

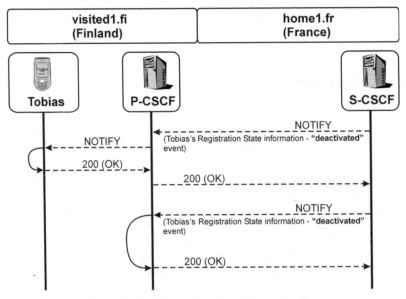

Figure 10.16   Network-initiated de-registration.

In both cases the S-CSCF will send NOTIFY requests to the P-CSCF and all other subscribers to Tobias's registration-state information, indicating that Tobias has been de-registered. By sending these NOTIFY requests the dialogs that were created during subscription to the registration-state event will also be terminated.

10.14.2   User-initiated de-registration

If Tobias decides to switch off his phone, the UE will send a REGISTER request to the network in order to de-register:

```
REGISTER sip:home1.fr SIP/2.0
Via: SIP/2.0/UDP [5555::1:2:3:4]:1357;comp=sigcomp;branch=99uetb
Route: sip:[5555::a:b:c:d]:7531;comp=sigcomp;lr
Max-Forwards: 70
From: <sip:tobias@home1.fr>;tag=ulkomaa
To: <sip:tobias@home1.fr>;tag=kotimaa
Authorization: Digest username="user1_private@home1.fr",
               realm="home1.fr",
               nonce=A34Cm+Fva37UYWpGNB34JP, algorithm=AKAv1-MD5,
               uri="sip:home1.fr",
               response="6629fae49393a05397450978507c4ef1",
               integrity-protected="yes"
               uri="sip:home1.fr",
Require: sec-agree
Proxy-Require: sec-agree
Security-Verify: tls ;q=0.2, IPsec-3gpp ;q=0.1 ;alg=hmac-sha-1-96
                ;spi-c=98765434 ;spi-s=87654322
                ;port-c=8644 ;port-s=7533
Security-Client: digest, IPsec-3gpp ;alg=hmac-sha-1-96
                ;spi-c=23456790 ;spi-s=12345679
                ;port-c=2472 ;port-s=1357
Contact: <sip:[5555::1:2:3:4]:1357;comp=sigcomp>;expires=0
Call-ID: apb03a0s09dkjdfglkj49222
CSeq: 49 REGISTER
Content-Length: 0
```

This is principally the same information that we have already seen in the other REGISTER requests; the main difference is that the expires value is set to 0, which means that the user wants to de-register the binding between the public user identity (in the To header) and the IP address (in the Contact header).

This REGISTER request will be routed in exactly the same way as every other REGISTER request (i.e., it will not follow the stored Service-Route). Therefore, it will:

- traverse the P-CSCF – which checks for integrity protection and adds the integrity-protected=yes flag to the Authorization header;

- traverse the I-CSCF – which will ask the HSS for the S-CSCF address that was selected for the user; and
- finally, be received at the S-CSCF – where de-registration will take place.

The S-CSCF will immediately send back a 200 (OK) response to the UE, which will also include the expires header set to the value 0.

Afterwards, the S-CSCF will generate NOTIFY requests to all subscribers of the registration-state information of Tobias, including Tobias's UE. Each of these NOTIFY requests will include the Subscription-State header set to the value "terminated", which indicates that the subscription to the registration-state information of that user has been terminated. For example:

```
NOTIFY sip:[5555::1:2:3:4]:1357;comp=sigcomp SIP/2.0
Subscription-State: terminated
```

The body of these NOTIFY requests will include Tobias's registration-state information:

```
<?xml version="1.0"?>
<reginfo xmlns="urn:ietf:params:xml:ns:reginfo" version="3" state="partial">
```

Once again, this XML document includes a "partial"-state notification, as it does not explicitly list those public user identities that have not been registered (see Section 10.12.7):

```
<registration aor="sip:tobias@home1.fr" id="a1" state="active">
  <contact id="15" state="terminated" event="unregistered">
    sip:[5555::1:2:3:4]
  </contact>

  <contact id="20" state="active" event="registered">
    sip:[5555::171:171:172:173]
  </contact>
</registration>
```

The public user identity *sip:tobias@home1.fr* is still active, as it was registered by Tobias's pager (see Section 10.12.10). Only the contact address of the mobile phone was set to terminated:

```
<registration aor="tel:+44-123-456-789" id="a2" state="terminated">
  <contact id="16" state="terminated" event="unregistered">
    sip:[5555::1:2:3:4]
  </contact>
</registration>
```

Tobias's tel URL has been completely de-registered:

```
  <registration aor="sip:gameMaster@home1.fr" id="c1" state="active">
    <contact id="45" state="active" event="refreshed">
      sip:[5555::101:102:103:104]
    </contact>

    <contact id="19" state="terminated" event="unregistered">
      sip:[5555::1:2:3:4]
    </contact>
  </registration>
</reginfo>
```

Finally, the gaming URI *sip:gameMaster@home1.fr* also remains registered, as another UE is still actively using it. Only the contact that was explicitly de-registered ended up being removed.

10.14.3   Network-initiated de-registration

Whenever the network sees the need to de-register the user or some of the user's identities, the S-CSCF will generate NOTIFY requests in the same way as described in Section 10.14.2, only the content of the XML document will look different:

```
<?xml version="1.0"?>
<reginfo xmlns="urn:ietf:params:xml:ns:reginfo" version="3" state="partial">
  <registration aor="sip:tobias@home1.fr" id="a1" state="terminated">
    <contact id="15" state="terminated" event="deactivated">
      sip:[5555::1:2:3:4]
    </contact>

    <contact id="20" state="terminated" event="deactivated">
      sip:[5555::171:171:172:173]
    </contact>
  </registration>
  <registration aor="tel:+44-123-456-789" id="a2" state="terminated">
    <contact id="16" state="terminated" event="deactivated">
      sip:[5555::1:2:3:4]
    </contact>

  </registration>
  <registration aor="sip:gameMaster@home1.fr" id="c1" state="terminated">
    <contact id="45" state="terminated" event="deactivated">
      sip:[5555::101:102:103:104]
    </contact>

    <contact id="19" state="terminated" event="deactivated">
      sip:[5555::1:2:3:4]
    </contact>
  </registration>
</reginfo>
```

All public user identities are now set to "terminated", as the network consequently de-registered every registration that was active for Tobias, even those from other terminals. The event has changed to "de-activated", which indicates that it was the network that de-registered, not the user.

10.14.4   *Related standards*

Specifications relevant to Section 10.14 are:

RFC3265   Session Initiation Protocol (SIP)-specific Event Notification.
RFC3680   A Session Initiation Protocol (SIP) Event Package for Registrations.

10.15   Early IMS Secuirty

10.15.1   *Example IMS registration with fallback to early IMS security*

In Section 3.19 the general principles and requirements for early IMS security are described. Figure 10.17 shows how early IMS security works in principle when the network requires IP-based authentication and the UE supports both – i.e., full IMS security as well as early IMS security. There are other possible scenarios, which are listed in Section 10.15.2.

When the UE establishes an IMS signalling PDP context, the GGSN creates a RADIUS "Accounting-Request START" request towards the GGSN, in which it indicates the user's Mobile Subscriber Integrated Services Digital Network (MSISDN) number (i.e., the phone number) as well as the IP address for the IMS-specific PDP context.

After establishing a signalling PDP context the UE will send out an initial REGISTER request, as described in the previous sections, including the Authorization header, a Security-Client header as well as the "sec-agree" option tag in the Require and the Proxy-Require header:

```
REGISTER sip:home1.fr SIP/2.0
From: <sip:tobias@home1.fr>;tag=pohja
To: <sip:tobias@home1.fr>
Authorization: Digest username="tobias_private@home1.fr",
               realm="home1.fr", nonce="",
               uri="sip:home1.fr", response=""
Security-Client: digest, IPsec-3gpp; alg=hmac-sha-1-96
               ;spi-c=23456789 ;spi-s=12345678
               ;port-c=2468; port-s=1357
Require: sec-agree
Proxy-Require: sec-agree
Contact: <sip:[5555::1:2:3:4]>;expires=600000
```

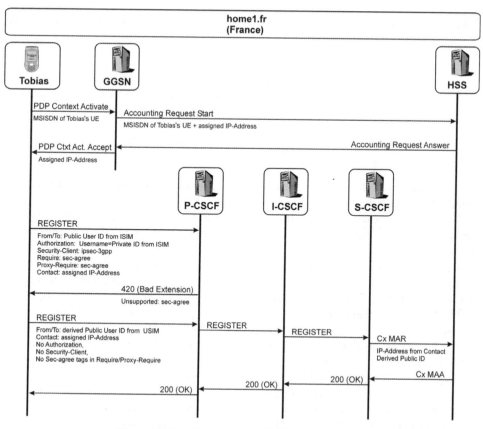

Figure 10.17   Example early IMS security flow.

When the P-CSCF, which in this example only supports early IMS security, receives this REGISTER request, it will reject it with a 420 (Unsupported) response, indicating that it does not support the Sip-Sec-Agree extension.

```
SIP/2.0 420 Unsupported
Unsupported: sec-agree
```

This rejection is used at the UE as an indication that early IMS security has to be applied.

It is for this reason that the UE constructs a new REGISTER message, which does not include the Authorization and Security-Client headers nor the sec-agree option tags in the Require and Proxy-Require header.

The identities used in the REGISTER request have now to be derived from the USIM appliction in the UE. This means that even if the UE is equipped with an ISIM application (in the case of early IMS security), it nevertheless needs to derive identities from the USIM (see Section 10.12.3). This is required as this second REGISTER request does not include the private user identity in the Authorization header. As a consequence, the user will always register a temporary public user identity in the case of early IMS security.

```
REGISTER sip:33.222.IMSI.3gppnetworks.org SIP/2.0
From: <sip:222330999999999@33.222.IMSI.3gppnetworks.org>;tag=t2
To: <sip:222330999999999@33.222.IMSI.3gppnetworks.org>

Contact: <sip:[5555::1:2:3:4]>;expires=600000
```

This second REGISTER request will now be routed through the network in exactly the same way as described in the previous sections of this chapter.

The P-CSCF, when receiving this REGISTER request, will detect that early IMS security is used, as the "sec-agree" option tag is not included in the Require and Proxy-Require headers.

The S-CSCF will detect that early IMS security is used, as no Authorization header is included in the received REGISTER request. It will, therefore, send the identity in the To header and the IP address from the Contact header to the HSS via the Cx interface.

The HSS checks whether the provided MSISDN is correlated with the IMSI provided in the user identity. Afterwards, it checks whether the IP address indicated by the S-CSCF was assigned by the GGSN to that specific MSISDN. As seen before, the HSS receives from the HSS the IP address that was assigned to that MSISDN during PDP context establishment procedures. Only if this check succeeds will the HSS give to the S-CSCF the indication that the authentication was successful. When receiving this indication from the HSS, the S-CSCF will respond to the received REGISTER request with a 200 (OK) response. On completion of this procedure, the user is authenticated by IMS.

In order to guarantee that SIP messages sent into the IMS network were really originated by the authenticated UE, the GGSN will check the IP addresses within these messages against the registered contact address.

10.15.2 Early IMS security scenarios

The scenarios given in Table 10.4 for early and full IMS security are possible. In the case when the UE only supports early IMS security but the network only supports full IMS security, the P-CSCF will reject the intitial REGISTER request with a 421 (Extension Required) response, indicating that it requires the Sip-Sec-Agree procedures to provide full IMS security.

```
SIP/2.0 421 Extension Required
Require: sec-agree
```

As the UE in this case does not support full IMS security, it will stop its attempt to register to the IMS network.

Table 10.4 Early IMS security registration scenarios.

UE supports	Network supports		
	Only early IMS security	Only full IMS security	Both
Only early IMS security	Early IMS security	P-CSCF rejects "early IMS" REGISTER request – no IMS registration possible	UE initiates with "early IMS" REGISTER – network accepts
Only full IMS security	P-CSCF rejects initial REGISTER request – no IMS registration possible	Full IMS security	Full IMS security
Both	P-CSCF rejects initial REGISTER request – UE sends "early IMS" REGISTER	Full IMS security	Full IMS security

Table 10.4 shows two cases in which IMS registration is not possible. This underlines the early nature of this alternative security mechanism. Only in networks that provide both mechanisms – early and full IMS security – will subscribers not suffer from service discontinuity.

11

An example IMS session

11.1 Overview

This chapter shows an example session between Tobias and his sister Theresa, who are both registered in their home networks and are both currently roaming in different countries (see Section 9.1).

The Session Initiation Protocol (SIP) and Session Description Protocol (SDP) are facilitated by the Internet Protocol Multimedia Subsystem (IMS) to ensure that Tobias and Theresa can talk to each other and see each other on the screens of their mobile phones. In order to achieve this within the wireless environment certain steps have to be taken:

- Tobias's UE needs to construct an INVITE request that includes a registered public user identity of Theresa in order to reach her – Section 11.2.4.
- All SIP messages must traverse the Proxy Call Session Control Functions (P-CSCFs) and the Serving-CSCFs (S-CSCFs) of both users – Section 11.3.
- All SIP messages are sent via the established IP security (IPsec) Security Associations (SAs) between User Equipment (UEs) and their P-CSCFs – Section 11.3.3.1.
- All SIP messages are sent compressed between UEs and their P-CSCFs – Section 11.4.
- The two UEs agree on the media streams that they will exchange. In the example case they will exchange a bidirectional audio stream, so that brother and sister can talk to each other, and a bidirectional media stream, so that they can also see each other – Section 11.5.
- The two UEs agree on a single codec for every media stream that they will exchange – Section 11.5.
- The networks will authorize the media for the session, so that the users can reserve the related resources – Section 11.7.2.
- Both UEs perform resource reservation (i.e., they set up the necessary media PDP contexts over which the media streams to and from the network will be transported) – Section 11.6.
- Theresa's UE will not get any indication that her brother is calling her before the resources for media sessions (i.e., the media PDP contexts) have been reserved at both ends, in order to be sure that media sessions can really be established – Section 11.6.4.

The IMS: IP Multimedia Concepts and Services Second Edition
Miikka Poikselkä, Georg Mayer, Hisham Khartabil and Aki Niemi © 2006 John Wiley & Sons, Ltd

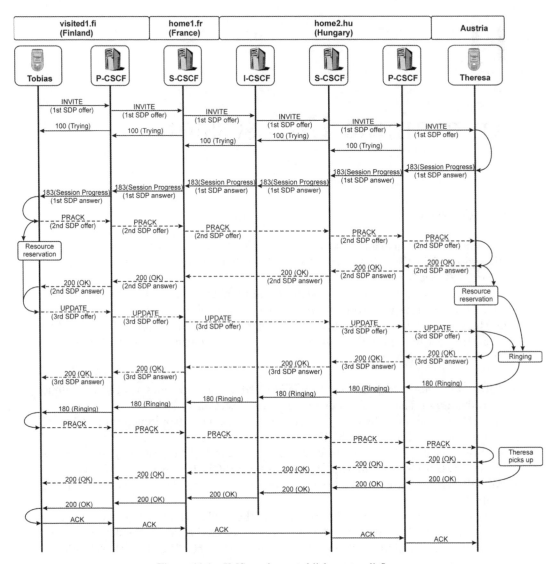

Figure 11.1   IMS session establishment call flow.

- The network elements will exchange charging information, so that media sessions can be billed correctly – Section 11.8.
- The S-CSCFs may initiate advanced services for their served users – Section 11.3.8.
- Theresa's UE will finally start to ring and Theresa will accept the session; this completes the session establishment phase.

After Tobias and Theresa have finished their call, they will hang up and one of their UEs will send a BYE request to the other UE – Section 11.9.1. The SIP message sequence for the example session will look like that shown in Figure 11.1.

11.2   Caller and callee identities

11.2.1   Overview

Section 10.12 described how an IMS user becomes aware during registration of the public user identities he can use and which of them are currently registered. Subsequently, the users – in our example Theresa and Tobias – will use these identities for different purposes. For every kind of dialog within the IMS – in this example the INVITE dialog – two identities are essential:

- A registered and authenticated public user identity of the calling user (Tobias) needs to be indicated in the request, in order to guarantee the identification of the user within his home network and for execution rights for extended services. This is provided in the P-Asserted-Identity header within the INVITE request.
- A registered and authenticated public user identity of the called user (Theresa) needs to be indicated in the request, in order to be able to contact the user and to execute services for her. This is provided in the request Uniform Resource Identifier (URI) of the INVITE request and in the P-Asserted-Identity header of the first response.

11.2.2   From and To headers

The INVITE request that Tobias's UE sends toward Theresa includes the following headers that are related to either his or her identity:

```
INVITE sip:theresa@home2.hu SIP/2.0
From: "Your Brother" <sip:tobi@brother.com>;tag=veli
To: "My beloved Sister" <sip:Theresa@sister.com>
P-Preferred-Identity: <sip:tobias@home1.fr>
Privacy: None
```

Obviously, the From and To headers can be set to any value the sender likes. We chose the wording in this example to clearly indicate that the values of these two headers in any request other than the REGISTER request have no influence on any IMS routing or security procedures – they can be freely set. The only information that is needed by the protocol itself are the tag parameters in these two headers.

Tobias's home network operator may have certain restrictions to some values that the To header can be set at. If this is the case, the home network can only reject the request if the setting of the From or To header does not fulfil the operator policy, because SIP does not allow any of these headers to be changed.

11.2.3   Identification of the calling user: P-Preferred-Identity and P-Asserted-Identity

11.2.3.1   Inclusion of the P-Preferred-Identity header by the originating UE

In the above example Tobias includes the P-Preferred-Identity header, which is optional. When present, it should include a registered public user identity of that user. In Section 10.12.5 we saw how Tobias became aware of all the public user identities that he can use. By means of the registration-state information to which his terminal subscribed, he also discovered which of these user identities he currently has registered.

If Tobias wanted to completely hide his identity from his sister, he would have needed to set the Privacy header to the "id" value. This value would force Theresa's P-CSCF to remove the P-Asserted-Identity header from the INVITE request, so that Theresa could only see the identity in the From header as the caller identification.

11.2.3.2   Originating P-CSCF includes the P-Asserted-Identity header

Tobias's UE will send out the INVITE request which is then first received by the P-CSCF. The P-CSCF checks whether the request was received over a valid IPsec SA. If the request was received unprotected (i.e., not over an SA), the P-CSCF will reject the request.

Afterwards, the P-CSCF inserts a P-Asserted-Identity header in the INVITE request, which replaces the received P-Preferred-Identity header, if one was received. The P-Asserted-Identity header is the only identity within an IMS dialog that is guaranteed to include a registered and authenticated public user identity of the user.

If a P-Prefered-Identity header is present, the P-CSCF will check whether the URI in the header is a currently registered public user identity of the user who sent in the request. It will discover whether it is a registered public user identity from the registration-state information it is subscribed to (see Section 10.12.6). The P-CSCF can ensure that a certain request was sent in by a specific user based on the SA it was received over (see Section 10.7). If both checks are successful, the P-CSCF will replace the P-Preferred-Identity header with a P-Asserted-Identity header that includes the same content.

If the P-Preferred-Identity header did not include a currently registered public user identity, then the P-CSCF will remove the header. In this case or when no P-Preferred-Identity header was received at all, the P-CSCF will add a P-Asserted-Identity header that includes the default public user identity of the user. How the default public user identity of the user is determined is described in Section 10.12.4:

```
INVITE sip:theresa@home2.hu SIP/2.0
From: "Your Brother" <sip:tobi@brother.com>;tag=veli
To: "My beloved Sister" <sip:Theresa@sister.com>
P-Asserted-Identity: <sip:tobias@home1.fr>
Privacy: None
```

11.2.3.3   Originating S-CSCF and P-Asserted-Identity header

On receiving this INVITE request the S-CSCF of Tobias's home network operator will identify Tobias by the information given in the P-Asserted-Identity header. The S-CSCF will also check the authentication and registration state of the public user identity indicated in the header. Because of these checks, the header serves as the main identification of the user for the whole dialog. Application Server (ASs) – see

Section 11.3.8 – base the identification and even the authentication of the user on this header as well.

Tobias's S-CSCF may add an additional URI to the P-Asserted-Identity header. In this example it adds the telephone Universal Resource Locator (tel URL) of Tobias to the header:

```
INVITE sip:theresa@home2.hu SIP/2.0
From: "Your Brother" <sip:tobi@brother.com>;tag=veli
To: "My beloved Sister" <sip:Theresa@sister.com>
P-Asserted-Identity: <sip:tobias@home1.fr>, <tel:+44123456789>
Privacy: None
```

Before the S-CSCF of Tobias's home network routes the request toward Theresa's home network, it will check whether that network is within its trust domain (see Section 3.8.4.2). If the S-CSCF and the home network of Theresa do not share the same trust domain, the S-CSCF will remove the P-Asserted-Identity header from the request, as long as the Privacy header is set to the value "id". For this example we assume that the two networks have a trust relationship that allows the header to be forwarded.

11.2.3.4   P-Asserted-Identity header at the terminating end

The P-CSCF of Theresa has to check the value of the Privacy header of the request. As it is not set to the value "id" it can send the P-Asserted-Identity header to Theresa's UE.

So, finally, the UE of Theresa receives the P-Asserted-Identity header. It can facilitate the information in the header by, say, displaying the "real name" of Theresa's caller.

11.2.4   Identification of the called user

11.2.4.1   The request URI

Let us look again at the INVITE message that Tobias sends. Its first line, the request URI, looks like:

```
INVITE sip:theresa@home2.hu SIP/2.0
```

The request URI is set to the final destination of the request (i.e., to Theresa's SIP URI). Section 11.3 explains how this URI is used for SIP and IMS routing procedures. But, this URI also identifies Theresa as the called user within her home network. This means that Theresa's S-CSCF will check whether this public user identity is currently registered and authenticated. If Theresa is currently not registered with this public user identity, the S-CSCF will return, say, a 404 (Not Found) response to the INVITE request and the call will fail or, based on the filter criteria for an unregistered user, will forward the INVITE to Theresa's voice mail box.

For our example we assume that Theresa has registered the public user identity that Tobias's UE put into the request URI.

11.2.4.2   The request URI and P-Called-Party-ID header

Another problem arises when this request is sent by Theresa's S-CSCF toward the terminating P-CSCF. The S-CSCF, which also acts as Theresa's SIP registrar, will re-write the request URI with the registered contact address of Theresa, in order to route the request to the UE at which Theresa is currently registered. Therefore, the public user identity in the request URI will be lost.

However, Theresa might have several public user identities and might want to know under which of them she receives a call: for example, she might have work-related user identities and others that relate to her private life. Maybe her UE even provides different ring tones for each of her user identities.

We already saw in Section 11.2.2 that Theresa cannot trust the To header in the request, as the originator can set it to any value – one that might be completely different from the public user identity in the request URI.

In order not to lose information about the public user identity that is used by Tobias to call his sister, the S-CSCF, when re-writing the request URI with the registered contact address, will add a P-Called-Party-ID header to the INVITE request. This P-Called-Party-ID header includes the public user identity that was received in the request URI:

```
INVITE sip:[5555::5:6:7:8]:1006 SIP/2.0
P-Called-Party-ID: sip:theresa@home2.hu
```

11.2.4.3   P-Asserted-Identity header

After receiving the INVITE request, Theresa's UE will send back a P-Preferred-Identity header in the first response to the INVITE request – the 183 (Session in Progress) response – which will include one of Theresa's public user identities:

```
SIP/2.0 183 Session in Progress
From: "Your Brother" <sip:tobi@brother.com>;tag=veli
To: "My beloved Sister" <sip:Theresa@sister.com>;tag=schwester
P-Preferred-Identity: <sip:theresa@home2.hu>
Privacy: None
```

The P-CSCF of Theresa will perform the same checks as described before for Tobias's P-CSCF (see Section 11.2.3) and will replace it by a P-Asserted-Identity header:

```
SIP/2.0 183 Session in Progress
From: "Your Brother" <sip:tobi@brother.com>;tag=veli
To: "My beloved Sister" <sip:Theresa@sister.com>;tag=schwester
P-Asserted-Identity: <sip:theresa@home2.hu>
Privacy: None
```

11.2.5   Related standards

Specifications relevant to Section 11.2 are:

RFC3323   A Privacy Mechanism for the Session Initiation Protocol (SIP).
RFC3325   Private Extensions to the Session Initiation Protocol (SIP) for Asserted
           Identity within Trusted Networks.
RFC3455   Private Header (P-Header) Extensions to the Session Initiation Protocol
           (SIP) for the 3rd-Generation Partnership Project (3GPP).

11.3   Routing

11.3.1   Overview

One of the most complex issues within IMS is the routing of requests, especially the
routing of initial requests. In our example, Tobias is sending the initial INVITE request
to Theresa. Consequently, a SIP dialog is created within which several subsequent
requests – such as ACK, PRACK, UPDATE and BYE – are sent.

Tobias's UE is unaware at the time of sending the INVITE request how Theresa's
UE can be reached. All it can provide is:

- The final destination of the INVITE request – which is the SIP URI of Theresa (one
  of her public user identities) that Tobias had to indicate (e.g., by selecting it from his
  phone book).
- The address of the P-CSCF – which is the outbound proxy of Tobias's UE and will
  be the first hop to route to. This address is obtained before SIP registration during
  the P-CSCF discovery procedures (see Section 10.3).
- The address of the S-CSCF – which was discovered during registration procedures by
  means of the Service-Route header (see Section 10.5.8).

Armed with this partial route information the INVITE request is sent on its way. It first
traverses the P-CSCF and then the S-CSCF that have been selected for Tobias.

Tobias's S-CSCF now has no further routing information available for the request
other than the final destination (i.e., the public user identity of Theresa, *sip:theresa@
home2.hu*). As Tobias's S-CSCF does not act as a registrar for Theresa, it can only
resolve the host part of the address: *home2.hu*. This domain name is sent to the Domain
Name System (DNS) server and the S-CSCF will receive back one or more Interrogating-CSCF (I-CSCF) addresses of Theresa's home network, will select one of them and
will send the INVITE request to it.

The I-CSCF just acts as the entry point to Theresa's home network. It asks the local
HSS for the address of the S-CSCF that was selected for Theresa and sends the INVITE
further on to the returned address.

Theresa's S-CSCF now acts as the registrar and replaces her SIP URI with the
contact address that she has registered. It does not send the request directly to Theresa's
UE, because it has not established an SA with it (see Section 10.7). The INVITE
request, therefore, is first sent to Theresa's P-CSCF. The S-CSCF knows the address

of the P-CSCF, as that was received within the Path header during Theresa's registration (see Section 10.5.9). The P-CSCF finally forwards the INVITE request to Theresa's UE over the IPsec SA.

This shows that for the initial request the route from Tobias to Theresa is put together piece by piece, as the originating UE and the CSCFs have only information about the next one or two hops that have to be traversed. In order to make further routing within the dialog easier, SIP routing mechanisms (see Section 12.12) will be used:

- All CSCFs put their addresses on top of the Via header – this allows all responses to the INVITE request to be sent back over exactly the same route as the request.
- All CSCFs, other than Theresa's I-CSCF, put their addresses on top of the Record-Route header – this allows all subsequent requests in the dialog to be sent over the CSCFs that put themselves in the Record-Route header. The I-CSCF in Theresa's home network fulfilled its routing task when it discovered the addresses of Theresa's S-CSCF; so, it is no longer needed on the route.

When sending out subsequent requests the UEs will include a list of Route entries, which will force the request to follow the recorded route (Figure 11.2). Routing issues related to the provision of services are explained in Section 11.3.8.

11.3.2 Session, dialog, transactions and branch

During session establishment and while the session is active, different types of signalling messages are exchanged and different kinds of relations between the two UEs are established.

The term "session" describes the media connections between the two users. Tobias wants to exchange audio and video media streams with his sister. This exchange of media is done on the so-called "bearer level": this means that Real-time Transport Protocol (RTP) packets are sent from the two items of UE to their Gateway GPRS Support Nodes (GGSNs) and the GGSNs exchange these packets between each other directly over the backbone. This session is established on the basis of SIP and SDP signalling that are exchanged via the "control plane".

A SIP dialog is the signalling relation between the two UEs which is needed to establish, modify and release the multimedia session. The dialog will be first established (with the INVITE request) and will exist as long as the related session is active. Every SIP dialog is identified by the value of the Call-ID header and by the tags in the To and the From headers of the SIP requests, which in our example look like:

```
From: "Your Brother" <sip:tobi@brother.com>;tag=veli
To: "My beloved Sister" <sip:theresa@home2.hu>;tag=schwester
Call-ID: apb03a0s09dkjdfglkj49555
```

The SIP dialog for the multimedia session between Tobias and Theresa starts with the INVITE request and ends with the 200 (OK) response for the BYE request.

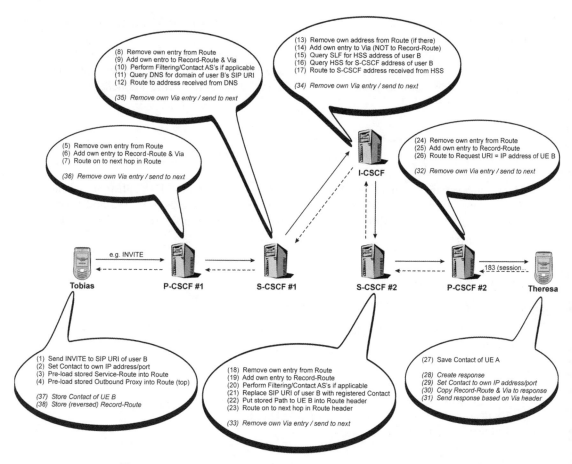

Figure 11.2   Routing an initial INVITE request and its responses.

A SIP transaction comprises a single SIP request and all the responses related to it. In order to establish the session, Tobias's UE sends an INVITE request to Theresa's UE. At the very outset it receives a 100 (Trying) response from the P-CSCF in response to the request. Afterwards, Theresa's UE responds with a 183 (Session in Progress), a 180 (Ringing) and, finally, a 200 (OK) response. All these five messages belong to the same dialog and have the same CSeq number:

```
From: "Your Brother" <sip:tobi@brother.com>;tag=veli
To: "My beloved Sister"
<sip:theresa@home2.hu>;tag=schwester
Call-ID: apb03a0s09dkjdfglkj49555
CSeq: 1112 INVITE
```

Every subsequent request sent from the same end (in this case from Tobias's UE) will have a higher CSeq number than the preceding request: this means that, for example, the first PRACK request includes CSeq 1113, the following UPDATE request CSeq 1114 and so forth.

Every entity – either UE or CSCF – will correlate the responses that are received for a sent request on the basis of the branch parameter that it added as a parameter to its Via header entry: for example, the P-CSCF of Tobias adds the following Via header to the INVITE request:

```
INVITE sip:theresa@home2.hu SIP/2.0
Via: SIP/2.0/UDP pcscf1.visited1.fi;branch=9pctb
```

The branch parameter identifies the INVITE transaction (i.e., the INVITE request and the responses to it) at the P-CSCF. It is constructed from the tags in the To and From headers, the Call-ID, the CSeq number and the information in the topmost Via header of the request.

11.3.3   Routing of the INVITE request

11.3.3.1   From Tobias's UE to the P-CSCF

Tobias's UE will include the following routing-related headers in the initial INVITE request:

```
INVITE sip:theresa@home2.hu SIP/2.0
Via: SIP/2.0/UDP [5555::1:2:3:4]:1357;branch=8uetb
Route: <sip:[5555::a:b:c:d]:7531;lr>
Route: <sip:orig@scscf1.home1.fr;lr>
Contact: <sip:[5555::1:2:3:4]:1357>
```

The destination of the request is Theresa's SIP URI, which is indicated in the request URI.

During registration the route between Tobias's UE and its S-CSCF in the home network was discovered by the Service-Route header (see Section 10.5.8). The UE pre-loads this first part of the route into the Route header and places the P-CSCF above it, because it always needs to contact its outbound proxy first.

Tobias's UE puts its IP address in the Contact header of the request, so that the remote UE B can directly reach it. It also adds its IP address to the Via header in order to receive responses to that request.

As the request is sent over established IPsec SAs (see Section 10.7), Tobias's UE puts:

- The protected server port of the UE (1357) as the port value in the Contact header, because it wants to receive all subsequent requests within this dialog via the established IPsec SA.
- The protected server port of the UE (1357) as the port value in the Via header, because it wants to receive all responses to the INVITE request via the established IPsec SA.

- The protected server port of the P-CSCF (7531) as the port value of the address of the P-CSCF in the Route header, because the P-CSCF must receive all requests from the UE via an established IPsec SA. The UE became aware of the P-CSCF's protected server port during SIP Security Mechanism Agreement procedures (see Sections 10.7.5 and 10.8).

The To and From headers are never used for routing purposes (see Section 11.2.2).

The INVITE request is now sent to the topmost entry in the Route header, which is the P-CSCF that serves Tobias.

11.3.3.2 From Tobias's P-CSCF to the S-CSCF

When receiving this request the P-CSCF:

- removes its own entry from the topmost Route header;
- checks that the request includes further routing information in accordance with the routing information it saved during registration (i.e., that the UE does not try to deviate from the Service-Route);
- puts its address at the top of the Via header, as it needs to receive all responses to the requests;
- adds the first Record-Route header and puts its own address there – this guarantees that all subsequent requests within this dialog will traverse the P-CSCF;
- does not include the protected server port number in both the Via and the Record-Route entry – the protected server port number identifies only the port over which the P-CSCF wants to receive SIP messages that are sent from the UE over the set of established IPsec SAs.

On completion of this, the P-CSCF again routes toward the topmost entry of the Route header, which in this case is the S-CSCF that serves Tobias:

```
INVITE sip:theresa@home2.hu SIP/2.0
Via: SIP/2.0/UDP pcscf1.visited1.fi;branch=9pctb
Via: SIP/2.0/UDP [5555::1:2:3:4]:1357;branch=8uetb
Record-Route: <sip:pcscf1.visited1.fi;lr>
Route: <sip:orig@scscf1.home1.fr;lr>
Contact: <sip:[5555::1:2:3:4]:1357>
```

11.3.3.3 From Tobias's S-CSCF to Theresa's home network (I-CSCF)

Tobias's S-CSCF removes its entry from the topmost Route header, which afterwards is empty and can be removed. It then adds its address at the top of the Record-Route and Via headers.

Afterwards, the S-CSCF will perform the procedures for service provisioning that are described in Section 11.3.8.

Having done this, it needs to route the request further. But, now there is a problem: there is no Route header left to point to the next hop. All the S-CSCF can do now is take the host part of the address of Theresa's public user identity that is indicated in the request URI (i.e., *home2.hu*) and resolve a SIP server in that domain from the DNS (see Chapter 15.1). In return, it receives one or more addresses of I-CSCFs that are located in the home network of Theresa. It takes one of them and sends the request there.

Note that the S-CSCF can only put the address of the I-CSCF into a Route header when it is aware that this I-CSCF is able to act as a loose router. In the example case the S-CSCF and the I-CSCF are in different networks and it is not assumed that the S-CSCF knows about the routing capabilities of the I-CSCF. Therefore, it sends the UDP packet that transports the initial INVITE to the I-CSCF address:

```
INVITE sip:theresa@home2.hu SIP/2.0
Via: SIP/2.0/UDP scscf1.home1.fr;branch=asctb
Via: SIP/2.0/UDP pcscf1.visited1.fi;branch=9pctb
Via: SIP/2.0/UDP [5555::1:2:3:4]:1357;branch=8uetb
Record-Route: <sip:scscf1.home1.fr;lr>
Record-Route: <sip:pcscf1.visited1.fi;lr>
Contact: <sip:[5555::1:2:3:4]:1357>
```

11.3.3.4 From the I-CSCF to Theresa's S-CSCF

The I-CSCF in Theresa's home network now needs to discover the address of the S-CSCF that is allocated to her. Even if Theresa is not currently registered, the I-CSCF may well be able to discover the address of a default S-CSCF as long as Theresa is subscribed to some services as an unregistered user. Information about the S-CSCF currently allocated to a user is stored in the Home Subscriber Server (HSS); as there are several HSSs within the network, the I-CSCF first has to query the Subscription Locator Function (SLF) to discover which HSS holds the data for Theresa. After the SLF returns the address of the HSS, the I-CSCF queries that HSS, which finally returns the address of the S-CSCF that serves Theresa. The I-CSCF now adds a Route entry at the top of the Route list and puts the received address of the S-CSCF into it. Furthermore, the I-CSCF:

- removes its entry from the topmost Route header, if one is present (in this example this is not the case);
- puts its address at the top of the Via list, in order to receive all responses for the INVITE request;
- does not put its address in the Record-Route, because it does not need to receive any subsequent requests in this dialog – the task of the I-CSCF is to find the S-CSCF of the called user, and, as this is done during initial request processing, there is no need for it to stay in the Route header.

The request once again goes toward the topmost entry in the Route header, which this time is set to Theresa's S-CSCF:

```
INVITE sip:theresa@home2.hu SIP/2.0
Via: SIP/2.0/UDP icscf1.home2.hu;branch=bicth
Via: SIP/2.0/UDP scscf1.home1.fr;branch=asctb
Via: SIP/2.0/UDP pcscf1.visited1.fi;branch=9pctb
Via: SIP/2.0/UDP [5555::1:2:3:4]:1357;branch=8uetb
Route: <sip:scscf2.home2.hu;lr>
Record-Route: <sip:scscf1.home1.fr;lr>
Record-Route: <sip:pcscf1.visited1.fi;lr>
Contact: <sip:[5555::1:2:3:4]:1357>
```

11.3.3.5 From Theresa's S-CSCF to the P-CSCF

Now Theresa's S-CSCF – her registrar – receives the INVITE request. Once again, it removes its entry from the Route header and puts itself in the Via and the Record-Route lists. Afterwards, it provides the services for Theresa as described in Section 11.3.8.

Having done so, the S-CSCF performs the actions of a registrar (i.e., it replaces the request URI, which is still set to Theresa's SIP URI, by her registered contact address). The registered contact address also includes the protected server port (1006) that is used to send requests from the P-CSCF to Theresa's UE via the established IPsec SA.

During Theresa's registration the S-CSCF received the Path header from the P-CSCF. It must now put the entries of the Path header in the Route header of the INVITE request. Were this not done, the request would immediately be sent to Theresa's UE, which could not accept the request as it had not established an IPsec SA with the S-CSCF.

The S-CSCF adds a new Route header, puts the P-CSCF address in it and, as this is now the topmost entry, sends the request to this address immediately:

```
INVITE sip:[5555::5:6:7:8]:1006 SIP/2.0
Via: SIP/2.0/UDP scscf2.home2.hu;branch=cscth
Via: SIP/2.0/UDP icscf1.home2.hu;branch=bicth
Via: SIP/2.0/UDP scscf1.home1.fr;branch=asctb
Via: SIP/2.0/UDP pcscf1.visited1.fi;branch=9pctb
Via: SIP/2.0/UDP [5555::1:2:3:4]:1357;branch=8uetb
Route: <sip:pcscf2.home2.hu;lr>
Record-Route: <sip:scscf2.home2.hu;lr>
Record-Route: <sip:scscf1.home1.fr;lr>
Record-Route: <sip:pcscf1.visited1.fi;lr>
Contact: <sip:[5555::1:2:3:4]:1357>
```

11.3.3.6 From the P-CSCF to Theresa's UE

The P-CSCF receives the request and does the usual: it removes the whole Route header, adds itself to the Record-Route and Via headers and then sends the request to the final destination indicated in the request URI – Theresa's UE:

```
INVITE sip:[5555::5:6:7:8]:1006 SIP/2.0
Via: SIP/2.0/UDP pcscf2.home2.hu:1511;branch=dpcth
Via: SIP/2.0/UDP scscf2.home2.hu;branch=cscth
Via: SIP/2.0/UDP icscf1.home2.hu;branch=bicth
Via: SIP/2.0/UDP scscf1.home1.fr;branch=asctb
Via: SIP/2.0/UDP pcscf1.visited1.fi;branch=9pctb
Via: SIP/2.0/UDP [5555::1:2:3:4]:1357;branch=8uetb
Route: <sip:pcscf2.home2.hu:1511;lr>
Record-Route: <sip:scscf2.home2.hu;lr>
Record-Route: <sip:scscf1.home1.fr;lr>
Record-Route: <sip:pcscf1.visited1.fi;lr>
Contact: <sip:[5555::1:2:3:4]:1357>
```

The entry of the P-CSCF in the Via header also includes the port number of the protected server port (1511), which was negotiated with Theresa's UE during the registration procedure in the same way as described for Tobias's registration in Section 10.7.5. This entry forces Theresa's UE to send all responses to this request over the established IPsec SA.

The selfsame protected server port value (1511) is put in the Record-Route header entry of the P-CSCF, where it expects to receive all subsequent requests from Theresa's UE that are sent in this dialog.

After Theresa's UE has received the INVITE request, it stores the received Contact value and the Record-Route header list, as it will route subsequent requests in the dialog based on them.

11.3.4   Routing of the first response

11.3.4.1   From Theresa's UE to the P-CSCF

Theresa's UE now creates a response to the received INVITE request, which is, due to the usage of preconditions (see Section 11.6), a 183 (Session in Progress) response.

The UE puts its own IP address in the Contact header to indicate the address it wants to use to receive subsequent requests in this dialog. The contact address also includes the protected server port of Theresa's UE (1006), which guarantees that all subsequent requests will be received via the established IPsec SA as well.

The Record-Route and Via headers of the INVITE request also go into the response. After doing so, Theresa's UE sends the response to the address and port number of the topmost entry in the Via header, which is the protected server port of the P-CSCF:

```
SIP/2.0 183 Session in Progress
Via: SIP/2.0/UDP pcscf2.home2.hu:1511;branch=dpcth
Via: SIP/2.0/UDP scscf2.home2.hu;branch=cscth
Via: SIP/2.0/UDP icscf1.home2.hu;branch=bicth
Via: SIP/2.0/UDP scscf1.home1.fr;branch=asctb
Via: SIP/2.0/UDP pcscf1.visited1.fi;branch=9pctb
Via: SIP/2.0/UDP [5555::1:2:3:4]:1357;branch=8uetb
```

```
Record-Route: <sip:pcscf2.home2.hu:1511;lr>
Record-Route: <sip:scscf2.home2.hu;lr>
Record-Route: <sip:scscf1.home1.fr;lr>
Record-Route: <sip:pcscf1.visited1.fi;lr>
Contact: <sip:[5555::5:6:7:8]:1006>
```

All other responses that are sent from Theresa's UE to this INVITE request will include the same Via header entries as the 183 (Session in Progress) response.

11.3.4.2   From Theresa's P-CSCF onward to Tobias's P-CSCF

The P-CSCF identifies the INVITE transaction the response belongs to by the branch parameter that it set in its own entry in the Via header. It then manipulates the routing information in the 183 (Session in Progress) response in the following way:

- it removes its own address from the the Via header;
- it re-writes its own Record-Route entry;
- it sends the request to the topmost entry in the Via header, which is the S-CSCF in Theresa's home network.

Why does the P-CSCF re-write its own Record-Route entry? Well, it does this to ensure that no other entity than Theresa's UE sends messages to the P-CSCF's protected server port that is used for the IPsec SA with the UE. If Theresa's S-CSCF were to send the next request (the PRACK) to the P-CSCF's protected server port (1511), the request would be dropped by the IPsec layer in the P-CSCF's protocol stack, as it had not been sent integrity-protected via the IPsec SA:

```
SIP/2.0 183 Session in Progress
Via: SIP/2.0/UDP scscf2.home2.hu;branch=cscth
Via: SIP/2.0/UDP icscf1.home2.hu;branch=bicth
Via: SIP/2.0/UDP scscf1.home1.fr;branch=asctb
Via: SIP/2.0/UDP pcscf1.visited1.fi;branch=9pctb
Via: SIP/2.0/UDP [5555::1:2:3:4]:1357;branch=8uetb
Record-Route: <sip:pcscf2.home2.hu;lr>
Record-Route: <sip:scscf2.home2.hu;lr>
Record-Route: <sip:scscf1.home1.fr;lr>
Record-Route: <sip:pcscf1.visited1.fi;lr>
Contact: <sip:[5555::5:6:7:8]:1006>
```

From then on, nothing of consequence happens to the response until it reaches Tobias's P-CSCF – every hop simply removes its own Via entry and sends the message toward the next entry in the Via. The Record-Route stays untouched. Note that other servers on the way back are permitted to re-write their Record-Route entries in order to distinguish requests received from different directions; however, this is not shown in this example, as it is an implementation option for a CSCF to carry out.

11.3.4.3 From Tobias's P-CSCF to his UE

When receiving the 183 (Session in Progress) response, Tobias's P-CSCF performs similar actions to Theresa's P-CSCF. It also re-writes its entry in the Record-Route header; but, instead of removing the protected server port value in its entry (as Theresa's P-CSCF did during the handling of the same response), it adds this port (7531). Consequently, it forces Tobias's UE to send all subsequent requests via the established IPsec SA.

As the P-CSCF routes the response on the basis of the Via header, it will send it to the protected server port of Tobias's UE (1357) – i.e., via the IPsec SA:

```
SIP/2.0 183 Session in Progress
Via: SIP/2.0/UDP [5555::1:2:3:4]:1357;branch=8uetb
Record-Route: <sip:pcscf2.home2.hu;lr>
Record-Route: <sip:scscf2.home2.hu;lr>
Record-Route: <sip:scscf1.home1.fr;lr>
Record-Route: <sip:pcscf1.visited1.fi:7531;lr>
Contact: <sip:[5555::5:6:7:8]:1006>
```

After receiving the response, Tobias's UE:

- stores the IP address of Theresa's UE, as received in the Contact header; and
- stores the Record-Route list after reversing the order of all entries in it.

11.3.5 Re-transmission of the INVITE request and the 100 (Trying) response

After having sent out the INVITE request, Tobias's UE waits for responses from Theresa's UE. It will wait until its timer T_1 – in IMS this is set to the value of 2 seconds – expires. Afterwards, it will re-transmit the INVITE request repeatedly until either a response to the request is received or until $128 \ (= 64 * T_1)$ seconds have elapsed; it will then indicate to Tobias that establishment of the session has failed.

As the INVITE request has to pass through several CSCFs all over Europe, it might take longer than 2 seconds for it to reach Theresa's UE, which has to construct the 183 (Session in Progress) response before once again travelling back to Finland.

To avoid frequent re-transmissions of the INVITE request from Tobias's UE, the P-CSCF sends back a 100 (Trying) response after it has received the INVITE request. This indicates that from now on the P-CSCF will take care of such re-transmissions.

The same is done by all other call-statefull SIP proxies on the route (see Figure 11.1). The 100 (Trying) is always stopped at the SIP proxy that was the latest to take over responsibility for re-transmission. For example, the S-CSCF of Theresa's home network sends back the 100 (Trying) response, which first reaches the I-CSCF. As the I-CSCF is not a call-statefull SIP proxy it just sends it on (based on the Via header). Next it reaches the S-CSCF of Tobias's home network. Tobias's S-CSCF has sent the 100 (Trying) response to the P-CSCF; consequently, it took over responsibility for the re-transmission of the INVITE request. Now the receipt of the 100 (Trying) response indicates that it no longer needs to re-transmit the INVITE request, as this responsibility is taken over by Theresa's S-CSCF.

11.3.6 Routing of subsequent requests in a dialog

When one of the two UEs needs to send a subsequent request within a dialog, it copies the stored Record-Route entries into the Route header of the new requests and the remote UE's IP address into the request URI.

The request then is routed toward the remote UE by strictly following the entries in the Route header (Figure 11.3). Every CSCF that is traversed puts itself in the Via header, in order to get all the responses to this request.

As the I-CSCF did not record any route in the beginning, it does not receive any subsequent request. For example, Tobias's UE has to send back a PRACK request to acknowledge the received 183 (Session in Progress) response (see Section 11.5.2). This PRACK request would include the following routing-related information:

```
PRACK sip:[5555::5:6:7:8]:1006 SIP/2.0
Via: SIP/2.0/UDP [5555::1:2:3:4]:1357;branch=82uetb
Route: <sip:pcscf1.visited1.fi:7531;lr>
Route: <sip:scscf1.home1.fr;lr>
Route: <sip:scscf2.home2.hu;lr>
Route: <sip:pcscf2.home2.hu;lr>
```

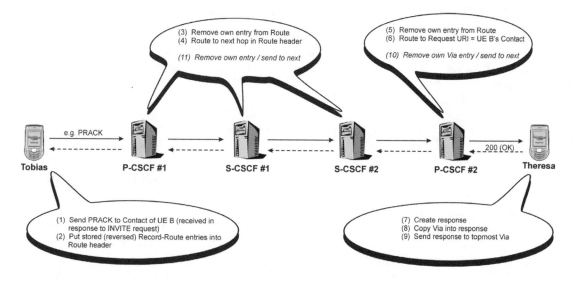

Figure 11.3 Routing of subsequent requests and their responses.

The PRACK request, therefore, will be routed:

- on the basis of the Route headers to Tobias's P-CSCF and his S-CSCF as well as Theresa's S-CSCF and her P-CSCF; and
- from Theresa's P-CSCF based on the address in the request URI, which Tobias's UE took from the received Contact header that was received in the 183 (Session in Progress) response to Theresa's UE over the IPsec SA.

A subsequent request within a dialog does not include a Contact header, as the addresses of the two UEs were already exchanged during the sending and receiving of the initial request and its first response. Furthermore, the CSCFs will not put any Record-Route headers in the request, because the route was already recorded during the initial request.

Theresa's UE will send back a 200 (OK) response to this PRACK request and will include the following routing information:

```
SIP/2.0 200 OK
Via: SIP/2.0/UDP scscf2.home2.hu;branch=c2scth
Via: SIP/2.0/UDP scscf1.home1.fr;branch=a2sctb
Via: SIP/2.0/UDP pcscf1.visited1.fi;branch=92pctb
Via: SIP/2.0/UDP [5555::1:2:3:4]:1357;branch=82uetb
```

This response will be routed back on the basis of the Via header entries. Record-Route headers are no longer returned.

11.3.7 *Standalone transactions from one UE to another*

For standalone transactions, such as MESSAGE or OPTIONS, the same routing procedures as those used for an initial request are performed, although record routing does not need to be done, because a standalone transaction does not create a dialog.

11.3.8 *Routing to and from ASs*

11.3.8.1 Filter criteria evaluation in the S-CSCF

Service provisioning in the IMS is achieved by ASs, which are contacted on the basis of initial filter criteria. When Tobias's or Theresa's S-CSCF receives an initial request, they will go through these filter criteria one by one and, if one or more of them matches, they will send the request toward the indicated AS. Filter criteria are downloaded by the S-CSCF from the HSS during registration and are part of Tobias's and Theresa's service profile; this is further described in Section 3.13.

In this example we assume that there are three ASs that have set filter criteria for requests that originate from Tobias (see Table 11.1). Tobias's S-CSCF will check these

Table 11.1 Filter criteria in Tobias's S-CSCF.

Element of filter criteria	Filter criteria 1	Filter criteria 2	Filter criteria 3
SPT: session case	Originating	Originating	Terminating
SPT: public user identity	*tel:+44-123-456-789*	*sip:tobias@hom1.fr* *tel:+44-123-456-789*	*sip:tobias@home1.fr*
SPT: SIP method	*	INVITE	SUBSCRIBE
Further SPT	—	—	SIP header: event: pres
Application server	*sip:as1.home1.fr;lr*	*sip:as2.home1.fr;lr*	*sip:as3.home1.fr;lr*

The asterisk signifies that any value is going to match.

filter criteria one by one against the information received in the INVITE request:

```
INVITE sip:theresa@home2.hu SIP/2.0
Via: SIP/2.0/UDP scscf1.home1.fr;branch=asctb
Via: SIP/2.0/UDP pcscf1.visited1.fi;branch=9pctb
Via: SIP/2.0/UDP [5555::1:2:3:4]:1357;branch=8uetb
Route: <sip:orig@scscf1.home1.fr;lr>
From: "Your Brother" <sip:tobi@brother.com>;tag=veli
To: "My beloved Sister" <sip:Theresa@sister.com>
P-Asserted-Identity: <sip:tobias@home1.fr>
Privacy: None
```

Filter criterion #1 does not match, because the P-Asserted-Identity header, which is checked against the Service Point Trigger (SPT) for the public user identity, does not include Tobias's tel URL.

Filter criterion #2 does match, because:

- the INVITE request is received from the originating user – the S-CSCF knows this from the user part it set in its Service-Route header entry (see Section 10.5.8) and which is now returned in the Route header;
- the P-Asserted-Identity is set to one of the public user identities that are filtered (*sip:tobias@home1.fr*);
- the SIP method is INVITE.

11.3.8.2 From the S-CSCF to the AS

Consequently, the S-CSCF now has to send the INVITE request to the AS (Figure 11.4) that is indicated in filter criterion #2. It also needs to take care that it receives the request again after the AS has fulfilled its actions, because the S-CSCF needs to evaluate filter criterion #3 and to send the request toward the home network of

Figure 11.4   Routing to an application server.

Theresa. To achieve this, the S-CSCF adds a set of routing-related headers by putting:

- its own address at the top of the Route headers, in order to receive the INVITE request back from the AS;
- the address of the AS at the top of the Route headers, in order to route the INVITE request to the AS as the next hop;
- its own address on top of the Record-Route headers, so that it stays on the route for subsequent requests as well;
- its own address on top of the Via headers, so that it receives all responses to the request.

In addition to this, the S-CSCF will add an implementation-specific dialog identifier to its own Route header entry, which it has just added. It sets this dialog identifier to a value that allows it to identify the dialog that is created with this INVITE. What is the purpose of this?

The AS (as described in Section 3.13) could decide to act as a Back-to-Back User Agent (B2BUA) and terminate the INVITE request locally. It would then send a new INVITE request with a new Call-ID toward the S-CSCF. As this AS would use the URI that is included in the Route header for routing to the next hop, the S-CSCF would also get back the dialog identifier. Consequently, it recognizes that the new Call-ID is in fact related to the previously received INVITE request. The S-CSCF would then return to the point where it stopped after sending out the INVITE request to the AS (i.e., it will evaluate the third Filter Criteria).

We will not further consider the scenario of an AS acting as a B2BUA in this example:

```
INVITE sip:theresa@home2.hu SIP/2.0
Via: SIP/2.0/UDP sip:scscf1.home1.fr;branch=9sc2as2tb
Via: SIP/2.0/UDP pcscf1.visited1.fi;branch=9pctb
Via: SIP/2.0/UDP [5555::1:2:3:4]:1357;branch=8uetb
Route: <sip:as2.home1.fr;lr>
Route: <sip:scscf1.home1.fr;lr>;dia-id=6574839201
Record-Route: <sip:scscf1.home1.fr;lr>
Record-Route: <sip:pcscf1.visited1.fi;lr>
```

11.3.8.3   From the AS back to the S-CSCF

When receiving the INVITE request, the AS:

- will remove the topmost entry in the Route header that is pointing to the AS;
- provide the service based on the information in the request;
- may modify the request in compliance with [RFC3261] (e.g., add another header);
- put its own address at the top of the Via list.
- decide whether it wants to receive subsequent requests within this dialog – if it wants to then it puts its own address at the top of the Record-Route list (in this example we assume that the AS wants to stay in the Route header);
- route the INVITE request based on the topmost Route header back to the S-CSCF.

Our INVITE request now looks like:

```
INVITE sip:theresa@home2.hu SIP/2.0
Via: SIP/2.0/UDP sip:as2.home1.fr;branch=vas2tb
Via: SIP/2.0/UDP sip:scscf1.home1.fr;branch=9sc2as2tb
Via: SIP/2.0/UDP pcscf1.visited1.fi;branch=9pctb
Via: SIP/2.0/UDP [5555::1:2:3:4]:1357;branch=8uetb
Route: <sip:scscf1.home1.fr;lr>;dia-id=6574839201
Record-Route: <sip:as2.home1.fr;lr>
Record-Route: <sip:scscf1.home1.fr;lr>
Record-Route: <sip:pcscf1.visited1.fi;lr>
```

11.3.8.4   Evaluation of further filter criteria at the S-CSCF

When it receives the INVITE request again, the S-CSCF will then evaluate filter criterion #3; this does not match, because the SIP method is not SUBSCRIBE (as indicated in the SPT). Consequently, the S-CSCF will continue with its normal routing procedures, as described in Section 11.3.3.3 (i.e., it will send the INVITE request to the I-CSCF of Theresa's home network).

Because service provisioning further complicates the routing, no further attention is paid to it throughout this example; the Via, Route and Record-Route headers added here will likewise not be shown in the rest of this example.

11.3.9 *Related standards*

The IMS service provisioning architecture is further described in:

3GPP TS 23.218   IP Multimedia (IM) session handling; IM call model; Stage 2.

11.4   Compression negotiation

11.4.1 *Overview*

The basic compression capabilities of the UE and the P-CSCF have already been negotiated during the registration procedures (see Section 10.9). Consequently, all requests and responses that are sent between the two sets of UE and their P-CSCFs will be compressed.

   In this example we only show how compression parameters are basically set during session establishment and concentrate only on the compression between Theresa's UE and her P-CSCF. The procedures for Tobias's end are identical.

11.4.2 *Compression of the initial request*

We assume that Theresa has registered a contact address that included the comp=SigComp parameter at her S-CSCF. Therefore, Theresa's S-CSCF will include this parameter when it acts as a SIP registrar and re-writes the request URI of the INVITE request (see Section 11.3.3.5).

```
INVITE sip:[5555::5:6:7:8]:1006;comp=SigComp SIP/2.0
```

When the P-CSCF receives this request, it will route it toward Theresa's UE based on the request URI and, as the comp=SigComp parameter is included, it will send it compressed. Furthermore, the P-CSCF will:

- add the comp=SigComp parameter to its entry in the Via header, so that Theresa will send all responses to the INVITE request compressed;
- add the comp=SigComp parameter to its entry in the Record-Route header, so that Theresa will send all subsequent requests in this dialog compressed.

Our INVITE request now looks like:

```
INVITE sip:[5555::5:6:7:8]:1006;comp=SigComp SIP/2.0
Via: SIP/2.0/UDP pcscf2.home2.hu:1511;comp=SigComp
Via: SIP/2.0/UDP scscf2.home2.hu
Via: SIP/2.0/UDP icscf1.home2.hu
Via: SIP/2.0/UDP scscf1.home1.fr
Via: SIP/2.0/UDP pcscf1.visited1.fi
Via: SIP/2.0/UDP [5555::1:2:3:4]:1357
Record-Route: <sip:pcscf2.home2.hu:1511;comp=SigComp;lr>
Record-Route: <sip:scscf2.home2.hu;lr>
Record-Route: <sip:scscf1.home1.fr;lr>
```

```
Record-Route: <sip:pcscf1.visited1.fi;lr>
Contact: <sip:[5555::1:2:3:4]:1357;comp=SigComp>
```

11.4.3 Compression of responses

When Theresa's UE constructs the 183 (Session in Progress) response to the INVITE request, it will add its IP address in the Contact header and will also include the comp=SigComp parameter there. Based on this entry all subsequent requests will be routed from Theresa's P-CSCF to her UE.

The Record-Route headers are stored by Theresa's UE: whenever the UE sends a subsequent request (e.g., a PRACK or BYE request) it will send it compressed due to the compression parameter being set in the topmost entry.

Theresa's UE will send the 183 (Session in Progress) response to the P-CSCF and, as that shows the comp=SigComp parameter in the Via header, it will also send this response compressed:

```
SIP/2.0 183 Session in Progress
Via: SIP/2.0/UDP pcscf2.home2.hu:1511;comp=SigComp;lr
Via: SIP/2.0/UDP scscf2.home2.hu
Via: SIP/2.0/UDP icscf1.home2.hu
Via: SIP/2.0/UDP scscf1.home1.fr,
Via: SIP/2.0/UDP pcscf1.visited1.fi
Via: SIP/2.0/UDP [5555::1:2:3:4]:1357
Record-Route: <sip:pcscf2.home2.hu:1511;comp=SigComp;lr>
Record-Route: <sip:scscf2.home2.hu;lr>
Record-Route: <sip:scscf1.home1.fr;lr>
Record-Route: <sip:pcscf1.visited1.fi;lr>
Contact: <sip:[5555::5:6:7:8]:1006;comp=SigComp>
```

We saw in Section 11.3.4.2 that the P-CSCF re-writes its entry in the Record-Route header to remove its protected server port number from it. When doing so, it also removes the compression parameter from it, because it wants to receive compressed requests from the UE, and not from the S-CSCF:

```
SIP/2.0 183 Session in Progress
Via: SIP/2.0/UDP scscf2.home2.hu
Via: SIP/2.0/UDP icscf1.home2.hu
Via: SIP/2.0/UDP scscf1.home1.fr,
Via: SIP/2.0/UDP pcscf1.visited1.fi
Via: SIP/2.0/UDP [5555::1:2:3:4]:1357
Record-Route: <sip:pcscf2.home2.hu;lr>
Record-Route: <sip:scscf2.home2.hu;lr>
Record-Route: <sip:scscf1.home1.fr;lr>
Record-Route: <sip:pcscf1.visited1.fi;lr>
Contact: <sip:[5555::5:6:7:8]:1006;comp=SigComp>
```

11.4.4   Compression of subsequent requests

After the 183 (Session in Progress) request has reached Tobias's UE, it will send a PRACK request in the same dialog. The request URI of this PRACK request will be set to the address received in the Contact header of the 183 (Session Progress) response (see Section 11.3.5), which includes the compression parameter:

```
PRACK sip:[5555::5:6:7:8]:1006;comp=SigComp SIP/2.0
```

When this PRACK request is received at Theresa's P-CSCF, it will again be routed to Theresa's UE based on the request URI and can be sent compressed as it includes the compression parameter. Once again, the P-CSCF will add the comp=SigComp parameter to the Via header of the PRACK, so that Theresa can send the 200 (OK) response to it compressed.

Following these procedures, all requests and responses within the dialog will be sent compressed between the UE and their P-CSCFs.

11.4.5   Related standards

The comp parameter is defined in:

[RFC3486]   Compressing the Session Initiation Protocol (SIP).

11.5   Media negotiation

11.5.1   Overview

Media negotiation and the handling of preconditions, which will be described in Section 11.6.4, are closely related concepts in IMS. Both are more concerned with the description of the session parameters in SDP. Nevertheless, they have a major influence on SIP signalling.

During media negotiation the two items of UE agree on the set of media they want to use for the session and the codes which will be used for the different media types. Therefore, the SDP offer/answer mechanism is used, which – in IMS – basically works in the following way (Figure 11.5):

1. The calling UE sends a first SDP offer in the INVITE request to the called UE. This SDP lists all media types (e.g., audio, video or certain applications like whiteboard or chat) the caller wants to use for this session and lists the different codecs that the caller supports for encoding these different media types.
2. The called UE responds with a first SDP answer, in which it may reject some of the proposed media types. It also reduces the list of codecs by dropping those that it does not support, such that only the codecs that are supported on both ends remain.
3. After receiving the first answer the caller has to make the final decision on the used codecs. It sends a second offer to the called user, which indicates a single codec for every media type that will be used during the session.
4. The called UE accepts the second offer and sends an answer back as confirmation.

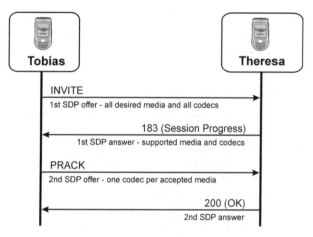

Figure 11.5   SDP offer/answer in IMS.

SIP allows media connections to be set up after just one offer/answer exchange. In IMS the selection of a single codec per media stream enforces a second exchange, if the first SDP answer includes more than one codec for any media type. This is done because both lots of UE must be prepared to receive any of the selected codecs and, therefore, would need to reserve resources on the air interface for the codec with the higher bandwidth, despite maybe using the codec with the lower bandwidth throughout the session.

Due to resource reservation, which is explained in Section 11.6, the two offer/answer exchanges must take place before the 200 (OK) for the INVITE is received. Consequently, the called UE needs to put the first answer in a 100-class response. We will also see in Section 11.6 that the first response is a 183 (Session in Progress) response. If this happens, two problems arise:

- the 183 (Session in Progress) response is – like all 100-class responses – a provisional response and, therefore, is not sent reliably, which means Theresa's UE cannot be sure that it will ever be received by the calling user;
- the calling end is no longer able to send a second offer back, as during a normal INVITE transaction there is no possibility for the calling UE to send any further SIP requests to the called UE besides the initial INVITE and the ACK at the very end of session establishment.

[RFC3262] solves both these problems by making the provisional 100-class responses reliable: this means that, when sending a provisional response back to Tobias's UE, Theresa's UE can indicate that it wants to send this response in a reliable way. Tobias's UE must then send back an acknowledgment (ACK) for the received provisional (PR) response: the PRACK request. As every request in SIP (besides the ACK) must be answered by a final response, Theresa's UE will send a 200 (OK) response back, after receiving the PRACK request.

With this addition to SIP, the first SDP answer in the 183 (Session in Progress) response can be sent reliably and the second SDP offer/answer exchange can be done in the PRACK request and in its 200 (OK) response.

11.5.2   Reliability of provisional responses

The 100-class responses in SIP are provisional: that is, the terminal that sends them out does not get any indication back whether these responses were ever received by the other end. As shown above, there are some cases that require provisional responses to be sent reliably: that is, that the UE that receives the response can explicitly acknowledge it. One of these cases in IMS is that the provisional response carries an SDP answer, which is obliged to be reliably delivered to the remote end.

The mechanism for sending provisional responses reliably is called, in short, "100rel" and its support is mandated for every UE that connects to IMS.

In order to indicate that it supports the 100rel mechanism, Tobias's UE includes a Supported header in the INVITE request, indicating the "100rel" option tag:

```
INVITE sip:theresa@home2.hu SIP/2.0
From: "Your Brother" <sip:tobi@brother.com>;tag=veli
To: "My beloved Sister" <sip:theresa@home2.hu>
Supported: 100rel
CSeq: 1112 INVITE
Call-ID: apb03a0s09dkjdfglkj49555
```

After receiving this, Theresa's UE can start sending provisional responses reliably, as it knows that Tobias's terminal is going to acknowledge them. So, when Theresa's UE sends the 183 (Session in Progress) response, it inserts two additional headers:

```
SIP/2.0 183 Session in Progress
From: "Your Brother" <sip:tobi@brother.com>;tag=veli
To: "My beloved Sister"
<sip:theresa@home2.hu>;tag=schwester
Require: 100rel
RSeq: 1971
CSeq: 1112 INVITE
Call-ID: apb03a0s09dkjdfglkj49555
```

The Require header indicates that the terminal that receives the provisional response must send a PRACK request for it. In order to distinguish between multiple provisional responses, the RSeq header is included.

Tobias's UE is now requested to send a PRACK request back, in order to acknowledge the provisional 183 (Session in Progress) response:

```
PRACK <sip:[5555::5:6:7:8]:1006 SIP/2.0
From: "Your Brother" <sip:tobi@brother.com>;tag=veli
To: "My beloved Sister"
```

```
<sip:theresa@home2.hu>;tag=schwester
RAck: 1971 1112 INVITE
CSeq: 1113 PRACK
Call-ID: apb03a0s09dkjdfglkj49555
```

This request is now sent to the IP address of Theresa's terminal, which was returned in the Contact header of the 183 (Session in Progress) response. In addition, the CSeq number is incremented by 1, as the PRACK is a subsequent request in the dialog that was created by the INVITE/183 (Session in Progress) exchange.

The provisional response that is explicitly acknowledged is identified in the RAck header, which includes the values of the RSeq and CSeq headers that were included in the received response. Both the RSeq and CSeq values are included in the RAck header in order to uniquely identify the 183 (Session in Progress) response.

As the PRACK is a normal SIP request, Theresa's terminal needs to return a final response to it:

```
SIP/2.0 200 OK
From: "Your Brother" <sip:tobi@brother.com>;tag=veli
To: "My beloved Sister"
<sip:theresa@home2.hu>;tag=schwester
CSeq: 1113 PRACK
Call-ID: apb03a0s09dkjdfglkj49555
```

The CSeq header in this 200 (OK) response indicates that this is a final response to the PRACK request (CSeq value "1113 PRACK") and not to the INVITE request (CSeq value "1112 INVITE"). Tobias's terminal is still waiting for the final response to the INVITE request and, as we are looking at an IMS case here, this will still take some time to be sent.

11.5.3   SDP offer/answer in IMS

The SDP offer/answer mechanism allows two users to agree on the media types and the codecs that they want to use for a specific session.

We assume that Tobias wants to use the following media types for his multimedia session with his sister:

- an audio stream – so that they can talk to each other;
- a first video stream – which is filmed by the camera built in to Tobias's UE, so that Theresa can see him and, if Theresa has a similar camera, he can see her;
- a second video stream – which is recorded by an external camera that is connected to the UE; this currently pictures a wooden house in Oulu in Finland.

Therefore, the INVITE will include an SDP body that will look like this (note that not all SDP parameters are shown here):

```
v=0
o=- 2987933615 2987933615 IN IP6 5555::1:2:3:4
s=-
c=IN IP6 5555::1:2:3:4
t=907165275 0
m=audio 3458 RTP/AVP 0 96 97 98
a=rtpmap:0 PCMU
a=rtpmap:96 G726-32/8000
a=rtpmap:97 AMR-WB
a=rtpmap:98 telephone-event
m=video 3400 RTP/AVP 98 99
a=rtpmap:98 MPV a=rtpmap:99 H.261
m=video 3456 RTP/AVP 98 99
a=sendonly
a=rtpmap:98 H.261
a=rtpmap:99 MPV
```

11.5.3.1 General SDP parameters

Let us have a closer look at this SDP information: first, there is the heading with five lines:

```
v=0
o=- 2987933615 2987933615 IN IP6 5555::1:2:3:4
s=-
c=IN IP6 5555::1:2:3:4
t=907165275 0
```

The v-line indicates the protocol version and is always set to 0.

The o-line holds parameters related to the owner of the session, who is in this case Tobias:

- the first parameter should include the name that Tobias wants indicated to the receiver of SDP; however, as he is already identified by the SIP From header in the INVITE request, this can be left out;
- the second parameter is Tobias's session identifier, which is a number that allows his UE to make a link between the session description and the media he wants to set up;
- the third parameter is the version of the session information that Tobias sends – in this case it is initialized with the same value as the session identifier, but it could have any other value;
- the subsequent parameters tell us that Tobias's UE uses the Internet (IN) and IPv6 addressing and that he is located at a terminal that has a certain address.

The s-line may include a subject for the session; but, again, this is already handled by SIP.

The c-line contains information about the connection that has to be established for the multimedia session: that is, this indicates the addresses used for the real media streams and not those for signalling. In this case Tobias's UE just indicates that he wants to receive all media for this session on the IP address that he is also using for SIP signalling: that is, the address that was assigned to it during activation of the signalling PDP context (see Section 10.2). Tobias's UE will also establish all secondary PDP contexts for media with the same IP address toward this access point.

Finally, there is the t-line, which indicates when the session was created and how long it is intended to last. There need be no time limitation to the session set in SDP, as SIP users already end a session by manually sending a BYE request. So, the second parameter of the t-line can safely be given as 0.

The number given as the time at which the session starts is based on definitions from the Network Time Protocol (NTP): this is a 64-bit, unsigned, fixed-point number representation of the seconds that have elapsed since 00 : 00 o'clock on 1 January 1900.

11.5.3.2   Media lines

Individual media lines, or m-lines, represent the three different media streams that Tobias wants to send:

```
m=audio 3458 RTP/AVP 0 96 97 98
m=video 3400 RTP/AVP 98 99
m=video 3456 RTP/AVP 98 99
```

The first line indicates that Tobias intends to use audio for this session and that his UE will send this on the local port 3458. RTP/AVP (Audio Video Profile) will be used as the transport protocol for audio-related media and the terminal seems to be capable of coding the audio in four different ways, because the m-line gives four different formats (0, 96, 97 and 98). Later on, we will see that only three formats are supported, the last one of which (98) points to Dual-Tone Multi-Frequency (DTMF).

The last two m-lines represent two different video streams, which are intended to be sent in parallel: one shows Tobias and the other the wooden house. One is sent over the local port 3400 of Tobias's terminal and the other over port 3456. Both will be transported by RTP/AVP. The terminal holds two formats for each of them.

11.5.3.3   Audio and video formats

The m-lines in SDP include one or more formats that indicate the codecs in which the media streams are encoded. [RFC3551] includes up to 35 formats or codecs that have statically assigned RTP/AVP payload-type numbers (0 to 35). Since that assignment, many more codecs have been defined, which can also be transported via RTP/AVP. These newer codecs can be dynamically assigned to a payload-type number between 96 and 127.

In our example the first media line includes four RTP/AVP payload types, which are further explained in the subsequent lines of the SDP information. The lines between two

SDP m-lines form a block that describes in more detail the media of the m-line under which they are placed:

```
m=audio 3458 RTP/AVP 0 96 97 98
a=rtpmap:0 PCMU
a=rtpmap:96 G726-32/8000
a=rtpmap:97 AMR-WB
a=rtpmap:98 telephone-event
```

The four format numbers that are indicated in the m-line for audio are now mapped to individual payload types or codecs.

Payload type 0 is statically assigned in [RFC3551] to the Pulse Code Modulation μ-Law (PCMU; ITU-T G.711) codec. The first a-line (attribute line) shows this relation again.

The following two attribute lines map (rtpmap) the next two dynamic RTP/AVP format numbers that are indicated in the media line (96 and 97) to specific codecs. In this case:

- payload type 96 gets mapped to G.726; and
- payload type 97 gets mapped to Adaptive Multi-Rate Wideband (AMR-WB).

The last payload type gets mapped to the telephone event representation of DTMF tones, as described in [RFC2833]. This means that the terminal is able to generate standardized tones whenever the user presses a specific key. These tones are well known from normal (Circuit-Switched, or CS) phones when, say, an automatic voice response unit asks questions that the user has to answer with certain keys (e.g., "if you want information in English, press 1, if you want information in German, press 2"). The telephone event defines a text-based representation of these DTMF tones and other telephone-related tones that can be transported over RTP.

The first three payload types (PCMU, G.726 and AMR-WB) represent alternatives – each of which can be used to encode the media indicated in the m-line above them. The telephone event payload type cannot be seen as an alternative: it will always be possible for the UE to send this information whenever a tone needs to be generated.

Consequently, it should now be easy to interpret the last two media blocks in the SDP description:

```
m=video 3400 RTP/AVP 98 99
a=rtpmap:98 MPV
a=rtpmap:99 H.261
m=video 3456 RTP/AVP 98 99
a=sendonly
a=rtpmap:98 H.261
a=rtpmap:99 MPV
```

The terminal is able to encode the two video streams using either MPV (dynamically assigned RTP/AVP payload type 98) or H.261 (dynamically assigned payload type 99).

The first video stream is sent from the camera that pictures Tobias; as nothing further is indicated this is deemed a send-and-receive stream, which means that Theresa can send video over the same stream toward Tobias. Video of the wooden house can be watched via the second video stream and, as it is assumed that Theresa will not be sending any video back, it is set to "sendonly".

11.5.3.4   The first SDP offer and answer exchange

The SDP information as shown above is the initial offer, and Tobias's UE sends it within the body of the INVITE request to Theresa. The SDP offer arrives at Theresa's UE due to SIP routing of the INVITE request. Theresa's UE afterwards generates an SDP answer for the received offer, which looks like this:

```
v=0
o=- 1357924 1357924 IN IP6 5555::5:6:7:8
s=-
c=IN IP6 5555::5:6:7:8
t=907165275 0
m=audio 4011 RTP/AVP 96 97 98
a=rtpmap:96 G726-32/8000
a=rtpmap:97 AMR-WB
a=rtpmap:98 telephone-event
m=video 4012 RTP/AVP 99
a=rtpmap:99 H.261
m=video 0 RTP/AVP 98
```

The SDP header includes information about Theresa's UE – i.e., the IP address in the o-line along with the session identification and the version, as well as the UE's IP address in the c-line.

From the m-lines we see that the terminal is capable of handling the audio stream (on port 4011) and the first video stream (on port 4012), but cannot deliver the second video stream to the user; therefore, the port number in the third m-line is set to zero and all related a-lines are simply dropped.

The SDP answer must repeat all the m-lines that are included in the SDP offer; therefore, the terminal cannot drop the last m-line from here: the only way to indicate that this video cannot be handled is to set the port number to 0.

Furthermore, the UE does not support the MPV codec for video: this RTP/AVP payload-type value was dropped from the first video m-line, and the related a-lines are omitted in the answer.

For audio the UE does not support the PCMU payload type, but both the G.726 and AMR-WB would work with it. The UE does not make a decision here: it just sends both supported codecs back to the caller.

Additionally, DTMF signals can be represented as telephone events.

11.5.3.5 The second SDP offer and answer exchange

After receiving the first SDP answer, Tobias's UE has to make the final decision about the media streams and the codecs used to encode them. As already mentioned, SIP and SDP do not insist on a user or a terminal choosing a single codec per media stream. As air interface resources are usually scarce, it is strongly recommended to make this decision when attached to the IMS via a wireless link.

The second SDP offer is, therefore, sent back in the PRACK request toward Theresa's terminal:

```
v=0
o=- 1357924 1357925 IN IP6 5555::1:2:3:4
s=-
c=IN IP6 5555::1:2:3:4
t=907165275 0
m=audio 3458 RTP/AVP 97 98
a=rtpmap:97 AMR-WB
a=rtpmap:98 telephone-event
m=video 3400 RTP/AVP 99
a=rtpmap:99 H.261
m=video 0 RTP/AVP 98
```

For the audio media the terminal chose AMR-WB as the desired codec. In parallel, DTMF signals can still be sent as telephone events. The first video stream, which was accepted by Theresa's terminal, will be sent in H.261 format, and, as the second video stream was not accepted by the remote end, Tobias's terminal also sets the port to the 0 value.

As the second offer now includes all media-related information, the answer does not include any new information; that is, it is sent back including the port numbers used by Theresa's terminal and its address information:

```
v=0
o=- 1357924 1357924 IN IP6 5555::5:6:7:8
s=-
c=IN IP6 5555::5:6:7:8
t=907165275 0
m=audio 4011 RTP/AVP 97 98
a=rtpmap:97 AMR-WB
a=rtpmap:98 telephone-event
m=video 4012 RTP/AVP 99
a=rtpmap:99 H.261
m=video 0 RTP/AVP 98
```

11.5.4 Related standards

Specifications relevant to Section 11.5 are:

RFC1305	Network Time Protocol (Version 3) Specification, Implementation and Analysis.
RFC3550	RTP: A Transport Protocol for Real-time Applications.
RFC3551	RTP Profile for Audio and Video Conferences with Minimal Control.
RFC2327	SDP: Session Description Protocol.
Draft-ietf-mmusic-sdp-new	SDP: Session Description Protocol.
RFC2833	RTP Payload for DTMF Digits, Telephony Tones and Telephony Signals.
RFC3262	Reliability of Provisional Responses in the Session Initiation Protocol (SIP).
RFC3264	An Offer/Answer Model with SDP.

11.6   Resource reservation

11.6.1   Overview

The media sessions between Tobias and Theresa are negotiated via SIP and SDP signalling. As in our example both UEs make use of a dedicated signalling PDP context for the transport of SIP signalling, they will have to establish one or more media PDP contexts for the transport of the media streams.

The procedure for establishing media PDP contexts is called "resource reservation". Both UEs perform these procedures completely independently of each other.

The establishment of media PDP contexts may consume some time and may even fail: when, say, insufficient resources are available over the wireless link. This means that, until the resources have been reserved, it cannot be guaranteed that the agreed media sessions can be established at all.

Therefore, Theresa's UE should not inform her about the incoming session request (INVITE): that is, it should not start to ring until it has confirmation that resource reservation has succeeded locally as well as at the calling user's end.

In order to achieve this, both UEs exchange preconditions during the SDP offer/answer negotiation, which basically instruct:

- Tobias's UE to send a SIP UPDATE request to Theresa's UE, when resource reservation has succeeded at Tobias's UE; and
- Theresa's UE not to ring until it receives the SIP UPDATE request from the remote end and has also successfully reserved its own/local resources.

Furthermore, the preconditions indicate what should happen with a session if a specific media stream could not be successfully reserved.

Figure 11.6 gives an overview of the relations between SIP, SDP offer/answer, resource reservation, preconditions, the SIP PRACK and the SIP UPDATE method during the establishment of a session.

Figure 11.6 SIP, SDP offer/answer and preconditions during session establishment.

11.6.2 The 183 (Session in Progress) response

As shown at the beginning of this chapter, the IMS session establishment attempt should only be indicated to Theresa (the called user) if resource reservation on the wireless link for both users has succeeded. Only after successful resource reservation will the SIP 180 (Ringing) response for the INVITE message be sent by Theresa's terminal.

On the other hand, Tobias's UE expects an early response from the called terminal, especially as SDP offer/answer negotiation in IMS needs to be done before resource reservation can start. Therefore, after receipt of the INVITE message, Theresa's UE will send back the 183 (Session in Progress) response, indicating that session establishment procedures have been started, but the called user has not been informed yet.

11.6.3 Are preconditions mandatorily supported?

Before sending out the first INVITE request, the calling terminal must also include a Require header with the value "precondition":

```
Require: sec-agree, precondition
```

This guarantees that the called terminal will reject this INVITE if it does not support the preconditions mechanism at all. But, this would mean that the session between Tobias and Theresa could not be established at all if Theresa's terminal does not support the precondition extension to SDP. Is that really what the IMS intended?

Let us assume that Theresa's terminal does not support preconditions but, nevertheless, goes ahead with session establishment (Figure 11.7). Assume further that resource reservation at Theresa's end is successful. Consequently, her terminal could immediately afterwards indicate the incoming call to her and start sending early media to Tobias.

At this point in time, Tobias would not even have started to reserve his resources over the wireless link, as he is still waiting for the SDP answer from Theresa's end. Even had he started reserving resources immediately after sending out the initial INVITE request,

Figure 11.7 SIP session establishment without preconditions.

it cannot be guaranteed that this process would have finished when Tobias receives the 180 (Ringing) response from Theresa's UE.

Even worse, Theresa could pick up the phone – which would result in her terminal sending a 200 (OK) response – and start talking, while Tobias is still in the process of resource reservation. If resource reservation at Tobias's end fails in this scenario, Theresa will be left with a call that was indicated as successful on the signalling level, but never had a chance to connect through the session on the media plane.

Therefore, if Theresa's UE does not support preconditions, it will, due to the received Require header in the INVITE request, send back a 420 (Bad Extension) response that includes an Unsupported header:

```
SIP/2.0 420 Bad Extension
Unsupported: preconditions
```

After Tobias's UE terminal has received this response, it can decide whether to send the INVITE request again, this time without preconditions, or not to bother to call the other end at all.

In IMS Release 5 it is assumed that both terminals support the preconditions mechanism; this practically made calls between an IMS terminal and a terminal outside the IMS, which does not support preconditions, impossible. For Release 6, interworking procedures have been specified that allow such scenarios. Nevertheless, the support of preconditions in IMS is mandated and necessary, due to resource reservation procedures over the wireless link.

11.6.4 Preconditions

[RFC3312] introduces the SDP preconditions mechanism that allows a UE to delay completion of a SIP session establishment until one or both ends have successfully completed their resource reservation. This extension to SIP and SDP is mandatorily supported by every UE that connects to IMS.

Up to now, Internet Engineering Task Force (IETF) has only defined the "qos" precondition type, but there might be more to come. The "qos" tag means that the precondition is set due to certain quality requirements of the related media stream, which is indicated in the m-line. The Quality of Service (QoS) depends mostly on the bandwidth that is reserved over the wireless link and the priority by which the routers in the network handle the packets that transport the "chunks" of speech and video data.

11.6.4.1 Preconditions in the first SDP offer

As explained in Section 11.5 for the SDP offer/answer mechanism, Tobias's UE proposes three different media types, which are offered in the first INVITE to Theresa's UE. In order to illustrate the preconditions mechanism, we take just one of these three media types as an example and extend it with precondition-specific indications (written in bold):

```
m=audio 3458 RTP/AVP 0 96 97 98
a=rtpmap:0 PCMU
a=rtpmap:96 G726-32/8000
a=rtpmap:97 AMR-WB
a=rtpmap:98 telephone-event
a=curr:qos local none
a=curr:qos remote none
a=des:qos mandatory local sendrecv
a=des:qos none remote sendrecv
```

Every m-line block in SDP needs to indicate a separate set of preconditions. Remember that the m=audio line is only taken as an example, the two remaining m-lines in the original SDP would also include precondition setting. First, we look at the penultimate line:

```
a=des:qos mandatory local sendrecv
```

This indicates the desired (des) quality of service (qos) precondition at the calling user (local) end. The resources for the calling user need to be reserved in both the sending and receiving directions (sendrecv), as the audio stream is bidirectional (i.e., both users can talk to each other and hear what the other says). It also states that the session will not take place if the indicated resources cannot be reserved by the calling user (mandatory).

The last line:

```
a=des:qos none remote sendrecv
```

indicates the desired (des) quality of service (qos)-related preconditions at the called user (remote) end. As the calling and the called terminal are not directly connected to each other, the calling terminal is unaware of how the remote terminal is attached to the network. It might be that Theresa's terminal does not need to perform any resource reservation, as it is connected via a CS telephony network. Therefore, the calling end can only indicate that, if the remote end needs to reserve resources, then they should be reserved in both the sending and receiving directions (sendrecv); however, the calling end is currently unaware that this is really required in order to get a media session established (none).

Up to now, the calling terminal has only expressed the desired (des) preconditions for each end, but it also needs to talk about the current status of resource reservation. This is the subject of the first two new lines:

```
a=curr:qos local none
a=curr:qos remote none
```

These two lines indicate that currently (curr) no (none) quality of service (qos)-related preconditions have been fulfilled by either the calling end (local) or the called end (remote).

The a=des lines make it possible to set preconditions for the local and the remote user, while the a=curr lines indicate the extent to which the set preconditions are already being fulfilled.

11.6.4.2   Preconditions in the first SDP answer

As we know, Theresa's terminal is also attached over a wireless link to IMS and supports the preconditions mechanism. Therefore, it will respond to the received SDP offer with a well-formed answer and will include its own preconditions. Once again, we only show here those preconditions for the first media type:

```
m=audio 4011 RTP/AVP 96 97 98
a=rtpmap:96 G726-32/8000
a=rtpmap:97 AMR-WB
a=rtpmap:98 telephone-event
a=curr:qos local none
a=curr:qos remote none
a=des:qos mandatory local sendrecv
a=des:qos mandatory remote sendrecv
a=conf:qos remote sendrecv
```

The important thing to note here is that the remote and local ends have changed, because from Theresa's point of view her terminal is local and Tobias's is remote. Her terminal now indicates its own preconditions line by line. Line 5:

```
a=curr:qos local none
```

it has currently not reserved any local qos-related resources. Line 6:

```
a=curr:qos remote none
```

it received information from the remote end (in the first offer) that no qos-related resources have been reserved at the moment. Line 7:

```
a=des:qos mandatory local sendrecv
```

it mandatorily requires that its own resources get reserved in both the sending and receiving directions, before the audio session can start. Note that the initial value "none", as set from the calling end, has changed to "mandatory" because Theresa's terminal is attached via the air interface and, therefore, is also mandated to reserve the resources locally before it can start sending media. Line 8:

```
a=des:qos mandatory remote sendrecv
```

it received information from the remote end (in the first offer) that resources are mandatorily reserved in both the sending and receiving directions. Line 9:

```
a=conf:qos remote sendrecv
```

the calling terminal (remote) should send a confirmation (conf) at the moment the resources (qos) have been reserved in the sending and receiving directions (sendrecv). This is a new line that the called end adds to SDP. It is a necessary addition because the called terminal is not intended to ring the called user or to start sending media until *both ends have reserved the resources.*

This SDP answer is now sent in the 183 (Session in Progress) response to the calling terminal.

11.6.4.3   Start of resource reservation

Subsequently, the second SDP offer/answer exchange takes place: that is, Tobias's terminal sends a PRACK request with the final selection of media types and codecs, which for the audio stream will look like:

```
m=audio 3458 RTP/AVP 97 98
a=rtpmap:97 AMR-WB
a=rtpmap:98 telephone-event
a=curr:qos local none
a=curr:qos remote none
a=des:qos mandatory local sendrecv
a=des:qos mandatory remote sendrecv
```

The received a=conf line is not repeated because the terminal knows that it needs to send a confirmation after it has reserved the resources.

At the instant Tobias's UE sends the PRACK request it can start resource reservation, because the QoS requirements for the different media streams and the specific codecs that will be used to encode these media streams are known.

It is also possible for Tobias's terminal to have already started resource reservation after the initial INVITE request was sent. All it would need do is to modify the reserved resources (i.e., the characteristics of the established media PDP context in the case of the GPRS) when it sends the PRACK request. This modification would be necessary because the initial reservation would have been made for media streams and codecs with the highest QoS requirements, while the agreed media and codecs could well have different QoS requirements.

Theresa's UE can start resource reservation after sending the 183 (Session in Progress) response, but it is assumed here that it starts after sending the 200 (OK) response for the PRACK request. This response will include the second SDP answer, which for the audio stream will look like:

```
m=audio 4011 RTP/AVP 97 98
a=rtpmap:97 AMR-WB
a=rtpmap:98 telephone-event
a=curr:qos local none
a=curr:qos remote none
a=des:qos mandatory local sendrecv
a=des:qos mandatory remote sendrecv
```

Once again, there is no need for the a=conf line to be repeated. As the status of resource reservation has not changed, the precondition-related information in the second answer does not include anything new.

11.6.4.4   Successful resource reservation

Both UEs now try to reserve the requested resources. In the General Packet Radio Service (GPRS) case this would mean that they establish one or more media PDP contexts as secondary PDP contexts (see Chapter 13). It is possible for either Tobias's or Theresa's UE to finish the reservation first. In either case Theresa's terminal must not start to indicate the incoming call to Theresa (i.e., it must not ring) before knowing that both ends have successfully reserved resources.

Let us assume that the called end finishes the reservation first: in this case it will simply wait until it gets the confirmation that it requested from the calling end in the 183 (Session in Progress) response answer by setting the a=conf line in the first SDP answer.

As soon as the calling end has reserved the required resources, it will confirm this to the called terminal by sending a SIP UPDATE request as a subsequent request in the dialog that was established with the INVITE request. This UPDATE request will include a third SDP offer in its body, which shows that the resources at the calling end have been reserved:

```
m=audio 3458 RTP/AVP 97 98
a=rtpmap:97 AMR-WB
a=rtpmap:98 telephone-event
a=curr:qos local sendrecv
a=curr:qos remote none
a=des:qos mandatory local sendrecv
a=des:qos mandatory remote sendrecv
```

The only difference here is that the first a=curr line has changed from "none" to "sendrecv". Consequently, Tobias's terminal indicates that the status of the local QoS-related resources for the audio stream has changed. They have now been success-fully reserved in both the sending and receiving directions.

As Theresa's terminal was the first to secure resource reservation, it can immediately start to ring after sending the 200 (OK) response for the UPDATE request, because it is now sure that both ends have sufficient resources reserved to send and receive the audio stream. When it starts to ring, it sends a 180 (Ringing) response to the INVITE request in parallel.

The 200 (OK) for the UPDATE will include a third SDP answer with the following SDP information about the audio stream:

```
m=audio 4011 RTP/AVP 97 98
a=rtpmap:97 AMR-WB
a=rtpmap:98 telephone-event
a=curr:qos local sendrecv
a=curr:qos remote sendrecv
a=des:qos mandatory local sendrecv
a=des:qos mandatory remote sendrecv
```

We can see from this that all the current states for resource reservation match the desired states; so, the preconditions negotiation has been successful and has finished.

Let us assume that Theresa's resource reservation takes longer than Tobias's: in this case Theresa's UE will receive the UPDATE request while it is still reserving resources. Although it will not start to ring, it will send back a 200 (OK) response with the following SDP information about the audio stream:

```
m=audio 4011 RTP/AVP 97 98
a=rtpmap:97 AMR-WB
a=rtpmap:98 telephone-event
a=curr:qos local none
a=curr:qos remote sendrecv
a=des:qos mandatory local sendrecv
a=des:qos mandatory remote sendrecv
```

The only difference here is that the status of local resource reservation is still set to "none". After local resources have been reserved, Theresa's UE will start to ring and send a 180 (Ringing) response that will *not* carry another SDP body. Nevertheless, Tobias's UE can interpret the 180 (Ringing) response as an indication that resources at the remote end (i.e., Theresa's) have been reserved successfully: this is the reason why the 180 (Ringing) response also has to be sent reliably, obliging Tobias's UE to answer it with a PRACK request (see Section 11.5.2).

Once both ends have reserved their resources, media can be exchanged between the two UEs.

11.6.5 Related standards

Specifications relevant to Section 11.6 are:

RFC3311   The Session Initiation Protocol (SIP) UPDATE Method.
RFC3312   Integration of Resource Management and Session Initiation Protocol.

11.7 Controlling the media

11.7.1 Overview

As each UE is usually attached to the IMS over a wireless link, bandwidth resources are scarce and need to be handled with care. In Section 11.5 we saw that both ends exchange their preferences on the media and codecs used during a session until they agree on a specific set of media streams, each media stream using a single codec.

As media streams will not be routed through any CSCF, the IMS network needs to authorize reservation of agreed resources. Consequently, the P-CSCF requests the Policy Decision Function (PDF) to generate a media authorization token (Figure 11.8) whenever the first SIP message for the dialog is sent toward the UE (see Section 3.11.1). This token is added to the INVITE request and sent to Theresa's UE. Tobias's UE will receive its local token in the 183 (Session in Progress) response.

During the SDP offer/answer process, the local P-CSCF can tell the UE to group specific media streams together within one media PDP context or to keep certain media streams in separate media PDP contexts: this is achieved by the SDP media line grouping mechanism.

In addition to all this, network operators might want to restrict the media types and codecs that each UE can use. To do so, the P-CSCFs and S-CSCFs check the SDP that is included in the INVITE request. If one of them determines that a certain media type or codec should not be used, this CSCF will send a rejection to the UE. In our example we do not assume that such a situation occurs. Nevertheless, the general procedure for media policing is described in Section 11.7.6.

Figure 11.8 Transport of media authorization information.

After completion of the SDP offer/answer process, each UE can begin resource reservation for the agreed media. In the case of the GPRS and when using a dedicated signalling PDP context, this means that each UE will now try to establish one or more secondary PDP contexts for the media. Within every request for media PDP context establishment, the UE will include the media authorization token that it received from the P-CSCF.

The GGSN receives the request for a new PDP context and first sends the token back to the PDF that created it. It also includes the requested session parameters in the message to the PDF. The PDF and the Policy Enforcement Point (PEP) perform checks about the resource reservation request. Subsequently, the GGSN establishes the requested media PDP context with the UE.

Similar procedures will exist for other types of access networks in the future: the GPRS is used here as an example.

11.7.2   Media authorization

When Theresa's P-CSCF receives the INVITE request it will request the PDF to generate a media authorization token. The P-CSCF will add a P-Media-Authorization header to the INVITE request, include the received token and send the request to Theresa's UE:

```
INVITE sip:[5555::5:6:7:8]:1006 SIP/2.0
P-Media-Authorization: example-auth-token2
```

Theresa's UE will use this token when establishing the media PDP context or contexts for the media streams that are negotiated between the two UEs.

Shortly after, Tobias's P-CSCF will receive the 183 (Session in Progress) response and will also request a media authorization token from its PDF. It includes the token in a P-Media-Authorization header and sends it with the response to Tobias's UE:

```
SIP/2.0 183 Session in Progress
P-Media-Authorization: example-auth-token1
```

As the reservation and authorization of resources is a local matter for each UE, the two tokens will be different.

Each UE will receive only one token for all media sessions that are created due to a single INVITE request. Regardless of how many PDP contexts the UE establishes, it can always blindly use the same token when requesting the resource. Consequently, the UE does not need to be at all aware of the content or the coding of the token, it just receives it in the SIP message and puts it into the ACTIVATE PDP CONTEXT REQUEST.

11.7.3   Grouping of media lines

The section on authorizing QoS resources (Section 3.11.1) already includes an example
of the grouping of media lines. For the example in this chapter we will assume the
grouping is different at both ends.

11.7.4   A single reservation flow

Theresa's P-CSCF will send to her UE the following information related to media line
grouping in the second SDP offer in the PRACK request:

```
v=0
o=- 1357924 1357924 IN IP6 5555::1:2:3:4
s=-
c=IN IP6 5555::1:2:3:4
t=907165275 0
a=group:SRF 1
m=audio 3458 RTP/AVP 97 98
a=mid: 1
a=rtpmap:97 AMR-WB
a=rtpmap:98 telephone-event
m=video 3400 RTP/AVP 99
a=mid: 1
a=rtpmap:99 H.261
m=video 0 RTP/AVP 98
```

The a-line in the SDP heading defines a Single Reservation Flow (SRF) group with the
number "1": that is, all media streams in this group should go into a single PDP
context.

Both media lines – one for audio and one for video – are followed by an a-line that
indicates a Media Stream Identification (MID) that is set in both cases to the above-
defined group number "1". This means that the P-CSCF instructs Theresa's UE to
group the two media streams together in one PDP context.

11.7.5   Separated flows

The P-CSCF in Finland sees the need for separated media streams. Therefore, it adds
the following media line grouping-related information to the SDP answer that is sent to
Tobias's UE in the 200 (OK) response to the PRACK request:

```
v=0
o=- 1357924 1357925 IN IP6 5555::5:6:7:8
s=-
c=IN IP6 5555::5:6:7:8
t=907165275 0
a=group:SRF 1
```

```
a=group:SRF 2
m=audio 4011 RTP/AVP 97 98
a=mid: 1
a=rtpmap:97 AMR-WB
a=rtpmap:98 telephone-event
m=video 4012 RTP/AVP 99
a=mid: 2
a=rtpmap:99 H.261
m=video 0 RTP/AVP 98
```

The a=group lines create two individual SRF groups with the numbers "1" and "2". As the audio stream indicates SRF group "1" and the video stream SRF group "2", Tobias's UE has to reserve separate media PDP contexts for each of the two media streams (Figure 11.9).

11.7.6 Media policing

The P-CSCF and S-CSCF can reject certain media types or codecs that are offered in SDP. This might be due to operator policy. One reason for this could be that an operator does not allow the use of any unknown media types or unknown codecs, as the network would not be able to charge for such media.

If a CSCF detects that an unsupported media type or codec is included in the SDP offer, it will reject the request with 488 (Not Acceptable Here) and will indicate the supported media types in the body of the response.

In the example scenario this is not assumed to happen. Figure 11.10 shows a worst case scenario when every CSCF on the route does not like certain media elements. Such extreme scenarios are unlikely to occur in reality.

11.7.7 Related standards

Specifications relevant to Section 11.7 are:

RFC3313   Private Session Initiation Protocol (SIP) Extensions for Media Authorization.

RFC3388   Grouping of Media Lines in the Session Description Protocol (SDP).

RFC3524   Mapping of Media Streams to Resource Reservation Flows.

11.8   Charging-related information for sessions

11.8.1   Overview

In the case of a GPRS access network, charging is usually done on the basis of the amount of data that is sent via a PDP context. The network operator can decide to charge specific PDP contexts differently: for example, a Multimedia Message Service

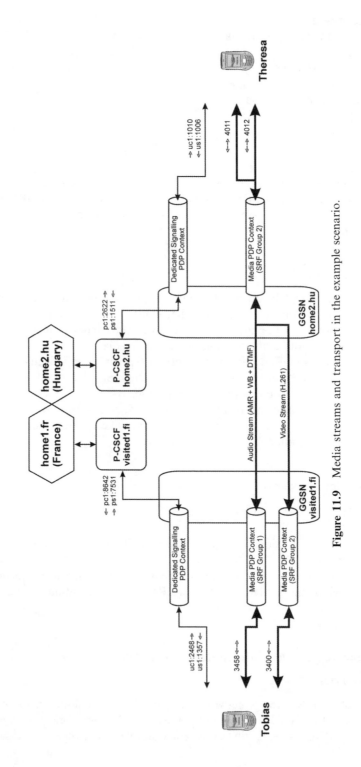

Figure 11.9 Media streams and transport in the example scenario.

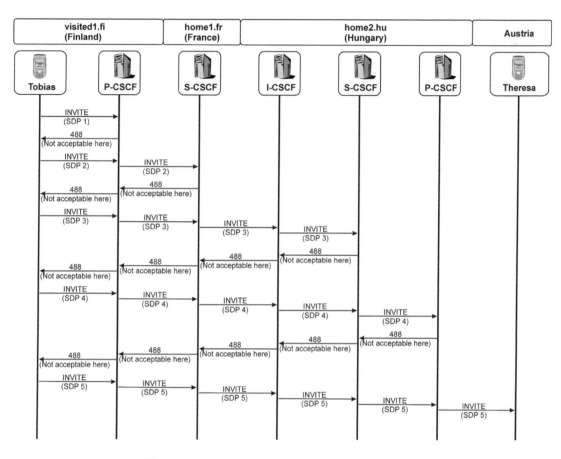

Figure 11.10 Worst case scenario for media policing.

(MMS) could be charged differently from "normal" Internet traffic. It is likely that operators will want to charge the signalling and media traffic that is related to the IMS in a different way than one based on the amount of transmitted data. To achieve this, IMS needs to correlate GPRS-specific charging records with IMS-specific charging information.

For the media session Tobias's UE needs to establish several media PDP contexts with the GGSN. The GGSN will start generating charging records whenever data (i.e., media) are sent over that PDP context. These charging records are based on the GPRS Charging ID (GCID) that is generated by the access network.

As described in Section 10.11, an IMS Charging ID (ICID) is created during initial registration by the P-CSCF. During a session the P-CSCF creates an additional ICID for the charging of media streams that are transported over established media PDP contexts.

The GGSN sends GCIDs of these media PDP contexts via the Go interface to the P-CSCF, which then creates an ICID and sends it toward the home network of the served user.

These procedures will take place at both ends (i.e., for Tobias and for Theresa). Which of them will get charged what will be decided by the charging entities in their home networks.

The S-CSCFs of the users will, furthermore, distribute within their home networks the addresses of the Charging Data Function (CDF) and the Event Charging Function (ECF) that will collect the charging records for the two users. This information is needed by all entities that create charging records for the session.

11.8.2 Exchange of ICID for a media session

When Tobias's P-CSCF receives the INVITE request for the session it will create a new ICID and include it in a P-Charging-Vector header that it adds to the INVITE request:

```
INVITE sip:theresa@home2.hu SIP/2.0
P-Charging-Vector: icid-value="AyretyU0dm+6O2IrT5tAFrbHLso=023551025"
```

Tobias's S-CSCF will add the originating Inter-Operator Identifier (IOI) to the P-Charging-Vector header. The IOI parameter makes the identity of Tobias's home network available to Theresa's home network for charging purposes and is usually set to the home domain of the user:

```
INVITE sip:theresa@home2.hu SIP/2.0
P-Charging-Vector: icid-value="AyretyU0dm+6O2IrT5tAFrbHLso=023551025"
 ;orig-ioi=home1.fr
```

The S-CSCF of Theresa's home network will store and remove the received orig-ioi parameter for the P-Charging-Vector header.

This P-Charging-Vector header will be sent further with the INVITE request until it reaches the P-CSCF of Theresa, where it will be removed. Charging-related headers are never sent toward a UE. All entities on the way that are involved in charging for this session will store the ICID.

When receiving the first response to the INVITE request – i.e., the 183 (Session in Progress) response – Theresa's P-CSCF will once again add the P-Charging-Vector header, including the same ICID value as received in the INVITE request:

```
SIP/2.0 183 Session in Progress
P-Charging-Vector: icid-value="AyretyU0dm+6O2IrT5tAFrbHLso=023551025"
```

Theresa's S-CSCF will add the terminating IOI information to the response, before sending it further:

```
SIP/2.0 183 Session in Progress
P-Charging-Vector: icid-value="AyretyU0dm+6O2IrT5tAFrbHLso=023551025"
 ;term-ioi=home2.hu
```

The terminating IOI information will again be stored and removed by Tobias's S-CSCF, and his P-CSCF will remove the P-Charging-Vector from the response.

11.8.3 Correlation of GCID and ICID

The media PDP context at Tobias's end is established after successful resource reservation, and, as we saw in Section 11.6, Tobias's UE will then immediately send an UPDATE request toward Theresa. As this is the first request that is received by Tobias's P-CSCF after the media PDP context has been established, it will include in the UPDATE request the P-Charging-Vector header with the following information:

- The ICID that it created for this media session.
- The charging parameters for every media PDP context that were established for this session:
  - ○ the GCID that was received from the GGSN;
  - ○ the address of the related GGSN;
  - ○ an indication ("pdp-sig") of whether the related PDP context is a signalling PDP context or not (in this case it is not a signalling PDP context);
  - ○ the flow identifier of the media stream; and
  - ○ the token (auth-token) that was used for media authorization (see Section 11.7.2).

The UPDATE request now looks like:

```
UPDATE sip:[5555::5:6:7:8]:1006 SIP/2.0
P-Charging-Vector: icid-value="AyretyU0dm+6O2IrT5tAFrbHLso=023551024"
  ;ggsn=[5555::4b4:3c3:2d2:1e1];
    ;pdp-sig=no; gcid=723084371
    ;auth-token=example-auth-token1
    ;flow-id=1
  ;ggsn=[5555::4b4:3c3:2d2:1e1]
    ;pdp-sig=no; gcid=723084372
    ;auth-token=example-auth-token1
    ;flow-id=2
```

As we saw in Section 11.7.3, Tobias has established two media PDP contexts; hence, two charging parameter lists are also included.

Tobias's S-CSCF will store and remove all data from the P-Charging-Vector header and add the originating and terminating IOI parameters, before sending the UPDATE request to Theresa's S-CSCF:

```
UPDATE sip:[5555::5:6:7:8]:1006 SIP/2.0
P-Charging-Vector: icid-value="AyretyU0dm+6O2IrT5tAFrbHLso=023551024"
  ;orig-ioi=home1.fr
  ;term-ioi=home2.hu
```

Once again, these two parameters (orig-ioi and term-ioi) will be removed by Theresa's S-CSCF before sending the UPDATE request to the P-CSCF, which removes the header before sending the request toward Theresa's UE.

In our example Theresa's UE has already finished resource reservation at the moment the UPDATE request is received from Tobias's end (see Section 11.6.4.4). Therefore, Theresa's P-CSCF will include the P-Charging-Vector header again in the 200 (OK) response to the UPDATE request, this time with all the information related to the reserved media PDP context:

```
SIP/2.0 200 OK
P-Charging-Vector: icid-value="AyretyU0dm+6O2IrT5tAFrbHLso=023551024"
  ;ggsn=[5555::802:53:58:6]
    ;pdp-sig=no
    ;gcid=306908949
    ;auth-token=example-auth-token2
    ;flow-id=2
```

As Theresa established only one media PDP context (see Section 11.7.3), her P-CSCF includes exactly one charging parameter list.

The S-CSCFs now behave in the same way as before:

- Theresa's S-CSCF will remove PDP context-related information from the P-Charging-Vector header and will add the saved orig-ioi and term-ioi parameters.
- Tobias's S-CSCF will remove the orig-ioi and term-ioi parameters.
- Finally, the P-CSCF will again remove the P-Charging-Vector header from the 200 (OK) response before sending it to Tobias's UE.

11.8.4   Distribution of charging function addresses

The addresses of the CCF and ECF are distributed within the home network of S-CSCFs.

Every S-CSCF adds a P-Charging-Function-Address header to the INVITE request, which is then sent along the route until it reaches the border of the home network. The CSCF at the border of the home network will remove this header.

In the case of Tobias's home network this means that his S-CSCF:

- Adds the P-Charging-Function-Address header when it first receives the INVITE request and then sends it to all ASs that belong to the home network and are contacted due to the filter criteria in Tobias's user profile (see Section 11.3.8). As the S-CSCF is the last entity in Tobias's home network, it will remove the P-Charging-Function-Address header before sending it toward Theresa's home network.
- Adds the P-Charging-Function-Address header again when it receives the 183 (Session in Progress) response, sends it once more to all ASs that belong to the home network and are contacted due to the filter criteria, and removes the header before sending the response toward the P-CSCF that is located in the visited network.

The P-CSCF in Tobias's visited network will need to discover its local charging function addresses by other means.

On the other hand, Theresa's S-CSCF:

- Adds the P-Charging-Function-Address when it receives the INVITE request and then sends it toward the ASs, as stated above. However, in this case the S-CSCF does not remove the header from the request before sending it to the P-CSCF, as Theresa's P-CSCF is located in her home network. She uses GPRS roaming, not IMS roaming, to access her home network's P-CSCF (see Section 9.1). The P-Charging-Function-Address header will, therefore, be removed from the INVITE request by the P-CSCF.
- Adds the P-Charging-Function-Address header again when it receives the 183 (Session in Progress) response and sends it once more with the response. However, this time the header will be removed by the I-CSCF of Theresa's home network, which will also receive the response.

The P-Charging-Function-Address header that is added by Tobias's S-CSCF in the initial INVITE request would look something like this:

```
INVITE sip:theresa@home2.hu SIP/2.0
P-Charging-Function-Addresses: ccf=[5555::b99:c88:d77:e66]
 ;ccf=[5555::a55:b44:c33:d22]
```

11.8.5   Related standards

IMS-specific SIP headers for these charging procedures are specified in:

[RFC3455]   Private Header (P-Header) Extensions to the Session Initiation Protocol (SIP) for the 3rd-Generation Partnership Project (3GPP).

11.9   Release of a session

11.9.1   User-initiated session release

Of course, Tobias and Theresa will at one point stop their conversation. Let us say that Theresa meets one of her friends at Stephansdom in Vienna and has to say goodbye to her brother (Figure 11.11). She will be the one who drops the call by pressing the red button on her mobile phone.

Consequently, her UE will generate a BYE request, which is sent to Tobias's UE in the same way as any other subsequent request. In parallel to this, her UE will also drop

Figure 11.11 Theresa releases the session.

the media PDP context that was established for the session:

```
BYE sip:[5555:1:2:3:4]:1357 SIP/2.0
Route: <sip:pcscf2.home2.hu:1511;lr>
Route: <sip:scscf2.home2.hu;lr>
Route: <sip:scscf1.home1.fr;lr>
Route: <sip:pcscf1.visited1.fi;lr>
To: "Your Brother" <sip:tobi@brother.com>;tag=veli
From: "My beloved Sister" <sip:Theresa@sister.com>;tag=schwester
```

We can see from this BYE request that the information in the To and From headers has been swapped, as this request is now sent from Theresa's end.

Tobias's UE will drop its PDP context immediately after it receives the BYE request. It will also respond to the request with a 200 (OK) response, which will be sent back toward Theresa. The four CSCFs and all ASs on the route will clear all dialog states and information related to the session.

11.9.2 P-CSCF performing network-initiated session release

There might be situations in which it is necessary for one of the CSCFs to release the session, rather than the user.

For example, Theresa's P-CSCF would need to release an ongoing session when it realizes that Theresa's UE has lost radio coverage and is no longer connected to the access network (Figure 11.12). In that case the P-CSCF would need to send a BYE

Figure 11.12 P-CSCF terminates a session.

request on behalf of Theresa:

```
BYE sip:[5555:1:2:3:4]:1357 SIP/2.0
Route: <sip:scscf2.home2.hu;lr>
Route: <sip:scscf1.home1.fr;lr>
Route: <sip:pcscf1.visited1.fi;lr>
To: "Your Brother" <sip:tobi@brother.com>;tag=veli
From: "My beloved Sister" <sip:Theresa@sister.com>;tag=schwester
```

11.9.3   S-CSCF performing network-initiated session release

There may be occasions when Tobias's S-CSCF needs to be shut down or Tobias may
be using a pre-paid card and runs out of money. In such cases Tobias's S-CSCF would
release the session (Figure 11.13) by issuing one BYE request toward Tobias's UE:

```
BYE sip:[5555:1:2:3:4]:1357 SIP/2.0
Route: <sip:pcscf1.visited1.fi;lr>
To: "Your Brother" <sip:tobi@brother.com>;tag=veli
From: "My beloved Sister" <sip:Theresa@sister.com>;tag=schwester
```

Figure 11.13   S-CSCF terminates a session.

and another BYE request toward Theresa's UE:

```
BYE sip:[5555:1:2:3:4]:1006 SIP/2.0
Route: <sip:scscf2.home2.hu;lr>
Route: <sip:pcscf2.home2.hu;lr>
From: "Your Brother" <sip:tobi@brother.com>;tag=veli
To: "My beloved Sister" <sip:Theresa@sister.com>;tag=schwester
```

To generate the BYE request with the correct set of Route headers, the P-CSCFs and
S-CSCFs need to keep track of all routing information that is collected during the
establishment of any dialog.

11.10   Alternative session establishment procedures

11.10.1   Introduction

The procedures as described in Chapter 11 so far explain the general mechanism used in
IMS for establishing multimedia sessions in a mobile environment. As shown there, the
procedures for reserving resources over the wireless link and authorization of media

add complexity and additional signalling to the SIP flows in IMS when compared with those in IETF.

When specifying 3GPP Release 6 and Release 7, IMS grew towards an open multimedia communication platform that is not only accessed from cellular access networks but in which resources are usually tight. UEs can attach to IMS from fixed networks as well as via Wireless Local Area Network (WLAN) based access technologies.

Also, certain services are not in need of specific resource reservation procedures. A text-based chat service, for example (see Section 11.10.2), has completely different resource and QoS requirements than an audio conversation.

In order to serve the requirements of new access technologies as well as of specific services, alternative session establishment mechanisms were introduced to IMS. Some of them are still in the process of being defined in detail as this book is being written.

Section 11.10.4 shows how session establishment can be performed between an IMS-attached terminal and a SIP terminal outside IMS.

All the examples in this section show the signalling flows directly between the two UEs, the networks inbetween are not depicted, as the involved network entities as well as the routing proceudres are mostly identical to those described so far in this chapter.

11.10.2   IMS session setup when no resource reservation is needed

There are several scenarios where sessions can be established and no prior resource reservation is needed.

IMS messaging sessions (see Chapter 5) are text-based chat sessions between two users. They have different real-time and bandwidth requirements than other multimedia services. A video stream, for example, requires the transport of all its media packets in real time, so that the receiving end can view the frames as a continuous stream to the user. Text messages are usually typed in first and then sent as a whole to the other end. A delay in the range of, say, 1 or 2 seconds does not diminish the user experience.

In 3GPP Release 6, procedures were defined to allow the re-use of an already established PDP context for media. A terminal or an application within the terminal may choose to keep an already established PDP context for IMS-related media, even though the related SIP dialog has already been ended – i.e., BYE/200 (OK) having been sent on the related dialog. It is the decision of the UE to determine how long such a PDP context is to be kept active. If a new session with the same or similar resource requirements is going to be established afterwards, the UE can re-use the pre-established PDP context.

In theory, there are two possible ways of establishing a session for which no resource reservation is needed:

• perform normal IETF session setup; or
• setting the preconditions to "met" when sending the initial INVITE (see Section 11.10.2.2).

The usage of preconditions usually lengthens the signalling flow (see Figure 11.1) and also extends the session setup time. It is for this reason that only the normal IETF session setup should be used in IMS when resource reservation is not needed.

Figure 11.14   IMS session setup with pre-reserved resources supported by both UEs.

11.10.2.1   IMS session setup with pre-reserved resources – supported by both UEs

Keeping the original example as described in Chapter 9, we assume that Tobias wants to invite Theresa to a chat session. For text-based chats the Message Session Relay Protocol (MSRP – see Chapter 5) is used in IMS at the bearer level. The chat application on Tobias's UE has pre-reserved a background PDP context for MSRP sessions; therefore, resources are already available at the originating end. At the terminating end we assume that Theresa is connected via a fixed terminal – e.g., a PC that is connected via Digital Subscriber Line (DSL) to the IMS of her home network provider. As a result of the fixed network connection, neither is there any resource reservation required at that end.

Tobias's UE sends an INVITE request and indicates MSRP in the first SDP Offer in the message body:

```
m=message 3404 msrp/tcp *
a=accept-types:message/cpim text/plain text/html
a=path:msrp://[5555::aaa:bbb:ccc:ddd]:3402/s111271;tcp
a=max-size:131072
```

As the resources are already available locally and are assumed to be available also at the remote end, Tobias UE does not set any precondition in the SDP Offer that it sends out.

This assumption is met at Theresa's end , her UE responding with a 180 (Ringing) to the received INVITE request. After Theresa has accepted the incoming chat request, a 200 (OK) response is sent, which includes the SDP Answer. After this, MSRP messages can be exchanged between the two UEs at the bearer level.

11.10.2.2   Fallback to full IMS session setup/setting preconditions to meet

In the above example we saw that the chat application on Tobias's UE assumed that resources are also available at Theresa's end. Let us assume that Theresa has signed in from her mobile device and did not pre-reserve any PDP context for an MSRP media stream. Tobias will send the initial INVITE as in the above example, but as resource reservation is needed at the terminating end, Theresa's UE will reject the INVITE with a 421 (Extension Required) response, indicating that the originating end is required to support preconditions:

```
SIP/2.0 421 Unsupported
Require: precondition
```

When receiving this response, Tobias's UE becomes aware that resource reservation is needed at the terminating end. It, therefore, constructs another INVITE request, which includes:

- The same media and codecs in the SDP Offer as in the first INVITE:
  - an indication that QoS-related preconditions are already met on the local link ("curr: qos local sendrecv" and "des: qos mandatory local sendrecv") – i.e., at Tobias's end;
  - an indication that Tobias's UE has no knowledge about the current status of the preconditions at the remote end ("curr: qos remote none") and that Theresa's UE can decide ("none") how to set them ("des: qos none remote sendrecv"):

```
m=message 3404 msrp/tcp *
a=accept-types:message/cpim text/plain text/html
a=path:msrp://[5555::aaa:bbb:ccc:ddd]:3402/s111271;tcp
a=max-size:131072
a=curr: qos local sendrecv
a=curr: qos remote none
a=des: qos mandatory local sendrecv
a=des: qos none remote sendrecv
```

As the preconditions on the local link are already met, the INVITE signalling flow is reduced in length. Theresa's UE no longer requires any confirmation from Tobias's UE as to whether preconditions are met or not; therefore, the "a=conf" line is not present in the SDP Answer that is sent from Theresa in the 183 (Session Progress) response. This means that the originating end will not generate an UPDATE request.

The session only makes use of one media type and one codec, so, therefore, there is not much to agree about in the SDP Offer/Answer process. The PRACK request that is sent to acknowledge the receipt of the 183 (Session Progress) and its 200 (OK) response do not carry any SDP-related information.

In this example it is also shown that there is no acknowlegement (PRACK) of the 180 (Ringing) response. As a consequence the session establishment procedure becomes rather short compared with the flow in Figure 11.1.

Figure 11.15 Fallback to full IMS session setup.

11.10.3 IMS session setup with resource reservation but without preconditions

The session setup procedures – as so far described in Chapter 11 – are too extensive and complicated for some scenarios. The PoC service (see Chapter 6), for example, makes use of shorter session setup procedures. Also, for networks that do not offer Service Based Local Policy (SBLP; see Section 3.9) or do not require IMS-specific charging for related media, shorter session setup signalling would be beneficial.

To satisfy these scenarios, alternative IMS session setup procedures have been defined which require resource reservation from both end points, but for which the preconditions mechanism is not used. These procedures will be introduced to IMS with 3GPP Release 7.

This new mechanism is based on the possibility of making media streams "inactive" in SDP. Inactive means that neither end is sending or waiting to receive media. This is identical to a scenario where both ends have put a call to mute or on hold. Usually, the SDP inactive flag is only used for call hold scenarios, but here it is applied with a slightly different background.

In this example we assume that Tobias again calls Theresa, this time wanting to invite her to a voice conversation. Instead of using preconditions, the UE sets the audio stream in the first SDP Offer to inactive. This SDP Offer is transported in the first INVITE request:

```
m=audio 3458 RTP/AVP 96
a=rtpmap: 96 AMR-WB
a=inactive
```

At Theresa's end we assume that these procedures are supported but resources are not reserved. As the terminating UE accepts the INVITE request, it responds to it with a 200 (OK), in which the media stream is also set to inactive. After sending/receiving this 200 (OK) response both ends start reserving resources. If media authorization (see Section 3.9) is required by one or both networks, the P-CSCF of Theresa would include the media authorization token in the first INVITE, whilst Tobias's P-CSCF would put the P-Media-Authorization header in the 200 (OK) response.

After Tobias's UE has reserved the related resources it constructs a re-INVITE request on the dialog that was established with the first INVITE request. This second INVITE request includes a second SDP Offer in which the media is set to active. This means that the originating end is ready to send and receive the related audio stream:

```
m=audio 3458 RTP/AVP 96
a=rtpmap: 96 AMR-WB
a=active
```

The networks will exchange charging-related information (see Section 11.8) within this second INVITE and its final 200 (OK) response, as at the time of sending them the resources at the bearer level (PDP contexts) and the dialogs at the signalling level can be correlated.

If Theresa's UE has not finished resource reservation when receiving the second INVITE request, it returns a 180 (Ringing) response, without indicating the call to Theresa yet. This scenario is shown in Figure 11.16. Only after local resources have been reserved, will the terminating UE ring Theresa and make her aware of the incoming call. After Theresa accepts the call, her UE returns a 200 (OK) response, in which the media streams are also set to active.

Figure 11.17 shows the scenario when resource reservation is finished at the terminating end first. In this case the terminating end waits for the re-INVITE request from the originating end.

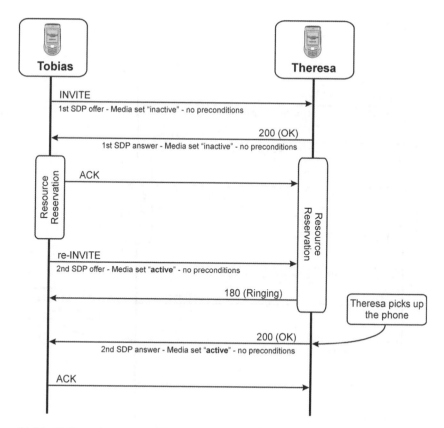

Figure 11.16   IMS session setup with resource reservation but without preconditions—originating end is first to finish resource reservation.

The two examples shown here leave several details out which, in reality, might further extend the call:

- if the terminating end sends back a list of codecs for one or more media streams in the first SDP Answer, it will include this SDP Answer in a 183 (Session Progress) response and request the originating end to send a PRACK, in order to allow the originating end to make the final codec selection (see Section 11.5.3.5);
- if Theresa's UE requires preconditions to be used, it would reject the first INVITE request and require the preconditions extension from the originating end, as shown in Section 11.10.2.2;
- if resource reservation fails at one of the two ends, that end would send a BYE request to terminate the established dialog;
- as the usage of inactive media streams during session establishment is only described for IMS terminals that require resource reservation it might happen that some UEs interpret the inactive indication in an initital INVITE request differently than expected. For example, a UE at the terminating end that does not require resource reservation might ring its user despite resources at the originating end have not being reserved yet. Prevention of such behaviour can only be guaranteed if both ends support and use the preconditions mechanism.

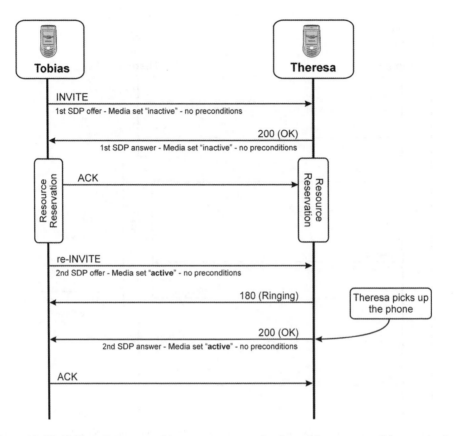

Figure 11.17 IMS session setup with resource reservation but without preconditions—terminating end is first to finish resource reservation.

11.10.4 Session setup between IMS and a SIP end point outside IMS

As IMS is not intended to be a closed system it also needs to interwork with other SIP-based Voice-over-IP networks. In most of these "non-IMS" networks, mechanisms such as resource reservation and media authorization are not needed and, therefore, not supported by the attached UEs. In order to make these mechanisms work porperly inside IMS and to guarantee seamless call establishment between any kinds of SIP-based UEs, interworking procedures are needed. These procedures were designed in a way that the non-IMS UE is not required to support any of the extensions that are needed within IMS.

11.10.4.1 Session originating from an IMS-attached terminal

In this example we assume Tobias tries to call his sister who is registered at a SIP terminal outside IMS. Tobias's UE will not be aware from Theresa's sip address that she is not registered to IMS and, therefore, construct a normal INVITE request,

Figure 11.18   IMS-attached UE originates session to non-IMS UE.

including preconditions and all IMS-related extensions (see Section 11.6.3):

```
INVITE sip:theresa@home2.hu SIP/2.0
Require: precondition
```

Theresa's UE does not support precondition mechanisms. As the precondition extension is indicated in the Require header of the INVITE request, Theresa's UE rejects this
INVITE request with a 420 (Bad Extension) response:

```
SIP/2.0 420 Bad Extension
Unsupported: precondition
```

When receiving this response, Tobias's UE constructs a new INVITE request, this time
avoiding the usage of preconditions. It falls back on the mechanism described in Section

11.10.2.1 – i.e., it sends a new initial INVITE request with the media set to inactive:

```
m=audio 3458 RTP/AVP 96
a=rtpmap: 96 AMR-WB
a=inactive
```

In the example flow here, it is assumed that Theresa's UE now indicates the call to Theresa and sends a 180 (Ringing) response. As Theresa's UE does not support the 100rel mechanism (see Section 11.5.2) it cannot send any SDP in a provisional response. After Theresa has accepted the call, her UE will send a 200 (OK) response with the media stream set to "inactive" as well. Although the terminating UE is ready to receive media, it must set the inactive flag in the SDP Answer as well, as it is only allowed to answer with "inactive" to an SDP Offer that included the same flag:

```
m=audio 3458 RTP/AVP 96
a=rtpmap: 96 AMR-WB
a=inactive
```

Afterwards, Tobias's UE starts reserving resources – the session establishment flow is basically identical to the one shown in Section 11.10.2.1.

11.10.4.2 Session originating from terminal outside IMS

Figure 11.19 shows the scenario when the originating UE is outside IMS. In this case Tobias calls Theresa, who is attached to her local IMS network. As Tobias's UE assumes that the remote end has the required resources for the media connection available it does not put any specific indication into the SDP Offer. This means that it has the audio stream available in both sending and receiving (sendrecv) directions. As sendrecv is the default flag for a media stream it can be surpressed:

```
m=audio 3458 RTP/AVP 96
a=rtpmap: 96 AMR-WB
```

Nevertheless, resource reservation is needed at Theresa's end and, therefore, her UE sets the media to inactive when accepting the incoming INVITE request.

```
m=audio 3458 RTP/AVP 96
a=rtpmap: 96 AMR-WB
a=inactive
```

The session setup flow is from there on identical to the one shown in Section 11.10.2.1.

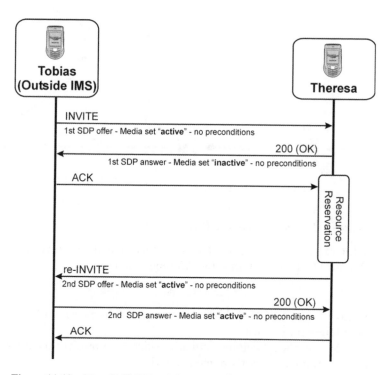

Figure 11.19   Non-IMS UE originates session to IMS UE-attached UE.

11.11   Routing of PSIs

The concept of a Public Service Identity (PSI; i.e., a URI that is not related to a user but to a service) is explained in Section 3.5.2.

This section is a basic introduction to the routing principles of PSIs, as they are quite different from those that are applied between two IMS users.

As PSIs are not registered, requests to and from them usually do not need to traverse any S-CSCF.

There are three scenarios for PSI routing.

11.11.1   Scenario 1: routing from a user to a PSI

This occurs, for example, when a user calls into a conference (Figure 11.20). In this case the request needs first to traverse the user's S-CSCF, which can then:

1. Either, resolve the PSI immediately and route it directly to the AS.
2. Or, is able to resolve the address of an I-CSCF in the home network of the PSI. The I-CSCF will then query the HSS – where routing information about the PSIs will be stored – and route the request directly to the hosting AS. Note that an S-CSCF can also be assigned for the PSI, in which case this would be contacted first. This scenario is not shown here.

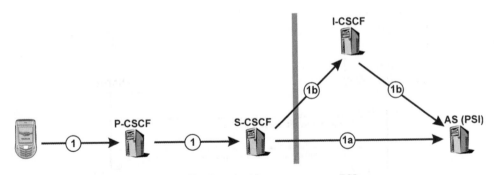

Figure 11.20   Routing from a user to a PSI.

11.11.2   Scenario 2: routing from a PSI to a user

This occurs, for example, when a conference server (the focus) invites a user to a conference (Figure 11.21). In this case the AS sends the request:

1. Either, directly to the I-CSCF of the terminating user's home network, if the AS can resolve this address on its own.
2. Or, to an S-CSCF in the ASs home network, which then resolves the address of the I-CSCF of the terminating user's home network.

Figure 11.21   Routing from a PSI to a user.

11.11.3   Scenario 3: routing from a PSI to another PSI

This occurs, for example, when two conferences are interconnected such that one conference server (focus) sends an INVITE request to another focus (Figure 11.22). For this case, several routing possibilities exist. The originating AS can route the

Figure 11.22   Routing from an AS to a PSI.

request:

1. Either, directly to the terminating AS that is hosting the second PSI, if the originating AS is able to resolve the PSI directly.
2. Or, if it cannot resolve the address of the second PSI directly, to an I-CSCF of the terminating AS, which then queries the HSS with the second PSI and sends the request directly to the terminating AS. Note that an S-CSCF can also be assigned for the PSI, in which case this would be contacted first. This scenario is not shown here.
3. Or, if it cannot resolve any part of the terminating address, to an S-CSCF in its own home network, which will then:
   ○ Either, resolve the address of the second PSI and send the request to the terminating AS directly.
   ○ Or, cannot resolve the address of the second PSI directly and, therefore, sends the request to an I-CSCF in the home network of the second PSI. This I-CSCF will then act in the same way as item (2) in Scenario 3.

Part IV

Protocols

12

SIP

This chapter does not provide a full Session Initiation Protocol (SIP) specification. Instead, it tries to point out the important aspects of SIP as they apply to the Internet Protocol Multimedia Subsystem (IMS). In particular, this chapter does not discuss how a SIP entity should behave using the maddr parameter in Uniform Resource Identifiers (URIs) nor does it explain how the SIP entity should behave in certain error conditions. For a full SIP specification please refer to [RFC3261].

12.1 Background

SIP is an application layer protocol that is used for establishing, modifying and terminating multimedia sessions in an Internet Protocol (IP) network. It is part of the multimedia architecture whose protocols are continuously being standardized by the Internet Engineering Task Force (IETF). Its applications include, but are not limited to, voice, video, gaming, messaging, call control and presence.

The idea of a session signalling protocol over IP dates back to 1992 where multicast conferencing was being considered. SIP itself originated in late 1996 as a component of the IETF Mbone (multicast backbone), an experimental multicast network on top of the public Internet. It was used by IETF for the distribution of multimedia content, including IETF meetings, seminars and conferences. Due to its simplicity and extensibility, SIP was later adopted as a Voice over IP (VoIP) signalling protocol, finally becoming an IETF-proposed standard in 1999 as [RFC2543]. SIP was further enhanced to take into account interoperability issues, better design and new features. The actual document was re-written entirely for clarity. The protocol remains mostly backward compatible with [RFC2543]. This newly created document became the proposed standard as [RFC3261] in 2002, making [RFC2543] obsolete.

12.2 Design principles

SIP, as part of the IETF process, is based on the Hypertext Transfer Protocol (HTTP) and the Simple Mail Transfer Protocol (SMTP). Figure 12.1 shows where SIP fits into a protocol stack.

The IMS: IP Multimedia Concepts and Services Second Edition
Miikka Poikselkä, Georg Mayer, Hisham Khartabil and Aki Niemi © 2006 John Wiley & Sons, Ltd

Figure 12.1 Protocol stack.

SIP was created with the following design goals in mind:

- transport protocol neutrality – able to run over reliable (TCP, SCTP) and unreliable (UDP) protocols;
- request routing – direct (performance) or proxy-routed (control);
- separation of signalling and media description – can add new applications or media;
- extensibility;
- personal mobility.

12.3   SIP architecture

Elements in SIP can be classified as User Agents (UAs) and intermediaries (servers). In an ideal world, communications between two end points (or UAs) happen without the need for intermediaries. But, this is not always the case as network administrators and service providers would like to keep track of traffic in their network.

Figure 12.2 depicts a typical network setup, which is referred to as the "SIP trapezoid".

A SIP UA or terminal is the end point of dialogs: it sends and receives SIP requests and responses, it is the end point of multimedia streams and it is, usually, the User Equipment (UE), which comprises an application in a terminal or a dedicated hardware appliance. The UA consists of two parts:

- User Agent Client (UAC) – the caller application that initiates requests;
- User Agent Server (UAS) – accepts, redirects, rejects requests and sends responses to incoming requests on behalf of the user.

Gateways are special cases of UAs.

Figure 12.2   SIP trapezoids.

SIP intermediaries are logical entities through which SIP messages pass on their way to their final destination. These intermediaries are used to route and redirect requests. These servers include:

- *Proxy server* – receives and forwards SIP requests. It can interpret or re-write certain parts of SIP messages that do not disturb the state of a request or dialog at the end points, including the body. A proxy server can also send a request to a number of locations at the same time. This entity is labelled a "forking proxy". Forking can be parallel or sequential. There are three proxy server types:
    - dialog-statefull proxy – a proxy is dialog-statefull if it retains the state for a dialog from the initiating request (INVITE request) right through to the terminating request (BYE request);
    - transaction-statefull proxy – a proxy that maintains client and server transaction-state machines during the processing of a request;
    - stateless proxy – a proxy that forwards every request it receives downstream and every response it receives upstream.
- *Redirect server* – maps the address of requests to new addresses. It redirects requests but does not participate in the transaction.
- *Location server* – keeps track of the location of users.
- *Registrar server* – a server that accepts REGISTER requests. It is used to store explicit binding between a user's address of record (SIP address) and the address of the host where the user is currently residing or wishes to receive requests.

Two more elements that are used to provide services for SIP users:

- *Application server* – an AS is an entity in the network that provides end-users with a service. Typical examples of such servers are presence and conferencing servers.

- *Back-to-back-user-agent* – as the name depicts, a B2BUA is where a UAS and a UAC are glued together. The UAS terminates the request just like a normal UAS. The UAC initiates a new request that is somehow related to requests received at the UAS end, but not in any protocol-specific link. This entity is almost like a proxy, but breaks all the rules that govern the way a proxy can modify a request.

12.4   Message format

As shown in Figure 12.3, the SIP message is made up of three parts: the start line, message headers and body.

The start line contents vary depending on whether the SIP message is a request or a response. For requests it is referred to as a "request line" and for responses it is referred to as a "status line".

An example SIP request looks like:

```
INVITE sip:bob.smith@nokia.com SIP/2.0
Via: SIP/2.0/UDP cscf1.example.com:5060;branch=z9hG4bK8542.1
Via: SIP/2.0/UDP [5555::1:2:3:4]:5060;branch=z9hG4bK45a35h76
Max-Forwards: 69
From: Alice <sip:alice@nokia.com>;tag=312345
To: Bob Smith <sip:bob.smith@nokia.com>
Call-ID: 105637921
CSeq: 1 INVITE Contact: sip:alice@[5555::1:2:3:4]
Content-Type: application/sdp
Content-Length: 159

[body]
```

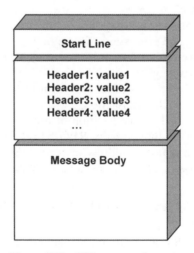

Figure 12.3   SIP message format.

12.4.1 *Requests*

SIP requests are distinguished from responses using the start line. As indicated earlier, the start line in the request is often referred to as the request line. It has three components: a method name, a request-URI and the protocol version. They appear in that order and are separated by a single space character. The request line itself terminates with a Carriage Return–Line Feed (CRLF) pair:

- *Method* – the method indicates the type of request. Six are defined in the base SIP specification [RFC3261]: the INVITE request, CANCEL request, ACK request and BYE request are used for session creation, modification and termination; the REGISTER request is used to register a certain user's contact information; and the OPTIONS request is used as a poll for querying servers and their capabilities. Other methods have been created as an extension to [RFC3261].
- *Request-URI* – the request-URI is a SIP or a Secure SIP (SIPS) URI that identifies a resource that the request is addressed to.
- *Protocol version* – the current SIP version is 2.0. All requests compliant with [RFC3261] must include this version in the request, in the form "SIP/2.0".

12.4.2 *Response*

SIP responses can be distinguished from requests by looking at the start line. As indicated earlier, the start line in the response is often referred to as the status line. It has three components: the protocol version, status code and reason phrase. They appear in that order and are separated by a single space character. The status line itself terminates with a CRLF pair:

- *Protocol version* – this is identical to the protocol version in the request line.
- *Status code* – the status code is a three-digit code that identifies the nature of the response. It indicates the outcome of the request.
- *Reason phrase* – this is a free text field providing a short description of the status code. It is mainly aimed at human users.

Status codes are classified in six classes (classes 2xx to 6xx are final responses):

- *1xx* – provisional/informational responses. They indicate that the request was received and the recipient is continuing to process the request.
- *2xx* – success responses. The request was successfully received, understood and accepted.
- *3xx* – re-direction responses. Further action needs to be taken by the requester in order to complete the request.
- *4xx* – client error responses. The request contains a syntax error. It can also indicate that the server cannot fulfil the request.
- *5xx* – server error responses. The server failed to fulfil a valid request. It is the fault of the server.

- *6xx* – global failure responses. The request cannot be fulfilled at any server. The server responding with this response class needs to have definitive information about the user.

The "xx" are two digits that indicate the exact nature of the response: for example, a "180" provisional response indicates ringing at the remote end, while a "181" provisional response indicates that a call is being forwarded.

12.4.3 Header fields

Header fields contain information related to the request: for example, the initiator of the request, the recipient of the request and call identifier. Header fields also indicate message body characteristics.

Header fields end with a CRLF pair. The headers section of a SIP message terminates with a CRLF.

The format of the header fields is as follows:

```
Header-name: header-value
```

Some headers are mandatory in every SIP request and response. These headers and their formats are listed below:

- To header   To: SIP-URI(;parameters)
- From header   From: SIP-URI(;parameters)
- Call-ID header   Call-ID: unique-id
- CSeq header   CSeq: digit method
- Via header   Via: SIP/2.0/[transport-protocol] sent-by(;parameters)
- Max-Forwards header   Max-Forwards: digit
- Contact header   Contact: SIP-URI(;parameters)

The Contact header is mandatory for requests that create dialogs, the Max-Forwards header is typically set to 70. Note that the brackets around parameters indicate that they are optional. The brackets are not part of the header syntax. Whenever (;parameters) appears it indicates that multiple parameters can appear in a header and that semicolons separate the parameters. The transport protocols used for the Via header are User Datagram Protocol (UDP), Transmission Control Protocol (TCP) or Transport Layer Security (TLS).

12.4.4 Body

The message body (payload) can carry any text-based information, while the request method and the response status code determine how the body should be interpreted.

When describing a session the SIP message body is typically a Session Description Protocol (SDP) message.

12.5   The SIP URI

The SIP URI follows the same form as an email address: `user@domain`. There are two URI schemes:

- *sip:bob.smith@nokia.com* is a SIP URI. This is the most common form and was introduced in [RFC2543].
- *sips:bob.smith@nokia.com* is a SIPS URI. This new scheme was introduced in [RFC3261] and requires TLS over TCP as transport for security.

There are two types of SIP and SIPS URIs:

- Address Of Record (AOR) – this is a SIP address that identifies a user. This address can be handed out to people in much the same way as a phone number: e.g., *sip:bob.smith@nokia.com* (needs DNS SRV records to locate SIP servers for the *nokia.com* domain).
- Fully Qualified Domain Name (FQDN) or IP address (identifies a device) of the host – e.g., *sip:bob.smith@127.233.4.12* or *sip:bob.smith@pc2.nmp. nokia.com* (which needs no resolution for routing).

The SIP URI has the form: *sip:userinfo@hostport[parameters][headers]*. The SIPS URI follows exactly the same syntax as the SIP URI:

- Userinfo – a user name or a telephone number.
- Hostport – the domain name or numeric network address and port.
- Parameters – defines specific URI parameters, such as transport, time to live, etc.
- Headers – another rarely used form that passes on extra information.

Below are some examples of SIP URIs:

- *sip:bob.smith@nokia.com*
- *sip:bob@nokia.com; transport=tcp*
- *sip:+1-212-555-1234@gw.com;user=phone*
- *sip:root@136.16.20.100:8001*
- *sip:bob.smith@registrar.com;method=REGISTER*

12.6   The tel URI

The telephone URI (tel URI) is used to identify resources using a telephone number. SIP allows requests to be sent to a tel URI. This means that the request-URI of a SIP request can contain a tel URI.

The tel URI can contain a global number or a local number. A global number follows the rules of E.164 numbers and starts with a "+", while a local number follows the rules of local private numbering plans. Local numbers need to have the phone-context parameter, which identifies the context (owner) of the local number and, therefore, the scope of the number. This makes the number globally unique. The context can be represented by a global number or a domain name: the former must contain a valid global number that is owned by the local number distributor, and the latter must contain a valid domain name that is under the authority of the owner distributing the local numbers. Here are some tel URI examples:

- a global number – *tel:+358-9-123-45678*
- a local number with a domain context – *tel:45678;phone-context=example.com*
- a local number with a global number context: *tel:45678;phone-context= +358-9-123*

Notice that the tel URI allows visual separators like hyphens "-" in the number to improve readability and the tel URI parameters are separated by semicolons ";". The full tel URI syntax can be found in [RFC3966].

12.7 SIP structure

SIP is a layered protocol that allows different modules within it to function independently with just a loose coupling between each layer. Figure 12.4 visualizes the layered approach taken.

12.7.1 Syntax and encoding layer

The first (bottommost) layer in the protocol is the syntax and encoding layer. Encoding makes use of augmented Backus-Naur Form (BNF) grammar, the complete description of which can be found in [RFC3261].

12.7.2 Transport layer

The second layer is the transport layer. As the name indicates, this is the layer that dictates how clients send requests and receive responses and how servers receive

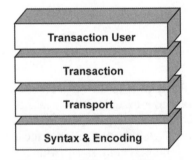

Figure 12.4 SIP protocol layers.

requests and send responses. The transport layer is closely related to the sockets layer of a SIP entity.

12.7.3   Transaction layer

The third layer is the transaction layer. A transaction, in SIP terms, is a request that is sent by a client to a server, along with all responses to that request sent from the server back to the client. The transaction layer handles the matching of responses to requests. Application-layer re-transmissions and application-layer transaction timeouts are also handled in this layer and are dependent on the transport protocol used. A client transaction sends requests and receives responses, while a server transaction receives requests and sends responses. The transaction layer uses the transport layer for sending and receiving requests and responses.

The transaction layer has four transaction-state machines. Each transaction-state machine has its own timers, re-transmission rules and termination rules:

- INVITE client transaction;
- non-INVITE client transaction;
- INVITE server transaction;
- non-INVITE server transaction.

12.7.4   TU layer

The fourth (topmost) layer is the Transaction User (TU) layer. This is the layer that creates client and server transactions. When a TU wishes to send a SIP request it creates a client transaction instance and sends the request along with the destination IP address, port and name of the transport protocol to use. TUs are defined to be UAC core and UAS core, or simply UAC and UAS. UACs create and send requests and receive responses using the transaction layer, while UASs receive requests and create and send responses using the transaction layer.

There are two factors that can affect TU behaviour: one is the method name in the SIP message and the other is the state of the request with regard to dialogs (dialogs are discussed in Section 12.9).

Other than these two factors, the TU behaves in a standard way. This is described in the following sections.

12.7.4.1   UAC behaviour

For requests that arrive outside a dialog, the steps that a UAC needs to take include populating the request-URI, the To header, the From header, the Call-ID header, the CSeq header and the Via header. Other headers like the Require header and Supported header that indicate any extension that the UAC requires or supports may also be added. A Contact header must be added if the request creates a dialog or if a registration binding is required. Any additional components can also be populated at this stage; this includes the message body. In the presence of a message body in the SIP request, the Content-type header and Content-length header must also be populated:

- The To header is populated with the target's AOR (an AOR is similar to a business card address).
- The From header is populated with the sender's AOR. It is also populated with a tag parameter. The tag is one of the components used to identify a dialog.
- The Call-ID header is populated with an identifier that is unique.
- The CSeq header is used to identify the order of transactions. The CSeq number is arbitrary for requests outside a dialog. It contains two parts, a CSeq and a method name separated by a space. The method part is populated with the same method as the one in the request line.
- The Max-Forwards header is used to limit the number of hops a request traverses and is used to avoid loops. It is typically set to 70 (indicating the number of hops). Each hop decrements the value by 1.
- The Via header contains two vital pieces of information: the transport protocol and the address where the response is to be sent. The protocol name and value are always set to SIP and 2.0, respectively. The Via header contains a branch parameter that identifies transactions and is used to match requests to responses. It must be unique. The branch inserted by an element compliant with this specification always begins with the characters "z9hG4bK".
- The Contact header is populated with a URI that is typically the address of the host where the request originated.
- The request-URI is normally populated with the value in the To header. REGISTER requests are special cases in which the request-URI is populated by the registrar address.

A UAC may have a pre-existing route set, which is a set of intermediaries to which the UAC wants requests to propagate before reaching their final destination, including an outbound proxy. This route set is represented in the request as Route headers. The request-URI population may differ in this case depending on whether the URI in the topmost Route header contains a loose-route parameter. Section 12.12.2 explains the concept of loose routing and how a remote-target should be populated in case the next hop is a loose router. The procedures in Section 12.12.2 are followed for populating the request-URI as the remote target.

The UAC must then route the request according to the rules defined in Section 12.12.

The UAC also handles the responses to requests it sends. These responses can be timeout error responses or SIP success or failure responses, including re-direction responses (3xx).

12.7.4.2   UAS behaviour

For requests that arrive outside a dialog, a UAS inspects the request method for recognition and inspects the request-URI and To header to determine whether this request is destined to it. If either of the two inspections fail, an error response is returned.

The UAS then decides whether any extensions are required and returns an error if it cannot satisfy them. If it can, the UAS continues processing the request by examining and processing the contents of the request (the message body).

If all the above is successful, the UAS can then apply any extensions that are supported by the UAC (as indicated in the Supported header). Processing of the request from this point on is method-specific: for REGISTER requests see Section 12.8 and for INVITE requests see Section 12.10.

Once the UAS has processed the request it generates a response, which can be provisional or final. Multiple provisional responses can be sent for one request, but only one final response must be sent. Typically, a provisional response is only sent in response to an INVITE request.

When generating a response, the From header, the Call-ID header, the CSeq-header, the Via header and the To header in the response are all copied from the request. The sequence of Via headers in the request must be maintained.

If a request contains a To header tag parameter in the request, then a new tag must not be created. However, if the To header in the request does not contain a tag, then the UAS must add a tag to the To header in the response. For "100" provisional responses, a To header tag is not mandated. This tag is used as one of the components that identify a dialog. It is also used by a forking proxy to identify the UAS.

12.8   Registration

SIP supports the concept of user mobility and discovery. A user can make herself available for communication by explicitly binding her AOR to a certain host address. This allows user mobility since the user can register from any device that supports SIP, including personal computers, wireless devices and cellular phones.

Discovery of the intended recipient of a SIP request is typically the function of SIP intermediary servers: for example, the user creates a binding to the registrar, which acts as a front end to a location server where all the bindings are stored, and then a proxy server, receiving a request that is destined to a domain it is responsible for, contacts that location server to retrieve the exact location of that intended recipient.

A user creates a binding by placing her AOR in the To header and the host address in the Contact header.

A user can be registered at many devices simultaneously by sending a REGISTER request from each device. Similarly, a user can create multiple bindings from the same device; this can be achieved by sending one REGISTER request with multiple bindings to the AOR. To do this, a user adds multiple contact headers in the REGISTER request.

A user can discover all the current bindings to her AOR using a process called "registration fetching". This is accomplished by sending a REGISTER request without a contact header. The registrar returns all the current bindings in the register response. Each binding has its own contact header in the response.

SIP registrations are, by nature, soft-state: this means that registration bindings must be periodically refreshed (updated). The expiration time of a binding is indicated by the registering entity using the expires parameter in a Contact header. If this parameter is not present, the registrar assumes an expiration time of 1 hour. If the UA does not refresh or, otherwise, explicitly remove the binding, the registrar silently removes it when the expiration time lapses. A UA can explicitly remove a binding by sending a

REGISTER request and adding a Contact header for the binding to be removed. This Contact header contains the expires parameter with a value of 0.

A registrar can be discussed using the procedures in Chapter 24.

12.9   Dialogs

A dialog is a SIP relationship between two collaborators. The dialog provides the necessary states required for the routing and sequencing of messages between those collaborators.

Dialogs are identified using Dialog-IDs and UAs use them to track messages sent within a dialog. A Dialog-ID consists of a Call-ID, a local tag and a remote tag. For a UAC the local tag is the tag that appears in the From header of the initial dialog-creating request and the remote tag is the tag that appears in the To header of the dialog-creating response. For a UAS the local tag is the tag that appears in the To header of the dialog-creating response and the remote tag is the tag that appears in the From header of the initial dialog-creating request. For subsequent requests using dialogs sent from either end, the local tag is placed in the From header and the remote tag is placed in the To header.

Note that a UAS needs to be prepared to receive a request without a tag in the From header, in which case the tag is considered to have a null value.

A dialog state is needed for creating, sending, receiving and processing of messages within a dialog. This state consists of the dialog-ID, a local sequence number, a remote sequence number, a local URI, a remote URI, a remote target, a Boolean flag called a "secure" flag and a route set.

When a dialog is in an "early" state it is referred to as an "early dialog". This occurs when a provisional response arrives at the UAC to an initial request, thus creating a dialog. A dialog moves to a "confirmed" state when a "2xx" success response arrives. If a final response other than a "2xx"-class response arrives or if no response arrives at all, the early dialog terminates.

A UAS responding to a request with a final response indicating success must copy all Record-Route headers that appear in the request into the response, maintaining the order they appear in. The UAS then stores these URIs in the Record-Route headers as the route set, maintaining the order. If no Record-Route headers are present, the route set is left empty. This route set, even if empty, is retained for the remainder of the dialog. This means that other Record-Route headers appearing in requests within dialogs do not override the already-existing route set.

Record-Route headers are added to a request by intermediaries that wish to remain on the signalling path of any subsequent requests sent from the UAC to the UAS, or vice versa, within a dialog.

A UAS must also add a Contact header in the response that indicates the address where subsequent requests within the dialog should be targeted.

The dialog state at the UAS is constructed as follows:

- If the request arrived over the TLS and the request-URI contained a SIPS URI, the "secure" flag is set to true; otherwise, it is set to false.

- The remote target is set to the URI from the Contact header of the request.
- The remote sequence number is set to the value of the sequence number in the CSeq header of the request.
- The local sequence number remains empty at this stage. It is populated when the remote end sends a request within a dialog.
- The remote URI is set to the URI in the From header.
- The local URI is set to the URI in the To header.
- The Dialog-ID is created as indicated above.
- The route set is set as indicated above.

The UAC must provide a Contact header in an initial request that creates a dialog. When the UAC receives a response that creates a dialog, it creates the dialog state at its end as follows:

- If the request was sent over TLS and the request-URI contained a SIPS URI, the "secure" flag is set to true; otherwise, it is set to false.
- The remote target is set to the URI from the Contact header of the response.
- The remote sequence number is set to empty at this stage. It is populated when the remote end sends a request within a dialog.
- The local sequence number is set to the value of the sequence number in the CSeq header of the request.
- The local URI is set to the URI in the From header.
- The remote URI is set to the URI in the To header.
- The Dialog-ID is created as indicated above.
- The route set is set using URIs in the Record-Route headers, but the order is reversed. If no Record-Route headers are present, the route set is left empty. This route set, even if empty, is retained for the remainder of the dialog. This means that other Record-Route headers appearing in requests within dialogs do not override the already-existing route set. If the route set was created using Record-Route headers in a provisional response, then the 2xx final response that confirms the dialog re-sets the route set using URIs in the Record-Route headers, but once again the order is reversed.

Requests within dialogs are populated using dialog states. The local CSeq header value is incremented by 1 for every new request created within a dialog. Requests within dialogs may update the remote target if they are target refresh requests: examples of target refresh requests are INVITE requests and UPDATE requests.

An early dialog is terminated when a non-2xx final response is returned to the request. Confirmed dialogs are terminated uniquely, depending on the method used.

12.10 Sessions

A multimedia session consists of a set of multimedia senders and receivers and the data streams that flow between them. Sessions use SIP dialogs and follow SIP rules for sending requests within dialogs.

The role SIP plays in establishing a multimedia session revolves around its ability to carry SDP media descriptions in its message bodies. SDP is used to describe the session, and the offer/answer model is employed [RFC3264]. SDP and the offer/answer model are described in Chapter 14. Section 12.10.1 describes the restrictions SIP has on such a model.

The session is initiated using the INVITE method, the request line and headers, which are populated with the UAC (see Section 12.7.4.1). The body is populated with an SDP offer. The answer may arrive in a provisional response or in the 2xx response.

INVITE requests follow a three-way handshake model: this means that the UAC, after receiving a final response to an INVITE request, must send an ACK request. The ACK request does not require a response; in fact, a response must never be sent to an ACK request.

If the UAC wants to cancel an invitation to a session after it sends the INVITE request, it sends a CANCEL request. The CANCEL request is constructed in a similar way to that in which the request-URI, the To header, the From header, the Call-ID header and the numeric part of the CSeq header are copied from the INVITE request. The method part of the CSeq header holds the CANCEL method. A UAS receiving a CANCEL responds to it with a 200 response and, then, follows it up with a "487 Request Terminated" response to the INVITE request. It is important to remember that all transactions must complete independently of each other: this is the reason the UAS must respond to the INVITE request.

If the UAC is not satisfied with the SDP answer that arrives in the 2xx response, it sends an ACK request followed by a BYE request to terminate the session. If the UAS is not satisfied with the SDP offer, it rejects the request with a 488 response.

INVITE requests can also be sent within dialogs to re-negotiate the session description.

A session is terminated by a BYE request. The BYE request is sent in exactly the same way as any other request within a dialog.

12.10.1 The SDP Offer/Answer model with SIP

Using basic SIP, Offer and Answer can only appear in INVITE requests, reliable responses to INVITE requests and ACK requests. However, Sections 12.13.4 and 12.13.5 describe further opportunities for offer/answer exchanges using SIP extensions.

If an INVITE request results in multiple dialogs, each dialog has its own offer/answer exchange.

The general rule for an offer/answer exchange is that an offer must not be sent unless a previously sent (received) offer, if any, has received (been sent) an answer. This rule restricts basic SIP to the following scenarios when an offer/answer exchange is possible:

- If the offer is in the INVITE request, then the answer must appear in the 2xx response of the INVITE.
- If the INVITE request did not contain an offer, then the 2xx response contains the offer and the ACK contains the answer.

12.11   Security

12.11.1   Threat models

SIP is susceptible to the following threats and attacks:

- *Denial of service* – the consequence of a DOS attack is that the entity attacked becomes unavailable. This includes scenarios like targeting a certain UA or proxy and flooding them with requests. Multicast requests are further examples.
- *Eavesdropping* – if messages are sent in clear text, malicious users can eavesdrop and get session information, making it easy for them to launch a variety of hijacking-style attacks.
- *Tearing down sessions* – an attacker can insert messages like a CANCEL request to stop a caller from communicating with someone else. He can also send a BYE request to terminate the session.
- *Registration hijacking* – an attacker can register on a user's behalf and direct all traffic destined to that user towards his own machine.
- *Session hijacking* – an attacker can send an INVITE request within dialog requests to modify requests en route to change session descriptions and direct media elsewhere. A session hijacker can also reply to a caller with a 3xx-class response, thereby redirecting a session establishment request to his own machine.
- *Impersonating a server* – someone else pretends to be the server and forges a response. The original message could be misrouted.
- *Man in the middle* – this attack is where attackers tamper with a message on its way to a recipient.

12.11.2   Security framework

There are six aspects to the SIP security framework:

- *Authentication* – this is a means of identifying another entity or user and making sure that the user is really who he claims to be. Typical methods involve user IDs, passwords or digitally signing a set of bytes using a keyed hash.
- *Authorization* – once the user is authenticated, he must be authorized. Authorization involves deciding whether the user with the authenticated identity should be granted access to the requested services. This is often achieved using Access Control Lists (ACLs).
- *Confidentiality* – this is where messages must remain confidential and only the intended recipient is allowed to see the contents of a message. This is usually achieved by means of encryption.
- *Integrity* – a user needs to be assured that a message was not tampered with en route. A message integrity check is a means of ensuring this.
- *Privacy* – anonymity of users is key. Users do not want others to know who they are, what they are communicating or with whom they are communicating.
- *Non-repudiation* – reverse protection.

12.11.3 Mechanisms and protocols

12.11.3.1 Hop-by-hop mechanism

Hop-by-hop authentication provides the user with total confidentiality. It involves a complex security infrastructure that requires each proxy to decrypt the message and, therefore, relies on trust relationships between hops. Two protocols are in use for SIP: IP security (IPsec: Chapter 22) and Transport Layer Security (TLS: Chapter 18).

SIP, TLS and SIPS URI

TLS provides authentication, integrity and confidentiality. As mentioned earlier, the use of TLS in SIP messages requires all SIP entities to use a SIPS URI. A UAC wishing to communicate securely places a SIPS URI in the To header. If the next hop URI or the request-URI of a SIP request contains a SIPS URI, the UAC must place a SIPS URI in the Contact header. If the request-URI contains a SIPS URI, any alternative destinations to the request must be contacted using TLS.

TLS-secured requests must be sent using a reliable transport protocol like TCP or the Stream Control Transmission Protocol (SCTP). The default port for sending TLS-secured requests and for sending TLS-secured responses is 5061.

When registering a binding, a UAC must create a SIPS URI in the Contact header unless it can guarantee that the host represented in the Contact header has other means of security.

A UAS responding to a request that creates a dialog with a dialog-creating response (non-failure response) places a SIPS URI in the Contact header of the response if the request-URI, top Record-Route header (if there is one) or the Contact header (if there are no Record-Route headers) has a SIPS URI. UASs must send responses using TLS if the request arrived on TLS (the Via header shows TLS as the transport).

UACs and UASs examine the "secure" flag in the dialog state when sending requests within dialogs. A "secure" flag value of true requires those entities to place a SIPS URI in the Contact headers.

For proxies inserting a Record-Route header, they must place a SIPS URI in the header if the request-URI or the topmost Route header (after post-processing the request) has a SIPS URI.

All SIP entities must use TLS if the next hop URI is a SIPS URI. Entities sending new requests using Contact headers in 3xx responses should not send a new request to a non-SIPS URI if the request-URI in the initial request contained a SIPS URI. Independently of which URI is being used as input to the procedures of discovering the next hop (Section 12.12), if the request-URI specifies a SIPS resource, the SIP entity making the discovery must follow the same procedures just as if the input URI was a SIPS URI.

A proxy – that is processing responses – changes the URI it places in a Record-Route header from a SIPS URI to a SIP URI if it receives the request from a non-TLS connection and forwards it to a TLS connection. Similarly, a proxy – that is processing responses – changes the URI it places in a Record-Route header from a SIP URI to a SIPS URI if it receives the request from a TLS connection and forwards it to a non-TLS connection.

The format of a SIPS URI is identical to that of a SIP URI except for the scheme: for

SIPS URIs it is "sips", while for SIP URIs it is "sip". A SIP URI and a SIPS URI are not equivalent.

IPsec
IPsec provides authentication, integrity and confidentiality by securing SIP messages at the IP layer. It supports both TCP and UDP (see Chapter 22 for more details).

12.11.3.2   User-to-user and proxy-to-user mechanisms

User-to-user (or end-to-end) and proxy-to-user security can be regarded as a more secure mechanism since only two entities are available for attack. Two protocols are used in SIP for this mechanism: SIP digest and Secure Multipurpose Internet Mail Extension (S/MIME) [RFC2633]. In addition, there is an extension to the digest framework – namely, digest AKA – which is used in Third Generation Partnership Project (3GPP) IMS.

Digest authentication
SIP digest authentication mostly makes use of the HTTP digest [RFC2617] authentication mechanism with a few minor modifications. Although digest only provides limited integrity protection, it does provide client authentication and replay protection. It also provides a form of mutual authentication that enables clients to authenticate servers.

Digest authentication requires a shared secret: this means that there is a need for a pre-existing relationship between all users and between all users and proxies. This is very problematic for public services.

Digest AKA Authentication
As described in Section 3.30.2, IMS authentication utilizes the Universal Mobile Telecommunications System (UMTS) Authentication and Key Agreement (AKA) protocol. This protocol needs to be transported within SIP signalling, which is the idea behind digest AKA: it integrates the AKA protocol and the digest authentication framework.

In practice, this means that an AKA authentication request is encapsulated in the WWW-Authenticate header field or the Proxy-Authenticate header field carried in the 401 "Unauthorized" and 407 "Proxy Authentication Required" responses, respectively. Similarly, the client's authentication response is encapsulated in the Authorization header field or the Proxy-Authorization header field of the request.

AKA parameters – namely, the random challenge (RAND) and the network authentication token (AUTN) – are concatenated and appended to the server nonce parameter. The response (RES) is calculated in the request digest by simply treating the RES parameter as the digest password. The normal authentication flow using digest AKA is illustrated in Figure 12.5.

The only exception to fully following the digest framework occurs when there is an AKA synchronization failure. Then, the synchronization failure parameter AUTS is included in an extension digest parameter, encoded in Base64: the reason for this is simply that there exists no other proper protocol element to carry the parameter (Figure 12.6).

Figure 12.5 Normal digest AKA message flow. Square brackets [ ] indicate that the element is optional and the message syntax is merely figurative.

Figure 12.6 Digest AKA message flow in a synchronization failure. Square brackets [ ] indicate that the element is optional and the message syntax is merely figurative.

S/MIME

Secure Multipurpose Internet Mail Extension (S/MIME) provides message integrity, confidentiality and authentication, which is achieved by protecting SIP headers in an encrypted and/or signed S/MIME SIP message body. It does not require a shared secret.

12.12 Routing requests and responses

12.12.1 Server discovery

The procedures indicated in this section are minimal and only show steps needed in the most basic scenarios (please refer to [RFC3263] for full procedures).

12.12.1.1 Sending requests

The TU layer performs the next hop discovery by using the URI it chose to be the next hop URI (see Section 12.12.5). The basic steps are as follows:

- If the URI contains a transport parameter, then that transport is used.
- If the URI contains a numeric IP address, but no transport parameter, then the response is sent using that IP address and the port indicated (or to the default port if no port is specified). The transport used is UDP for SIP URIs and TCP for SIPS URIs.
- If the sent-by field contains a domain name and a port, then an A (IPv4) or AAAA (IPv6) query is made. The resulting IP address and the available port are used to send the response. The transport used is UDP for SIP URIs and TCP for SIPS URIs.
- For URIs without numeric IP addresses or ports, a Domain Name System (DNS) server is queried for Naming Authority Pointer (NAPTR) records for the domain in the URI, which are then examined.
- The TU chooses a transport protocol, if more than one record different transport protocol is available, depending on external factors (e.g., configuration).
- Using the chosen transport protocol and the NAPTR records, the TU then queries the DNS server for service (SRV) records. The server may return one or more SRV records.
- These records are then tested in sequence, one by one, by performing an A or AAAA DNS query on each, so that an IP address and port can be found to send on the request. Note that an A or AAAA query on one SRV record might result in more than one IP address being returned. Each one of these IP addresses should be tried in turn and, only after all have failed, should the next A or AAAA query be performed on the next SRV record.

12.12.1.2 Sending responses

This procedure is followed when a UAS fails to send a request using the source IP address and the port in the "sent-by" field. The procedure is as follows:

- If the sent-by field contains a numeric IP address, then the response is sent there using the port indicated (or to the default port if no port is specified).
- If the sent-by field contains a domain name and a port, then an A or AAAA query is made. The resulting IP address and the available port are used to forward the response.

- If the sent-by field contains a domain name, but no port, then an SRV record query is performed on the domain name. These records are then tested in sequence, one by one, by performing an A or AAAA DNS query on each, so that an IP address and port can be found to send on the request. Note that an A or AAAA query on one SRV record might result in more than one IP address being returned. Each one of these IP addresses should be tried in turn and, only after all have failed, should the next A or AAAA query be performed on the next SRV record.

12.12.2   The loose routing concept

Loose routing was introduced in [RFC3261]. It offers a more robust way of forwarding messages to hops and provides a means for the request-URI to remain unchanged throughout the request's journey to the proxy that is responsible for servicing it (i.e., the proxy responsible for the domain present in the request-URI).

A SIP URI carrying the loose router parameter indicates that the owner (typically an intermediary) of this SIP URI is [RFC3261]-compliant and supports loose routing. An example of such a URI is: *sip:proxy.example.com;lr*. The lr parameter identifies the entity as a loose router. The absence of such a parameter indicates that the next hop is a strict router.

12.12.3   Proxy behaviour

For all new requests, including those with unknown methods, a statefull proxy performs the following:

- validates the request;
- pre-processes routing information;
- determines target(s) for the request;
- forwards the request to each target;
- processes all responses.

[RFC3261] describes each step in detail. For the purpose of completing routing analysis, the steps taken by a proxy to route a request are listed below:

- If the request-URI contains a URI that the proxy has previously placed in the Record-Route header of that request, then the proxy replaces it with the URI that populated the last Route header. The proxy then removes that Route header.
- If the first Route header contains a URI representing this proxy, then the proxy removes that Route header.
- If the domain name of the request-URI indicates a domain that this proxy is not responsible for, then the proxy proceeds with the task of forwarding the request and only places the request-URI in the target set of addresses where the request will be proxied.
- If the domain name of the request-URI indicates a domain that this proxy is responsible for, then the proxy uses any mechanism with which it has been configured to determine the target set. Taking the first target in the target set, the proxy places it

in the request-URI. The proxy places a Record-Route header as the topmost Record-Route header, if it wishes to remain on the path of any subsequent requests within a dialog. The URI in the Record-Route header must contain an lr parameter and must be a SIP or a SIPS URI.

- If a proxy server has a set of entities it wants a SIP request to pass through before it arrives at its final destination, it needs to insert Route headers with the addresses of these entities above any other Route headers, if present. The proxy needs to ensure that these entities are loose routers.
- If the request contains Route headers, then the proxy examines the topmost Route header and if it does not contain the lr parameter, then the proxy places the request-URI in a Route header as the last Route header. It then places the URI in the topmost Route header in the request-URI and removes that Route header. The proxy then sends the request using the steps in Section 12.12.5 and the calculated first target in the target set as the entry in the route set.

The proxy server may try each address in the target set serially or in parallel: a concept referred to as "forking".

12.12.4 Populating the request-URI

The UAC uses the remote target and the route set to populate the request-URI as follows:

- If the route set is empty, then the remote target is placed in the request-URI.
- If the route set is not empty and the topmost URI is a loose router, then the remote target is placed in the request-URI. Route headers are then built using the route set.
- If the route set is not empty and the topmost URI is a strict router, then the topmost URI is placed in the request-URI. Route headers are then built using the route set, but excluding the topmost URI in that route set. The remote target is then placed as the last Route header.

12.12.5 Sending requests and receiving responses

The procedures outlined in Section 12.12.1.1 are used by the TU layer to send the request as follows:

- If the topmost URI in the route set indicates that the next hop is a strict router and results in forming the request as described in Section 8.12.4, then the procedures are applied to the request-URI.
- If the topmost URI in the route set indicates that the next hop is a loose router, then these procedures are applied to the topmost URI in the route set.
- If there is no route set, then these procedures are applied to the request-URI.

The TU then creates a transaction instance, passes the request, the IP address and the port to it, and indicates the transport protocol to use. The transaction layer passes this information to the transport layer, which sends the request as follows:

- If the request is within 200 bytes of the Maximum Transfer Unit (MTU) en route to the destination, then the request must be sent using a congestion-safe protocol (TCP or SCTP).
- If the MTU is unknown and the request is longer than 1,300 bytes, then the request must be sent using a congestion-safe protocol (TCP or SCTP).
- If the transport protocol indicated in the Via header needs changing after the above steps, then it is changed. The Via header's sent-by field is populated by an IP address and a port (or preferably an FQDN).

Received responses are matched to requests using the branch parameter in the Via header.

12.12.6   Receiving requests and sending responses

If the request was received from a different IP address than the one indicated in the sent-by field of the Via header in the request, then the transaction layer adds a "received" parameter in the Via header and populates it with the IP address from which it received the request. The request is then matched to a server transaction or (if no match is found) is passed to the TU, which may choose to create a new server transaction instance.

Once the TU has completed processing the request and has generated a response, it passes the response to the transaction instance from which it received the request. The transaction layer forwards the response to the transport layer, which performs the following:

- if the request was received on a connection-oriented protocol, then the response is sent on the same connection;
- if the connection is no longer open, then the received parameter and port in the sent-by field (or the default port, if no port is specified) are used to open a new connection and send the request;
- if no received parameter is present, then the procedures in Section 12.12.1.2 are followed.

12.13   SIP extensions

12.13.1   Event notification framework

SIP has been the extension used for the purpose of event notification. A user or resource subscribes to another resource that has an event of interest and receives notifications of the state and any changes in such an event.

The SIP SUBSCRIBE method is used for subscription while the NOTIFY method is used to deliver notifications of any changes to an event.

[RFC3265] is the IETF paper that documents this extension. It is a framework that describes subscriptions and notifications in a generic manner and provides rules for creating SUBSCRIBE requests and NOTIFY requests. It also describes the behaviour of subscribers when sending and receiving subscription requests, as well as notifiers' behaviour when receiving subscription requests and sending notifications.

The event notification framework also introduces new SIP headers and response codes, along with the SUBSCRIBE and NOTIFY methods:

- *Event header* – this identifies the event to which a subscriber is subscribing for notifications.
- *Allow-Events header* – this indicates to the receiver that the sender of the header understands the event notification framework. The tokens present in the header indicate the event packages that it supports.
- *Subscription-State header* – this indicates the status of a subscription. "Active", "pending" and "terminated" are the three defined subscription states. This header also carries the reason for a subscription state: "deactivated", "probation", "rejected", "timeout", "giveup" and "noresource". Extensions are possible for subscription-state and reason values.
- *"202 Accepted" response* – this indicates that the subscription request has been preliminarily accepted, but is still pending a final decision, which will be indicated in the NOTIFY request.
- *"489 Bad Event" response* – this response is returned when the notifier does not understand an event as described in the Event header.

SUBSCRIBE requests are dialog-establishing requests. A dialog is created when a 2xx response or a NOTIFY request arrives for the SUBSCRIBE request. Subsequent SUBSCRIBE and NOTIFY requests are sent within the created dialog.

The request-URI in an initial SUBSCRIBE request addresses the resource about which the subscriber wishes to receive state information. The Event header identifies the event related to the subscription.

Much like registration, subscriptions are in soft state and need refreshing. The duration of a subscription is indicated in an Expires header. The default value is 1 hour if the header is not present in the SUBSCRIBE request. A subscription is terminated when not refreshed and can be explicitly terminated by sending a SUBSCRIBE request within the dialog and setting the Expires header value to 0.

The event notification framework also introduces the concept of an event package: an extension to the framework. Each event package created introduces a new use case for the event notification framework.

The NOTIFY request payload (body) is used to carry state information. Each event package defines its own MIME type for carrying such information.

The event template package is a special event package and is associated with other event packages, including itself. Template packages define states that can be applied to other event packages. A subscription to a template package is indicated in the Event header by appending a period (full stop) to an event package, followed by the template package name: for example, Event: *presence.winfo*.

12.13.2   State publication (the PUBLISH method)

The event notification framework specifies how to subscribe to the state of an event and how to get notification of changes to the state of an event. It does not specify how the state can be published. However, the SIP extension for state publication specification [RFC3903] allows a client to publish its event state to the state agent, which acts as the compiler of such state and generating notifications. This is achieved using the PUBLISH method.

PUBLISH requests are in soft state and need to be refreshed. The Event header defined in [RFC3265] (and in Section 12.13.1) is used by the publisher to identify the event whose state it is publishing. The request-URI is used to identify the resource whose state is being published. An entity tag is used by the client and is supplied by the server to enable the client to update a state using the PUBLISH method. The state of a resource is carried in the body of the PUBLISH request.

12.13.3   SIP for instant messaging

For a more detailed description of instant messaging, please refer to Chapter 5.

SIP is extended for instant messaging by the introduction of the MESSAGE method in [RFC3428].

There are two modes of instant message exchange: page mode and session mode. The MESSAGE method is used in page mode. Page mode is a one-shot instant message where a subsequent instant message is not related, at the protocol level, to the preceding one. It is used when a conversation or interaction is not unexpected.

The request-URI in a MESSAGE request carries the resource where the request will be sent. The MESSAGE request body carries the actual contents of an instant message, which, again, is a MIME type. The most common MIME body uses the "text/plain" MIME type. For interoperability with non-SIP instant-messaging clients using an IETF standard, the MIME type "message/cpim" as defined in [RFC3862] "Common Presence and Instant Messaging: Message Format" is used.

Session-based instant messaging uses SIP for signalling and Message Session Relay Protocol (MSRP) for carrying the data (instant messages) after the session has been established (see Section 5.4 for details about how an instant-messaging session can be established using SIP and SDP, as well as how the MSRP operates).

12.13.4   Reliability of provisional responses

In basic SIP, provisional responses are transmitted unreliably, unlike the 2xx responses for INVITE requests. It was later discovered that reliability in the transmission of provisional responses in some cases was both important and useful: therefore, [RFC3262] was created. In 3GPP, reliable provisional responses and their acknowledgments are used to exchange additional SDP Offer/Answer messages.

The reliability of the provisional responses extension only applies to INVITE requests.

A UAC generating an INVITE request and wishing to indicate its support for a reliable provisional responses extension includes the Supported header in the

INVITE request with the option tag "100rel". A UAC requiring the UAS to support a reliable provisional responses extension indicates this by placing the option tag "100rel" in a Require header of the INVITE request.

A UAS receiving an INVITE request with a Require header carrying the option tag "100rel" responds with an error response "420 Bad Extension" if it does not support such an extension. If the option tag appears in a Supported header, the UAS may ignore it.

A UAS supporting this extension must transmit the provisional response reliably if the "100rel" option tag is present in a Require header and may transmit it reliably if the option tag is present in a Supported header.

A UAS sending a reliable provisional response indicates this by placing the extension header RSeq in the provisional response and the option tag "100rel" in a Require header. The RSeq carries an integer value between 1 and $2^{31} - 1$.

Multiple reliable provisional responses can be sent to the same INVITE request. Any subsequent reliable provisional response must carry an RSeq value that is 1 higher than the previous value in the same transaction space (within the same INVITE transaction).

A UAC receiving a reliable provisional response responds to it with a provisional response acknowledgment (PRACK) request. The PRACK request carries a RACK header, which reflects the value in the RSeq header and that in the CSeq header, in that order.

Just like any other request, the PRACK must be responded to. The UAS holds off sending a 2xx to an INVITE request until it has received PRACK requests for all the reliable provisional responses it has sent. Examples:

```
Require: 100rel.
RSeq: 12345.
RACK: 12345 1 INVITE.
```

"100 Trying" responses are not sent reliably.

12.13.4.1 The SDP offer/answer model using PRACK requests

Section 12.10.1 described the offer/answer restrictions that occur when using basic SIP. This section offers additional opportunities for offer/answer exchange:

- if an INVITE contains an offer, then an answer may be present in a reliable provisional response;
- if an INVITE does not carry an offer, then the offer must be present in the first reliable provisional response;
- if a reliable provisional response carries the first offer, then the answer must be in the PRACK;
- if a reliable provisional response carries an answer, then the PRACK may carry an additional offer;
- if a PRACK carries an offer, then the 2xx of the PRACK must carry an answer.

12.13.5   The UPDATE method

The UPDATE method is an extension to SIP that enables UAs to update a session description without having any impact on a dialog. The UPDATE method can be sent within early and confirmed dialogs, but must not be sent if a dialog is not created (i.e., before a dialog-creating provisional response is sent/received). It is constructed like any other request within a dialog.

An UPDATE request can be sent by the caller (the INVITE UAC) as well as the callee (the INVITE UAS) and is only used for sessions: this means UPDATE requests are only used for dialogs created using INVITE requests.

A UAC wishing to indicate its support for an UPDATE method extension does so by including an Allow header in the INVITE request, listing UPDATE as a method. A UAS wishing to indicate its support for an UPDATE method does so by including an Allow header, using the UPDATE method to deliver a provisional response.

An UPDATE request is a target refresh request: that is, it can update the remote target of a dialog.

12.13.5.1   The SDP offer/answer model using UPDATE requests

An UPDATE request always carries the offer, while the 2xx response carries the answer.

UPDATE requests sent within early or confirmed dialogs can only carry an offer if answers have been returned to all the offers communicated by either side. Of course, an UPDATE with an offer cannot be sent if no offer/answer exchanges have yet taken place.

12.13.6   Integration of resource management and SIP (preconditions)

The inability to reserve network resources for a session is a serious drawback to session establishment. To minimize session failure once a session has started, it is necessary to reserve resources before the callee is alerted (i.e., before the phone rings).

To reserve resources the network needs to know the callee's IP address, port and session parameters. This is not possible without offer/answer exchanges. The problem is that a session can only be established after an offer/answer exchange and, typically, a user is only alerted after session establishment has occurred. To solve this problem the concept of preconditions was introduced. In this concept the caller indicates by means of an SDP offer a set of constraints about the session. The answerer responds to the offer with an SDP answer; however, it neither establishes a session nor does it alert the user. Session establishment occurs when the caller and callee learn that the preconditions have been met by local events and by the caller sending a new offer confirming that the preconditions have been met and the answerer sending an answer confirming the same. In IMS this new offer/answer exchange is carried in an UPDATE request and response.

This SDP extension is defined in [RFC3312], where preconditions for Quality of Service (QoS) are also defined. Three new SDP attributes are introduced:

- *Current status* – carries the current resource reservation status in the network for a particular media stream.
- *Desired status* – carries the preconditions for resource reservation in the network in order to proceed with session establishment.
- *Confirmation status* – the confirmation status attribute carries threshold conditions for a media stream. It is used so that one end point can indicate to the other end point that it needs confirmation that the preconditions at the remote end have been met.

Each of these three attributes have the following four fields, which appear in the same order (the current status attribute is an exception as it does not have a strength tag):

- *Precondition type* – at the moment only the QoS precondition type has been defined (extensions are possible).
- *Strength tag* – this indicates the strength of the preconditions. Defined values are "mandatory", "optional", "none", "failure" and "unknown".
- *Status type* – the end-to-end (or "e2e") status type indicates the status of end-to-end resource reservation. The "local" and "remote" status types indicate that a preconditions status is needed for local and remote networks, respectively. They are used when each end performs its own resource reservation locally.
- *Direction tag* – this indicates the current direction of resource reservation: desired or confirmed.

Here are some a-line attribute examples:

```
a=curr:qos e2e send
a=des:qos optional e2e send
a=conf:qos e2e recv
```

A new SIP extension option tag "precondition" was introduced to deal with preconditions. It appears in a Require header if one of the strength tags that appear in an offer is "mandatory". If all the strength tags are either "optional" or "none", then the option tag can appear in a Supported header. Here is an example of how the option tag is used:

```
Require: 100rel, precondition
```

12.13.7   The SIP REFER method

The REFER method is standardized in [RFC3515].

The sender of a REFER request refers the recipient to a resource that is identified in the REFER request itself. A REFER request also implicitly creates a subscription (implicit subscription means creating a subscription state as described in Section 12.13.1 without explicitly sending a SUBSCRIBE request) where senders of the REFER request can receive notifications about the outcome of such a referral (see Section 12.13.1 for details about event notification). A REFER request, therefore, is a dialog-creating request.

One application of the REFER method is call transfer. For instance, a secretary receives a call from an associate asking for the manager. The secretary then transfers the call to the manager by sending the associate a REFER request with the manager's contact information. The associate's UA then calls the manager by sending an INVITE.

The REFER standard also introduces a SIP extension header, the Refer-To header, which provides a URI. Its syntax follows that of the Contact header (with the exception of STAR "*", which cannot appear in a Refer-to header). The Refer-to header can only appear in a REFER request, but must not appear more than once.

The REFER request is constructed in the same way as any other request outside a dialog (defined in Section 12.7.4.1).

The receiver of a REFER request typically responds to it with a "202 Accepted" response. Its recipient also creates a subscription and sends notifications of the outcome. According to the event notification framework, a NOTIFY request is generated and sent immediately after accepting the REFER request. NOTIFY requests are constructed in the same way as any other request within a dialog. Every NOTIFY request contains the Event header with the tag "refer".

Unlike a SUBSCRIBE request, the REFER request does not carry an expiration time. The implicit subscription duration is determined by the referred party and is indicated to the referrer in the first NOTIFY request, using the Subscription-State header. A referrer wanting to extend subscription duration can do so by sending a SUBSCRIBE request within a dialog for that event.

The NOTIFY request also contains a body of MIME type "message/sipfrag" (described in Section 12.13.8). The body, at minimum, contains a SIP response status line.

12.13.8   The "message/sipfrag" MIME type

The "message/sipfrag" MIME type allows a fragment of a SIP message to be represented. If a start line, header or body of a SIP message is represented, then it must be complete according to the BNF of such a representation. If a SIP message body is represented, then the fragment must also carry appropriate SIP message headers that describe the body.

12.13.9   SIP extension header for registering non-adjacent contacts (the Path header)

Intermediaries, such as SIP proxy servers, could exist in a network topology between a UA and a registrar. Because of this scenario a SIP extension introduced the Path header, which is used to record the path taken (proxies traversed) by a REGISTER request from the UA to the registrar. This Path header is similar to the Record-Route header used for dialog-creating requests, but is only used by the home network of a user to send requests destined to that user.

The Path header syntax is similar to that of the Record-Route header. A UA sending a REGISTER request includes a Supported header in the request with the "path" option tag. Path header values returned in the 2xx response of a REGISTER request can be ignored by the UA.

An intermediate proxy wanting to remain on the path of any requests sent to the UA inserts a Path header value with a URI pointing to itself as long as the REGISTER request contained a Supported header with an option tag "path". The Path header value or a new Path header must appear as the topmost header. If a proxy requires the registrar to support the path extension, then it inserts a Require header with an option tag "path".

Proxies may add a Path header value with a URI pointing to other proxies, if it is aware of the network topology.

If the UA does not indicate support for the path extension and a Path header appears in the REGISTER request, it is recommended the registrar reject the registration request with a "420 Bad Extension" response.

If the registrar accepts the request, then it stores the Path header values preserving the order and associates them with the AOR and the Contact header. It then copies the Path header values into the 2xx response.

Typically, a home proxy is associated with the registrar of a user. If a home proxy receives a request that is destined to a user whose AOR is registered, then the home proxy determines whether there are Path headers stored with that registration. If there are, then the home proxy inserts those Path headers in the request as Route headers. The Path extension is defined in [RFC3327].

12.13.10   Private SIP extensions for asserted identity within trusted networks

[RFC3325] provides a private extension to SIP that enables a network of trusted SIP servers to assert the identity of end-users or end-systems. It also provides a mechanism whereby an end-user can indicate requested privacy. Entities are responsible for withholding asserted identity information outside the trusted domain when privacy is requested.

If a proxy server receiving a SIP request from an untrusted entity authenticates a user, it then inserts the P-Asserted-Identity extension header into a request before forwarding it to a trusted next hop. If any P-Asserted-Identity headers are already present in the request when it arrives from the untrusted entity, the proxy removes or replaces them. If a proxy receives a SIP request from a trusted entity, then it can safely rely on the identity information in already-existing P-Asserted-Identity headers.

In the case of multiple public user identities the user can give a proxy some idea of the preferred public identity to be inserted in the request. This is done using the P-Preferred-Identity extension header. A proxy receiving a request from an untrusted entity uses that header as a notion of what to insert in the P-Asserted-Identity header after it has authenticated the user. If no such identity exists, then the proxy may either reject the request or insert a P-Asserted-Identity header and ignore the P-Preferred-Identity header.

A proxy forwarding a request to an untrusted next hop removes all P-Asserted-Identity headers if the end-user has requested privacy. A user wishing to request privacy inserts the privacy token "id" into a Privacy header (defined in Section 3.6).

The syntax of the P-Asserted-Identity header and the P-Preferred-Identity header are similar to that of the From header.

12.13.11   Security mechanism agreement for SIP

12.13.11.1   Introduction

The SIP Security Agreement (Sip-Sec-Agree) [RFC3329] is an extension to SIP that allows a UA and its first hop server to exchange and agree security mechanisms that they will use in future communications. The most important aspect of Sip-Sec-Agree is that this negotiation is in itself secure. The protocol has mechanisms in place that make it impossible for an attacker to bid down on the exchanged security mechanism in an attempt to force an intentionally weak security mechanism on a UA and its first hop server. This makes systems future-proof: new stronger security mechanisms can be added and weaker ones deleted at a later date.

12.13.11.2   Operation

The Sip-Sec-Agree handshake contains the following stages (see Figure 12.7):

1. On receiving a request from the UA the first hop SIP server initiates the security agreement by challenging the UA. This challenge contains a list of security mechanisms supported by the server. Note that the client may pre-emptively include its supported security mechanisms in any request. However, the server always employs

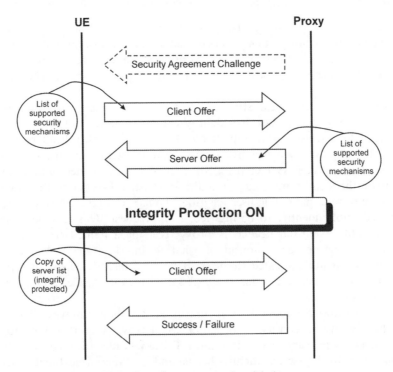

Figure 12.7   Security agreement handshake message.

its own full list of supported security mechanisms when making a challenge, irrespective of the contents of the client list.[1]

2. The UA inspects the server's security mechanism list, which is a list of mechanisms, ordered by preference, and chooses the security mechanism with the highest preference (i.e., most commonly supported). It then proceeds to switch on the chosen security mechanism (e.g., it establishes manually keyed IPsec SAs or starts a TLS handshake with the server). The UA then re-submits its original request, which is now protected by the security mechanism used. It also returns the server's list of supported security mechanisms.

3. The server receives a security mechanism list that is supposed to be a mirror image of the list it originally sent with the Sip-Sec-Agree challenge. The server verifies that the mirrored list is, in fact, identical to the list it had previously sent out. If the lists are not identical, then it means an attack is in progress,[2] in which case the server would immediately abort and reject the request. If everything checks out, then the request is forwarded to the next hop.

12.13.12 Private SIP extensions for media authorization

This SIP extension links the QoS applied to media in the bearer network with session signalling, which helps to guard against DOS attacks. Use of this extension is only feasible in administrative domains with trust relationships. A SIP intermediary (typically a proxy server) authorizes the QoS and the policy control function – or Policy Decision Function (PDF) – provides the QoS. This mechanism prohibits any end-to-end encryption of message bodies that describe media sessions; it must only be used in specialized SIP networks like 3GPP IMS.

This extension assumes that a UA wishing to obtain QoS for a session is connected to a QoS-enabled proxy and a Policy Enforcement Point (PEP). A QoS-enabled proxy is a proxy that has a link to a PDF.

The solution utilizes something called a token (i.e., a string of hexadecimal digits). The UA acquires a media authorization token from a QoS-enabled proxy server, which itself acquires a media authorization token from the PDF and, then, passes it to the UA. The proxy server passes this token to the UA by means of a SIP extension header. The UA then presents this token to the PEP in order to get the bearer with the allocated QoS.

The extension header is named P-Media-Authorization; it can carry multiple media authorization tokens separated by commas. The P-Media-Authorization header itself can only be carried in requests and responses that carry an SDP offer or answer. As already mentioned, a media authorization token is a string of hexadecimal digits.

When a SIP UA sends an INVITE request to a QoS-enabled proxy server (typically referred to as the "originating proxy server"), the proxy includes one or more media authorization tokens in all unreliable provisional responses (except 100), the first

---

[1] The reason for this behaviour is simply that there is no security applied to the client list; it cannot be trusted. It is merely a tool to relay certain parameters or ideas to the server.
[2] This aspect of Sip-Sec-Agree requires that any mechanism negotiated using it must at a minimum provide data integrity protection.

reliable 1xx or 2xx response and all re-transmissions of that reliable response. This is done for each dialog that is created as a result of the INVITE request. When the UA requests QoS from the bearer network it includes media authorization tokens with the resource reservation request.

When a SIP UA receives an INVITE request from a QoS-enabled proxy (typically referred to as the "destination proxy server"), it examines the request to see what media authorization tokens the proxy may have included. When the UA requests QoS from the bearer network it includes media authorization tokens with the resource reservation request.

The media authorization extension is defined in [RFC3313].

12.13.13   SIP extension header for service route discovery during registration

The UA is free to include a set of Route headers – the pre-loaded route – in an initial request to force that request to visit and, potentially, be serviced by one or more proxies. This allows a UA to request services from a specific set of proxy servers, which are typically located in the user's home network.

[RFC3608] defines a header called "Service-Route", which is used by a UA to learn the service route (or pre-loaded route). A registrar wanting to inform a UA of a service route set uses this Service-Route header in a 2xx response to a REGISTER request. If used by the UA, the service route set will provide services from a set of one or more proxies associated with that registrar.

A UA choosing to use the service route set provided by the registrar places the contents that appear in the Service-Route headers in the Route headers, maintaining the order in which they appear.

The mechanism that the registrar uses to construct the header value is not described in [RFC3608]. It is a registrar-local policy.

12.13.14   Private header extensions to SIP for 3GPP

[RFC3455] defines the following 3GPP IMS-specific headers for SIP:

- *P-Charging-Vector* – this transports the IMS Charging ID (ICID) and correlated access network (e.g., GPRS) related charging information between IMS network entities.
- *P-Charging-Function-Address* – this transports the addresses of the charging functions between the IMS network entities of the user's home network.
- *P-Visited-Network-ID* – this transports the identification string of the visited network to the home network of the user during registration, thus allowing the home network to discover details about the roaming agreements between the two networks.
- *P-Access-Network-Info* – this transports information about the access network technology and the user's location (GPRS Cell-ID) from the visited network to the home network.
- *P-Called-Party-ID* – this is included in an initial request by the registrar of the terminating user. The registrar re-writes the request-URI with the registered contact address of the terminating user. Consequently, the URI originally indicated

in the request-URI would be lost; therefore, the URI originally indicated in the request-URI is saved to the P-Called-Party-ID header and sent along with the request.

- *P-Associated-URI* – this is included in the 200 (OK) response of a REGISTER request. It includes additional URIs that are associated with the user. The registrar may implicitly register some of these additional URIs.

12.13.15   *Compressing SIP*

[RFC3486] provides a SIP mechanism to indicate whether signalling compression is supported. Two parameters are defined to enable a SIP entity to show that it supports signalling compression. The parameters, when present, also indicate the entity's willingness to receive compressed messages.

The first parameter is defined as a SIP URI parameter and is named "comp". Currently, only one value is defined for this parameter: "SigComp". When sending requests the SIP entity (a UA or an intermediary) inspects the next hop URI and, if the URI carries the parameter "comp=SigComp", the UAC sends the request compressed according to the signalling compression specified in [RFC3320].

The second parameter is defined as a Via header parameter. It is identical to that of the URI, but is used by the SIP entity forwarding the response to learn the upstream entity's ability and willingness to receive compressed responses. Hence, if that parameter is present in the Via header of a request, then the response is sent compressed.

Typically, a UA sending a compressed request inserts the "comp=SigComp" parameter in the Via header. If a Contact header is present, the parameter is also added to the Contact header URI. Proxy servers forwarding the compressed request also insert the parameter in their Via headers. If a proxy is record routing, then it adds the parameter in the URI of the inserted Record-Route header.

13

SDP

The Session Description Protocol (SDP) is an application-layer protocol intended to describe multimedia sessions. It is a text-based protocol. When describing a session the caller and callee indicate their respective "receive" capabilities, media formats and receive address/port.

Capability exchange can be performed during session setup or during the session itself – i.e., while the session is in progress.

At the time of writing, a new SDP specification is in the process of being finalized. The current version is [Draft-ietf-mmusic-sdp-new] and the earlier version was specified in [RFC2327].

13.1 SDP message contents

An SDP message contains three levels of information:

- Session-level description – this includes the session identifier and other session-level parameters, such as the IP address, subject, contact info about the session and/or creator.
- Timing description – start and stop times, repeat times, one or more media-level descriptions.
- Media type and format – transport protocol and port number, other media-level parameters. Note that the media address may be different from the signalling address.

The three levels of information must appear in the order described above. The SDP message is a collection of SDP lines.

13.1.1 Session description

Table 13.1 lists all the session-level description lines and indicates their mandate and the letter used as the line name.

The IMS: IP Multimedia Concepts and Services Second Edition
Miikka Poikselkä, Georg Mayer, Hisham Khartabil and Aki Niemi © 2006 John Wiley & Sons, Ltd

Table 13.1 Session-level description SDP lines.

Field	Description	Mandate
v	Protocol version	m
o	Origin and session ID	m
s	Session name	m
i	Session information	o
u	URI for sesssion	o
e	Email address	o
p	Phone number	o
c	Connection information	m*
b	Bandwidth information	m
z	Time zone correction	o
k	Encryption key	o
a	Attribute lines	o

* Not required if present in every media line.

13.1.2 Time description

Table 13.2 lists all the time description lines and indicates their mandate and the letter used as the line name.

Table 13.2 Time-level description SDP lines.

Field	Description	Mandate
t	Time session is active	m
r	Repeat times	o

13.1.3 Media description

Table 13.3 lists all the media-level description lines and indicates their mandate and the letter used as the line name.

Table 13.3 Media-level description SDP lines.

Field	Description	Mandate
m	Media and transport	o
i	Media title	o
c	Connection information	o*
b	Bandwidth information	o
k	Encryption key	o
a	Media attributes	o

*Not required if present at the session level.

13.2   SDP message format

The SDP syntax is very strict and all lines follow the same format. Every SDP line has the format:

```
<character>=<value>
```

No spaces are allowed on either side of the "=" sign. The <value> part of the line contains one or more parameters and there is exactly one space between each parameter:

```
value=parameter1 parameter2 ... parameterN
```

Each SDP line ends with a Carriage Return–Line Feed (CRLF) and each line has a defined number of parameters.

13.3   Selected SDP lines

13.3.1   Protocol version line

The SDP protocol version is 0 and, therefore, the v-line in an SDP message must always be set to 0:

```
v=0
```

13.3.2   Connection information line

The c-line must be either present at the session level or media level. It must be present at the media level if it is not present at the session level. If it is present at both levels, then media-level connection information overrides session-level information:

```
c=<network type> <address type> <network address>
```

The c-line has three parameters:

- Network type – the only currently defined network type is the Internet. The value appears as "IN".
- Address type – there are two address types, IP4 or IP6.
- Network address – this parameter identifies the Internet Protocol (IP) address or domain name where media are received.

13.3.3   Media line

The m-line carries information about the media, including transport information. The syntax is as follows:

```
m=<media> <port> <transport> <format-list>
```

The m-line has four parameters:

- *Media* – the type of media (e.g., audio, video, game).
- *Port* – contains the port number where these media can be received.
- *Transport* – the transport protocol to use, either the User Datagram Protocol (UDP) or Real-time Transport Protocol Audio and Video Profile (RTP/AVP) (RTP is explained in Chapter 15).
- *Format-list* – contains more information about the media, usually payload types defined in RTP/AVPs (see Section 15.3 for details).

If the transport is RTP/AVP, then the port number for the RTP Control Protocol (RTCP) = RTP port + 1. RTCP is assumed to be sent whenever RTP is carrying the media. The RTP port number must always be an even number and, therefore, the RTCP port is an odd number.

13.3.4   Attribute line

The a-line defines the attributes of the media. It is used to extend SDP: in fact, it is the only way of doing so. Attributes can be session-level attributes, media-level attributes or both. Attribute interpretation depends on the media tool being invoked. Its syntax is as follows:

```
a=<attribute field> [":"<attribute value>]
```

The attribute field contains the name of the attribute. The attribute value is optional, but if present is separated from the attribute field by a colon.

Table 13.4 shows a list of the most commonly used attributes that are defined in [Draft-ietf-mmusic-sdp-new]. This list is not exhaustive since extensions have also been defined in separate Internet Engineering Task Force (IETF) documents (see Section 12.13.6 on preconditions as an example of how SDP can be extended).

13.3.5   The rtpmap attribute

For RTP-transported media, SDP can be used to bind a media-encoding codec to the media's RTP payload type. This is done using a payload-type number.

For static payload types (see Section 15.4.1 for the definitions of static and dynamic payload types) the payload-type number is sufficient for binding, but for dynamic payload types the payload-type number is insufficient and additional encoding information is needed. This is achieved using the rtpmap attribute. The payload-type number is carried in the format-list parameter of the media line. The syntax for the rtpmap attribute is:

```
a=rtpmap:<payload type> <encoding name>/<clock rate>[/<encoding parameters>]
```

Table 13.4 Most common SDP attribute lines.

Attribute name	Syntax	Description	Session or media-level attribute
Packet time	`a=ptime:<packet time>`	It represents how long media are allowed to remain in one packet, in milliseconds. It is just a recommendation. It is not mandatory to use it when decoding RTP	Media
Maximum packet time	`A=maxptime:<max packet time>`	The maximum number of milliseconds of media to be included in a packet	Media

RTP map	`a=rtpmap:<payload type>`
	`<encoding name>/<clock rate>`
	`[/<encoding parameters>]`
		See 13.3.5 for details	Both

Receive only	`a=recvonly`	The sender of the media description only wants to receive the media stream (or streams, if in session level)	Both
Send only	`a=sendonly`	The sender of the media description only wants to send the media stream (or streams, if in session level)	Both
Send and receive	`a=sendrecv`	The sender of the media description wants to send and/ or receive the media stream (or streams, if in session level). This is the default value if none of "sendonly", "recvonly", "sendrecv" or "inactive" are present	Both
Inactive	`a=inactive`	In this mode the media are neither sent nor received. It can be used to put media streams on hold. Note that in the case of RTP, RTCP continues to be sent even if the stream is set to inactive	Both
Frame rate	`a=framerate:<frame rate>`	It indicates the maximum video frame rate in frames per second	Media
Formats	`a=fmtp:<format-specific parameters>`	Allows parameters for specific formats to be conveyed. The formats are the encoding parameters in the rtpmap attribute	Media

The rtpmap attribute consists of four parameters:

- *Payload type* – carries the payload-type number as indicated in the m-line.
- *Encoding name* – the codec name.
- *Clock rate* – bits per second.
- *Encoding parameters* – media-specific parameters, including the number of channels, but not codec-specific parameters.

14

The Offer/Answer Model with SDP

The Session Description Protocol's (SDP) Offer/Answer Model is used by two entities to reach agreement on session description, such as which media streams are in the session, the codecs, etc. The offerer indicates the desired session description in the offer. The answerer replies to the offer by indicating the desired session description from the answerer's viewpoint. The Offer/Answer Model can be used to create sessions or to modify an existing session. There are, however, restrictions on what can appear in an offer and answer. The following sections describe how offers and answers are created. The Offer/Answer Model is described in more detail in [RFC3264].

14.1   The offer

An entity wanting to create a session or currently in a session can generate an SDP session description that constitutes the offer. This offer contains the set of media streams and the set of codecs the offerer wishes to use, as well as the Internet Protocol (IP) addresses and ports the offerer would like to use to receive the media.

The session line, or s-line, is always set to "-" or a single space in the offer/answer.

Since signalling protocols, such as SIP, are used to create unicast sessions, the start and stop times in an SDP offer/answer exchange is set to 0 and 0, respectively.

14.2   The answer

The answerer generates an answer, which is an SDP answer that responds to an offer made to it. The answer indicates:

- whether a media stream is accepted;
- the codecs that will be used;
- the IP addresses and ports that the answerer wants to use to receive media.

The IMS: IP Multimedia Concepts and Services Second Edition
Miikka Poikselkä, Georg Mayer, Hisham Khartabil and Aki Niemi © 2006 John Wiley & Sons, Ltd

14.3   Offer/answer processing

There are certain restrictions to when an offer or answer can be generated:

- the client that received an offer cannot generate a new offer until it has responded with an answer to the offer or until it has rejected that offer;
- the client that generated an offer cannot generate a new offer until it has received an answer to the original offer or until that original offer is rejected.

There is also a constraint on the number of m-lines an answer can have. The number of m-lines in an answer must equal the number of m-lines in the offer.

If multiple media streams of different types are present, this means that the offerer wishes to use these streams at the same time. Rejecting media can be achieved by setting the port in a media line to 0, but not by removing the media line because, as stated earlier, the number of m-lines in an answer must equal that in the offer.

If the offerer wishes to send (and/or receive) multiple streams of the same type at the same time, then it places multiple media streams of that type in an offer. Each stream may use different encoding.

The offerer sends media after receiving an answer (it cannot do so before it receives an answer since it does not know the IP address and port to which to direct media). It uses the media format listed in the answer and, typically, uses the first media format listed in the answer: it is recommended that formats appear in an SDP offer/answer in such an order that the most preferred appears at the top of the list and the least appears last.

14.3.1   Modifying a session description

At any point during the session, either participant can issue a new offer to modify the characteristics of the session. Each stream in the newly created offer must have a matching media stream in the previous offer/answer exchange, and the order of the media streams must not change. As a result of these two requirements, the number of m-lines in a stream never decreases: it either stays the same or increases. Below are some examples to illustrate this (note that the syntax is not shown correctly): First offer:

```
m=a
m=b
```

First answer:

```
m=a
m=b
```

Correct second offer:

```
m=a
m=b
m=c
```

Correct second answer:

```
m=a
m=b
m=c
```

Incorrect second offer (order changed):

```
m=a
m=c
m=b
```

Incorrect second offer (number of lines decreased):

```
m=a
```

As happens when creating an initial offer/answer, by setting the port value to 0 we can disable a media stream. Also, new media streams that are added can re-use a media line whose stream was disabled with a port-0 value by an earlier offer/answer exchange.

14.3.2 Putting the media stream on hold

A client that wishes to instruct the remote end to stop sending media on that stream, but does not wish to completely disable the stream, can do so in two different ways which depend on the stream direction that was negotiated in an earlier offer/answer exchange. A stream that was earlier set with a sendrecv attribute (media for that stream is sent and received) is placed on hold by assigning it an attribute with sendonly in the new offer/answer. A stream that was earlier set with a recvonly attribute is placed on hold by assigning it an attribute with inactive in the new offer/answer.

15

RTP

[RFC3550] defines the Real-time Transport Protocol (RTP) as a protocol for end-to-end delivery for real-time data. It also contains end-to-end delivery services for real-time data: payload-type (codec) identification, sequence numbering, timestamping and delivery monitoring. RTP does not provide Quality of Service (QoS); it does, however provide QoS monitoring using the RTP Control Protocol (RTCP). RTCP also conveys information about media session participants.

15.1 RTP for real-time data delivery

15.1.1 RTP fixed header fields

The RTP fields that are present in an RTP packet (Figure 15.1) represent the following:

- *Version (V)* – the version of RTP. It is always set to "2" if the implementation is [RFC3550]-compliant.
- *Padding (P)* – if set, then the RTP packet contains padding octets. Padding may be needed by some encryption algorithms with fixed block sizes or for carrying multiple RTP packets in a lower layer protocol packet.
- *Extension (X)* – if set, then there is an RTP header extension.
- *CSRC count (CC)* – the number of contributing source (CSRC) IDs that follow the fixed header.
- *Marker (M)* – its interpretation is defined by a profile.
- *Payload type (PT)* – Identifies the payload format (codec).
- *Sequence number* – this increments by 1 for each RTP data packet. This sequence number is used by the receiver to re-order packets that arrive out of sequence. It also allows the receiver to determine whether packets have been lost.
- *Timestamp* – indicates the time when the first octet in the payload was sampled. This field is used along with the sequence number to remove jitter (see Section 15.1.2).
- *Synchronization source (SSRC)* – RTP is not dependent on the underlying Internet Protocol (IP) and, therefore, uses this field to identify the source of RTP packets.
- *Contributing source (CSRC)* – this field carries a list of SSRCs indicating the sources that have contributed to a mixed media stream, if a mixer has been involved. It allows

The IMS: IP Multimedia Concepts and Services Second Edition
Miikka Poikselkä, Georg Mayer, Hisham Khartabil and Aki Niemi © 2006 John Wiley & Sons, Ltd

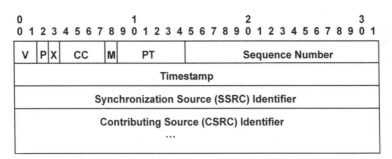

Figure 15.1   RTP packet format.

between 0 and 15 SSRCs. If there are more than 15 sources of media, then only the first 15 are identified. An audio conference mixer is a good example.

15.1.2   *What is jitter?*

The recipient may receive packets out of sequence and with delay, causing breaks and distortion in the media: this distortion is named "jitter". Figure 15.2 shows an example of how RTP packets may arrive out of sequence and be delayed. This is true of any Transmission Control Protocol over IP (TCP/IP) network where congestion is a factor. In RTP a sequence number is used to re-order out-of-sequence packets. The timestamp is used to allow the receiver to play packets at the right time and, therefore, minimize distortion.

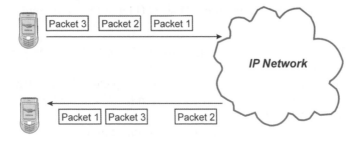

Figure 15.2   Packet jitter.

15.2   RTCP

RTP Control Protocol (RTCP) packets are transmitted periodically to all participants in a session. There are four RTCP functions:

- To provide feedback on the QoS of real-time data distribution.
- To carry a persistent identifier of the RTP source (called a CNAME).
- To permit an adjustable RTCP packet distribution interval (the report interval).
- To convey session control information.

15.2.1 RTCP packet types

There are five types of RTCP packets:

- *SR* – a sender report, providing transmission and reception statistics, sent by active media senders.
- *RR* – a receiver report, providing reception statistics, sent by non-active senders.
- *SDES* – source description items, such as CNAME.
- *BYE* – indicates end of participation.
- *APP* – application-specific functions (defined by a profile).

15.2.2 RTCP report transmission interval

Every participant is required to send RTCP packets. If there are many participants (say, a conference), then there is a scalability problem as further users join in. To control the scalability problem, the rate at which RTCP packets are sent must be scaled down by dynamically calculating the interval between RTCP packet transmissions.

A certain percentage of session bandwidth should be dedicated to RTCP; this percentage and the interval between report transmissions are profile-defined.

15.3 RTP profile and payload format specifications

When conventional protocols are designed, they are generalized to accommodate the additional functionality of their applications. For RTP, this is achieved through modifications and additions to the existing headers as needed by an application. Therefore, RTP for a particular application, like audio, requires one or more companion documents: namely, a profile specification document and a payload format specification document.

15.3.1 Profile specification

In [RFC3550] the following fields have been identified as possible items to be defined by an application profile specification: RTP data header, payload types (codecs), RTP data header additions, RTP data header extensions, RTCP application-specific packet types, RTCP report interval (see Section 15.2.2), SR/RR extensions, SDES use, security services and algorithm, string (password) to encryption key mapping, congestion control behaviour, underlying protocol, RTP and RTCP mapping to transport-level addresses and encapsulation of multiple RTP packets in a lower layer packet.

15.3.2 Payload format specification

A payload format specification for an RTP application defines how a particular payload is carried in RTP.

15.4 RTP profile and payload format specification for audio and video (RTP/AVP)

As defined in Section 15.3, an application using RTP must specify a profile and the payload formats used for it. In this section the profile and payload formats for audio and video applications are identified (the full specifications are defined in a combined document in [RFC3551]). Table 15.1 contains a summary of the profile specified for audio and video application and Table 15.2 contains a sample of the payload formats specified for audio and video.

15.4.1 Static and dynamic payload types

A static payload type is defined with a fixed identification number, a clock rate and a number of channels, when applicable. A dynamic payload type is defined without a number, but a number is dynamically allocated. Dynamic payload types are assigned numbers between 96 and 127.

Table 15.1 RTP/AVP-specific profile.

Profile item	RTP/AVP-specific
RTP data header	Fixed RTP data header is used (1 marker bit)
Payload types	Defined below
RTP header additions	None
RTP header extensions	None
RTCP extension packet types	None
RTCP report interval	As recommended in [RFC3550]
SR/RR extensions	None
SDES use	As defined in [RFC3550]
Security	RTP default
String-to-key mapping	None
Congestion	[RFC1633] and [RFC2475] can be used, as can best effort services
Underlying protocol	UDP, TCP
Transport mapping	Standard mapping of RTP/RTCP to transport-level addresses
Encapsulation	Not specified

Table 15.2 Sample payload formats for audio and video.

PT	Encoding name	Media type	Clock rate (Hz)	Channels
0	PCMU	Audio	8,000	1
3	GSM	Audio	8,000	1
4	G723	Audio	8,000	1
Dynamic	L8	Audio	Variable	Variable
26	JPEG	Video	90,000	—
31	H261	Video	90,000	—
34	H263	Video	90,000	—
Dynamic	H263-1998	Video	90,000	—
96–127	Dynamic	A/V	?	—

In the SDP a static payload type does not need the rtpmap attribute. For example, the PCMU payload type can be represented as:

```
m=audio 49170 RTP/AVP 0
```

or

```
m=audio 49170 RTP/AVP 0
a=rtpmap:0 PCMU/8000
```

An example of a dynamic payload type is:

```
m=video 49172 RTP/AVP 96
a=rtpmap:96 H263-1998/90000
```

16

DNS

16.1 DNS resource records

On a large network, such as the Internet, it would be impractical to identify each system solely by its numerical Internet Protocol (IP) address. Names are more convenient; they help users and administrators locate network resources more easily. The Domain Name Service (DNS) is a distributed database holding the alphanumeric names and their corresponding IP addresses (and more) of every registered system on a Transmission Control Protocol (TCP)/IP network, such as the Internet or the IP Multimedia Subsystem (IMS). Each entry is referred to as a Resource Record (RR).

The alphanumeric names, better known as domain names, are hierarchical in nature where country, company, department and even a host (machine) name can be identified. Each step in the hierarchy is identified as a zone. The domain name below identifies a machine named pc27 in the engineering department of an Australian company called Foobar:

```
pc27.engineering.foobar.com.au
```

An entry in a DNS server that maps a domain name to an IP address is referred to as an address record, or A record. For IPv6 these records are referred to as AAAA records.

16.2 The Naming Authority Pointer (NAPTR) DNS RR

NAPTR RRs are used to replace compact, regular expressions with a replacement field that may well be a pointer to another rule. [RFC2915] defines the NAPTR; however, it does not define why or how replacement fields are used. It is up to the applications using NAPTR to do this. The DNS-type code for NAPTR is 35.

The format of a NAPTR record is as follows (Table 16.1 describes each field in detail):

```
Domain TTL Class Type Order Preference Flag Service Regexp Replacement
```

The IMS: IP Multimedia Concepts and Services Second Edition
Miikka Poikselkä, Georg Mayer, Hisham Khartabil and Aki Niemi © 2006 John Wiley & Sons, Ltd

Table 16.1 NAPTR RR fields.

Field	Description
Domain	The key for an entry (the domain name of the RR)
TTL	Time to live
Class	Class of record (NAPTR records are of class IN, or Internet)
Type	DNS-type code (for NAPTR records it is 35)
Order	The order in which NAPTR records need to be processed. This allows NAPTR records to specify a complete rule in an incremental fashion. Records are processed from lower order numbers to higher order numbers
Preference	Specifies the order in which NAPTR records with the same "order" values need to be processed. Records are processed from lower preference numbers to higher preference numbers
Flags	Indicate what happens next after this look-up. The "S" flag indicates that the next look-up should be an SRV look-up. The replacement field carries an SRV reference. The "U" flag indicates that the next step is not a DNS look-up. The regexp (see below) carries a URI. The "A" flag indicates that the next step is a DNS A, AAAA or an A6 look-up. The "P" flag indicates that the remainder of the algorithm (or look-ups) is defined by the application that uses NAPTR records
Service	Specifies the services available in this domain. The replacement field is used to get to this service. It can also specify the protocol used to communicate with the server that offers this service. In SIP, three services are defined along with their resolution services (resolution services are defined after the "+" sign): "SIPS + D2T": Secure SIP, TLS over TCP "SIP + D2T": SIP over TCP "SIPS + D2S": Secure SIP, TLS over SCTP "SIP + D2S": SIP over SCTP "SIP + D2U": SIP over UDP
Regexp	Carries a substitution expression that is applied to the original domain name in order to construct a new domain name for the next look-up
Replacement	The next name used to query a DNS. This could be another NAPTR record, an SRV record or an address record. In SIP the replacement fields are SRV RRs and, hence, the flag field is set to "S"

16.2.1 NAPTR example

This example uses the Session Initiation Protocol (SIP) service. A NAPTR RR looks like:

Domain	TTL	Class	Type	Order	Preference	Flag	Service	Regexp	Replacement
example.com	7200	IN	35	50	50	"S"	"SIPS + D2T"	""	*_sips._tcp.example.com*
	7200	IN	35	90	50	"S"	"SIP + D2T"	""	*_sip._tcp.example.com*
	7200	IN	35	100	50	"S"	"SIPxD2U"	""	*_sip._udp.example.com*

In this example a client of the SIP service that does not support the Transport Layer Security (TLS) performs a NAPTR RR look-up for the "example.com" domain and selects the replacement:

```
_sip._tcp.example.com
```

16.3 ENUM – the E.164 to Universal Resource Identifier (URI) Dynamic Delegation Discovery System (DDDS) application

ENUM is described as the use of the DNS for storage of international public tele-communication numbers using international format (E.164) numbers [RFC3761]. This is used to discover what services are available for a specific domain.

Imagine that user A has a Public Switched Telephone Network (PSTN) phone and would like to call her friend user B who uses Voice over IP (VoIP). User A cannot enter a SIP address into her PSTN phone and, therefore, enters user B's E.164 number. The call reaches a gateway that performs a DNS ENUM look-up and returns user B's SIP address (see Figure 16.1). IP-to-IP calls are also feasible using E.164 numbers. The steps involved in this procedure are as follows:

1. User A dials the E.164 number, *+135812345678*.
2. The Circuit-Switched (CS) domain contacts a gateway.
3. The gateway formats the E.164 number into a Fully Qualified Domain Name (FQDN), *8.7.6.5.4.3.2.1.8.5.3.e164.arpa*, and looks up the FQDN in DNS.
4. DNS returns the NAPTR record, *sip:userB@example.com*

Figure 16.1 CS-to-IP call example.

5. The gateway contacts DNS for the IP address of the SIP server in the domain "*example.com*", following the procedures in Section 16.12.1.1.
6. DNS returns the IP address of the SIP server.
7. The gateway routes the call to the SIP server.
8. The SIP server contacts user B and the call is established.

The solution uses a NAPTR RR. First, the client turns the E.164 telephone number into a domain name. The domain "*e164.arpa*" is used to provide DNS with the infrastructure. The number is pre-pended to the domain by reversing the numbers, numbers are separated with a dot ".", each number becoming a zone. The NAPTR query returns a re-write rule using the regexp field. The NAPTR service parameter for ENUM is "E2U < enumservice > ".

16.3.1 ENUM service registration for SIP Addresses Of Record (AORs)

[RFC3761] defines an enumservice for SIP that allows users to map a telephone number to a SIP AOR. The enumservice for SIP is "E2U + sip". An example NAPTR record looks like:

Domain	TTL	Class	Type	Order	Preference	Flag	Service	Regexp
8.7.6.5.4.3.2.1.8.5.3.e164.arpa	7200	IN	35	50	50	"U"	"E2U + sip"	"*!^.*$!sip:userB@example.com!*"

16.4 Service records (SRV RRs)

Before service records (SRV RRs) are introduced an entity must know the exact address of a server to contact for a service. With the introduction of SRV RRs in [RFC2782], DNS clients can query a domain for a service or protocol and get back the names of servers. Administrators can use SRV RRs to distribute the load by using several servers to service a single domain. The DNS-type code for SRVs is 33.

The format of an SRV record is as follows (Table 16.2 describes each field in detail):

```
_Service._Proto.Name TTL Class SRV Priority Weight Port Target
```

16.4.1 SRV example

Continuing with the example, the SRV RR looks like:

Service, Proto, Name	TTL	Class	Record type	Priority	Weight	Port	Target
_sip._tcp.example.com	7200	IN	SRV	0	1	5060	*server1.example.com*
	7200	IN	SRV	0	3	5060	*server2.example.com*

In this example the client of the SIP service in the "*example.com*" domain performs an SRV look-up of "*_sip._tcp.example.com*" followed by an A record look-up for "*server2.example.com*".

Table 16.2 SRV RR fields.

Field	Description
Service	Service identifier. An underscore "_" is pre-pended
Proto	Desired protocol. An underscore "_" is pre-pended
Name	The domain of the SRV record
TTL	Time to live
Class	Class of record (SRV records are of class IN, or Internet)
SRV	Identifies the entry as an SRV RR
Priority	The priority of the target host in this record. Clients with lowest numbered priority are tried first. Weight (see below) is used if more than one record carries the same priority
Weight	Denotes a relative weight for entries with the same priority. Entries with larger weights typically have a higher probability of being selected
Port	The port number for the host identified in this record
Target	The domain name of the host targeted by this record

17

GPRS

17.1 Overview

The General Packet Radio Service (GPRS) is the Packet-Switched (PS) domain of the
Global System for Mobile communications (GSM) and the Universal Mobile
Telecommunications System (UMTS) network. It provides Internet Protocol (IP) connectivity to attached User Equipment (UE) via so-called Packet Data Protocol (PDP)
contexts. As expressed in the name, it is a logical connection (context) that is related to
a specific packet-based protocol.

   The UE will be able to send IP packets over the air interface after it has established a
PDP context.

   This chapter concentrates only on those parts of GPRS that are necessary for a UE
to access an IP Multimedia Subsystem (IMS) network. The detailed procedures within
the Gateway GPRS Support Node (GGSN) or the Serving GPRS Support Node
(SGSN), as well as detailed message coding and flows are not discussed here. The
aim of this chapter is to provide the reader with a short overview of the basic principles
that lie behind GPRS and its PDP contexts.

17.2 Packet Data Protocol (PDP)

A Packet Data Protocol (PDP) context offers a packet data connection over which the
UE and the network can exchange IP packets. Usage of these packet data connections
is restricted to specific services. These services can be accessed via so-called access
points.

17.2.1 Primary PDP context activation

This procedure is used to establish a logical connection with the Quality of Service
(QoS) functionality through the network from the UE to the GGSN. PDP context
activation is initiated by the UE and changes the session management state to active,
creates the PDP context, receives the IP address and reserves radio resources. After a
PDP context activation the UE is able to send IP packets over the air interface. The UE
can have up to 11 PDP contexts active concurrently.

The IMS: IP Multimedia Concepts and Services Second Edition
Miikka Poikselkä, Georg Mayer, Hisham Khartabil and Aki Niemi © 2006 John Wiley & Sons, Ltd

17.2.2   Secondary PDP context activation

A secondary PDP context activation allows the subscriber to establish a second PDP context with the same IP address as the primary PDP context. The two contexts may have different QoS profiles, which makes the feature useful for applications that have different QoS requirements (e.g., IP multimedia). The access point name, though, will be the same for the primary and secondary PDP contexts.

17.2.3   PDP context modification

The UE, the SGSN or the GGSN initiate this procedure for updating the corresponding PDP context. Additionally, the radio access network is able to request a PDP context modification from the SGSN (e.g., when coverage to the UE has been lost). The procedures modify parameters that were negotiated during an activation procedure for one or several PDP contexts.

17.2.4   PDP context deactivation

This procedure is used to delete a particular logical connection between the UE and the GGSN. The initiative to deactivate a PDP context can come from the UE, the SGSN, the Home Location Register (HLR) or the GGSN.

17.3   Access points

Access points can be understood as IP routers that provide the connection between the UE and the selected service. Examples of such services are:

- Multimedia Messaging Service (MMS);
- Wireless Application Protocol (WAP);
- direct Internet access;
- IP Multimedia Subsystem (IMS).

Depending on the operator of the network, more than one of these services might be provided by the same access point. The UE needs to be aware of an Access Point Name (APN) – the address of a GGSN – which gives access to the service-providing entity (e.g., an MMSC, the Internet or the P-CSCF). One GGSN may provide different services that can be accessed by different APNs.

When establishing a primary PDP context with an APN the UE receives an IP address or – in the case of IPv6 – an IPv6 prefix that it has to use when communicating over that PDP context. This means that when a UE has established several connections to different APNs the UE will have different IP addresses for each of the provided services.

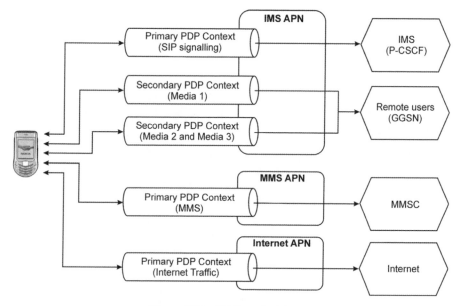

Figure 17.1   PDP context types.

17.4   PDP context types

From the viewpoint of the IMS, it is important to distinguish between the PDP context types shown in Figure 17.1. A primary PDP context is used whenever the first connection with a specific APN is established. If further connections with the same APN are needed, secondary PDP contexts are established.

IMS signalling can be transported either in a dedicated signalling PDP context or a general purpose PDP context. If a dedicated signalling PDP context is established, all IMS-related media streams must be put into one or more separate secondary PDP contexts.

Over a general IMS PDP context the UE may send signalling and media streams together. Nevertheless, it still has the option to create separate PDP contexts for media in this case as well.

18

TLS

This chapter is an overview of the Transport Layer Security (TLS) [RFC2246].

18.1 Introduction

TLS provides transport layer security for Internet applications. It provides for confidentiality and data integrity over a connection between two end points. TLS operates on a reliable transport, such as TCP, and is itself layered into the TLS Record Protocol and the TLS Handshake Protocol.

One advantage of TLS is that applications can use it transparently to securely communicate with each other. Another is that TLS is visible to applications, making them aware of the cipher suites and authentication certificates negotiated during the setup phases of a TLS session; whereas with Internet Protocol Security (IPsec) security policies are usually not visible to each application individually, which makes it difficult to assess whether there is adequate security in place.

TLS allows for a variety of cipher suites to be negotiated, for the use of compression and for a TLS session to span multiple connections. This reduces the overhead of having to perform an expensive TLS handshake for each new parallel connection between applications. It is also possible to resume a session: this means that the client and server can agree to use a previously negotiated session – if one exists in their session cache – instead of performing the full TLS handshake.

18.2 TLS Record Protocol

The TLS Record Protocol layers on top of a reliable connection-oriented transport, such as TCP. The Record Protocol provides data confidentiality using symmetric key cryptography and data integrity using a keyed Message Authentication Checksum (MAC). The keys are generated uniquely for each session based on the security parameters agreed during the TLS handshake. The Record Protocol is also used for encapsulating various upper layer protocols – most notably the TLS Handshake Protocol – in which case it can be used without encryption or message authentication. Other protocols encapsulated in the Record Protocol are the Alert Protocol and the Change Cipher Spec Protocol.

The IMS: IP Multimedia Concepts and Services Second Edition
Miikka Poikselkä, Georg Mayer, Hisham Khartabil and Aki Niemi © 2006 John Wiley & Sons, Ltd

The basic operation of the TLS Record Protocol is as follows:

1. Read messages for transmit.
2. Fragment messages into manageable chunks of data.
3. Compress the data, if compression is required and enabled.
4. Calculate a MAC.
5. Encrypt the data.
6. Transmit the resulting data to the peer.

At the opposite end of the TLS connection, the basic operation of the sender is replicated, but in the reverse order:

1. Read received data from the peer.
2. Decrypt the data.
3. Verify the MAC.
4. Decompress the data, if compression is required and enabled.
5. Re-assemble the message fragments.
6. Deliver the message to upper protocol layers.

18.3   TLS Handshake Protocol

The TLS Handshake Protocol is layered on top of the TLS Record Protocol. It is used to authenticate the client and the server, to exchange cryptographic keys and to negotiate the used encryption and data integrity algorithms before the applications start to communicate with each other. Figure 18.1 illustrates the actual handshake message flow. First, the client and server exchange Hello messages. The client sends a ClientHello message, which is followed by the server sending a ServerHello message. These two messages establish the TLS protocol version, the compression mechanism used, the cipher suite used, and possibly the TLS session ID. Additionally, both a random client nonce and a random server nonce are exchanged, both of which are used in the handshake later on.

Then, the server may send any messages associated with the ServerHello. Depending on the selected cipher suite, it will send its certificate for authentication. The server may also send a key exchange message (e.g., if the server certificate is for signing only) and a certificate request message to the client, depending on the selected cipher suite. To mark the end of the ServerHello and the Hello message exchange, the server sends a ServerHelloDone message.

Next, if requested, the client will send its certificate to the server. In any case, the client will then send a key exchange message that sets the pre-master secret between the client and the server. The pre-master secret is used along with previously exchanged random nonce values to create a master secret for the session. Optionally, the client may also send a CertificateVerify message to explicitly verify the certificate that the server requested.

Then, both the client and the server send the ChangeCipherSpec messages and enable the newly negotiated cipher spec: in other words, the record layer is provided with the

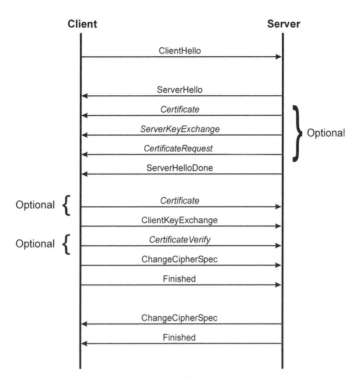

Figure 18.1 The TLS handshake.

necessary security parameters resulting from the handshake. The first message passed in each direction using the new algorithms, keys and secrets is the Finished message, which includes a digest[1] of all the handshake messages. Each end inspects the Finished message to verify that the handshake was not tampered with.

As mentioned earlier, it is also possible to resume a TLS session. To do this, the client suggests in its ClientHello message a session ID used in the TLS session to be resumed. If the server is willing to resume the session and is able to find a matching session ID in its session cache, the handshake will be an abbreviated one, resulting in ChangeCipherSpec and Finished messages immediately after the ServerHello message.

18.4 Summary

The TLS protocol provides transport layer security for Internet applications and confidentiality using symmetric key cryptography and data integrity using a keyed MAC. It also includes functionality for client and server authentication using public key cryptography.

---

[1] The results of applying a one-way hash function to handshake messages.

19

Diameter

This chapter is an overview of the Diameter Protocol. It covers the basic components, protocol properties and protocol operation of Diameter. The services provided by Diameter are given a thorough inspection. This chapter also includes a short introduction to Diameter accounting.

19.1   Introduction

Diameter is an Authentication, Authorization and Accounting (AAA) protocol developed by the Internet Engineering Task Force (IETF). Diameter is used to provide AAA services for a range of access technologies. Instead of building the protocol from scratch, Diameter is loosely based on the Remote Authentication Dial In User Service (RADIUS)[1] [RFC2865], which has previously been used to provide AAA services, at least for dial-up and terminal server access environments. As the basis for Diameter work, the AAA Working Group first gathered requirements for AAA services as they apply to network access from different interest groups:

- IP Routing for Wireless/Mobile Hosts WG (MOBILEIP) [RFC2977].
- Network Access Server Requirements WG (NASREQ) [RFC3169].
- Roaming Operations WG (ROAMOPS) [RFC2477].
- Telecommunications Industry Association (TIA).

The final Diameter protocol is actually split into two parts: Diameter Base Protocol and Diameter applications. The base protocol is needed for delivering Diameter data units, negotiating capabilities, handling errors and providing for extensibility. A Diameter application defines application-specific functions and data units. Each Diameter application is specified separately. Currently, in addition to the base protocol [RFC3588], a few Diameter applications have been defined and some are in the process of being defined: namely, Mobile IP [Draft-ietf-aaa-diameter-mobileip], NASREQ [Draft-ietf-aaa-diameter-nasreq], Extensible Authentication Protocol (EAP) [Draft-ietf-aaa-eap], Diameter credit control [Draft-ietf-aaa-diameter-cc] and Diameter SIP application

---

[1] The name is derived from geometry, as Diameter = 2 ∗ Radius.

The IMS: IP Multimedia Concepts and Services Second Edition
Miikka Poikselkä, Georg Mayer, Hisham Khartabil and Aki Niemi © 2006 John Wiley & Sons, Ltd

[Draft-ietf-aaa-diameter-sip-app]. In addition, the AAA transport profile [RFC3539] includes discussions and recommendations on the use of transports by AAA protocols. Third Generation Partnership Project (3GPP) Release 5 has also been allocated a specific set of Diameter command codes [RFC3589].

The Diameter Base Protocol uses both the Transmission Control Protocol (TCP) [RFC0793] and the Stream Control Transmission Protocol (SCTP) [RFC2960] as transport. However, SCTP is the preferred choice, mostly due to the connection-oriented relationship that exists between Diameter peers. It is beneficial to be able to categorize several independent streams to a single SCTP association, instead of keeping all streams open as independent TCP connections. Both Internet Protocol Security (IPsec) [RFC2401] and Transport Layer Security (TLS) [RFC2246] are used to secure connections.

19.2 Protocol components

Diameter is a peer-to-peer protocol, since any Diameter node can initiate a request. Diameter has three different types of network nodes: clients, servers and agents. Clients are generally the edge devices of a network that perform access control. A Diameter agent provides either relay, proxy, redirect or translation services. A Diameter server handles the AAA requests for a particular domain, or realm. Diameter messages are routed according to the Network Access Identifier (NAI) [RFC2486] of a particular user.

19.3 Message processing

Each Diameter node maintains a Diameter peer table, which contains a list of known peers and their corresponding properties. Each peer table entry is associated with an identity and can either be statically or dynamically assigned. Dynamically assigned peer entries have an expiration time associated with them, within which they are either refreshed or discarded. Each peer entry also includes a relative priority setting, which specifies the role of the peer as either primary, secondary or alternative. The status of the peer relates to a specific configuration of a Finite State Machine (FSM) of a peer connection, called the Diameter Peer State Machine. Each peer entry also specifies whether a peer supports TLS[2] and, optionally, includes other security information (e.g., cryptographic keys and certificates).

Diameter peer table entries are referenced by a Diameter realm-routing table, as part of the message-routing process. All realm-based routing look-ups are performed against a realm-routing table. The realm-routing table lists the supported realms (with each route entry containing certain routing information). Each route entry is either statically assigned or dynamically discovered. Dynamic entries are always associated with an expiry time. A route entry is also associated with an application identifier, which

---

[2] Support for IPsec is mandatory for all nodes in Diameter, whereas support for TLS is optional for clients.

Table 19.1 Diameter local action entries.

Local action field	Description
Local	Messages whose destination en route resolves to a route entry with the local directive set can be processed locally and need not be routed to another Diameter peer
Relay	Messages whose destination en route resolves to a relay route entry will be routed to a next-hop peer without modification
Proxy	Messages whose destination en route resolves to a proxy route entry will be routed to a next-hop peer, possibly after local policies are first applied
Redirect	Messages whose destination en route resolves to a redirect route entry will result in a redirection of a downstream[a] Diameter peer to the correct next-hop peer

[a] "Downstream" is a term used to identify the direction of a message (e.g., toward the originator of a request).

enables route entries to have a different destination depending on the Diameter application. The destination of a route entry corresponds to one or more peer entries in a Diameter peer table. As message processing can follow different procedures depending on the message, the local action field of a route entry can have one of four processing directives (see Table 19.1).

Each processing directive corresponds to a functional entity called a Diameter agent, as discussed in Section 19.5. All Diameter nodes maintain a transaction state (i.e., they keep track of request and answer pairs). A Diameter session is always bound to a particular service, defined by the corresponding Diameter application. A statefull agent maintains a session state (i.e., it stores information related to all of its authorized active sessions). Active sessions are either explicitly torn down or deactivated by expiration.

Local state can also be added to Diameter requests. This feature enables an otherwise stateless agent to encode its local state as part of the request, so that it gets reflected back in the corresponding answer message. Overall, maintaining the session state is necessary for some Diameter agents, such as translation agents and proxy agents.

19.4 Diameter clients and servers

Diameter clients originate AAA requests. The Diameter server[3] services the client's requests within a realm, according to the used Diameter application. Diameter agents provide value-added services for clients and servers.

Typically, the Diameter client is a Network Access Server (NAS) that performs AAA services for a particular access technology. The NAS needs to authenticate the terminals that are attached to a network before allocating network resources for them: for

---

[3] Often referred to as the home server.

example, a Wireless Local Area Network (WLAN) access point may need to ascertain the user's identity before allowing access.[4]

19.5   Diameter agents

Diameter agents can be deployed in a network to perform load balancing, distribute system administration and maintenance, concentrate requests and perform additional message processing. Diameter relay agents are protocol-transparent (i.e., they support the Diameter Base Protocol), and are also agnostic about Diameter applications. They simply accept requests and route messages based on the supported realms and known peers. Relay agents are usually used to reduce the configuration load on Diameter clients and servers: for example, NASs of a specific geographical area can be grouped behind a common relay agent, thus avoiding the need to configure each NAS with the information needed to communicate with other realms. Similarly, a Diameter server may have a relay agent that handles the configuration load caused by the addition and removal of individual NASs in the network.

Diameter redirect agents perform realm-to-server resolution. They use the Diameter Realm Routing Table to determine the next-hop Diameter peer to a particular request message. Instead of routing a request by itself, the redirect agent returns a special answer message that includes the identity of the next-hop peer. The originator of the request then contacts the next-hop peer directly. Since redirect agents operate in a stateless fashion, they provide scalability and centralized message routing. However, centralized message routing is always a compromise between ease of configuration and fail-over resilience.

Diameter proxy agents carry out value-added message processing on the requests and answers. They are similar to relay agents in that message routing is based on a Diameter routing table, but different in that they also modify the messages by implementing policy enforcement (e.g., resource usage enforcement, admission control or provisioning). Since policy enforcement always requires an understanding of the offered service, proxy agents need to support specific Diameter applications as well as the Diameter Base Protocol.

Diameter translation agents perform protocol translation services between Diameter and another AAA protocol. Translation agents are used to allow legacy systems to communicate with the Diameter infrastructure. Sometimes, migration from an installed equipment base that is not easy to upgrade (e.g., embedded devices) takes a considerable amount of time.

Perhaps the most common legacy system in this case is RADIUS, to which Diameter has extensive support with the NASREQ application.

19.6   Message structure

Diameter messages consist of a Diameter header, followed by a certain number of Diameter Attribute Value Pairs (AVPs). The Diameter header comprises binary data

---

[4] Usually, there exists a prior (billing) relationship between the network provider and the user.

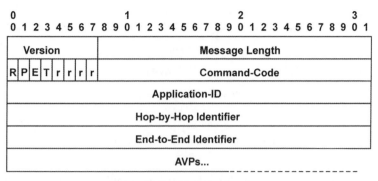

Figure 19.1 Diameter header.

and as such is similar to an Internet Protocol (IP) header [RFC0791] or a TCP header [RFC0793]. The format of the Diameter header is shown in Figure 19.1.

The command flags specify the type of Diameter message. A request message has the "R" bit (or the request bit) set, while an answer message has it cleared. The "P" bit (or the proxiable bit) indicates whether the message can be proxied or must instead be processed locally. The "E" bit (or the error bit) indicates that the answer message is an error message. The "T" bit indicates that the message is possibly a re-transmission and is set by Diameter agents in fail-over situations to aid detection of duplicate messages. The "r" bits are unused flags that are reserved for future use. Together with the Command-Code, the command flags specify the semantics associated with the particular Diameter message.

The Command-Code field indicates the command associated with the Diameter message. The Command-Code field is 24 bits long and includes values from 0 to 255 that are reserved for RADIUS backward compatibility, as well as values 16777214 and 16777215 that are reserved for experimental use. The Command-Code namespace is maintained by the Internet Assigned Numbers Authority (IANA) and includes either codes that are used by the Diameter Base Protocol or specific codes used by the Diameter applications.

The Application-ID field is used to identify the application for which the message is targeted. An Application-ID indicates either a standard application or a vendor-specific application. The Application-ID namespace is also maintained by IANA.

The Hop-by-Hop identifier helps in matching requests and responses and is unique within a connection at any given time. The End-to-End identifier is used end to end to detect duplicate messages.

AVPs contain authentication, authorization and accounting information elements, as well as routing, security and configuration information elements that are relevant to the particular Diameter request or answer message. Each AVP contains an AVP header and some AVP-specific data. The AVP header is shown in Figure 19.2.

The AVP-Code field uniquely identifies the attribute together with the Vendor-ID field. The first 256 AVP numbers, or codes, are reserved for backward compatibility with RADIUS, whereas numbers above 256 belong to Diameter attributes. AVP-Code field values are maintained by IANA.

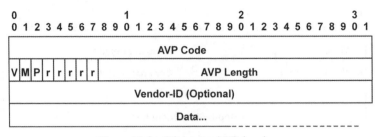

Figure 19.2   Diameter AVP header.

AVP flags carry information on how the attribute should be handled by the receiving end. The "V" bit (or the vendor-specific bit) indicates that the AVP-Code belongs to a vendor-specific address space, denoted by the otherwise optional Vendor-ID field. The "M" bit (or mandatory bit) mandates the support for a particular AVP. Any message with an unrecognized AVP carrying the "M" bit is always rejected by the receiver. AVPs that do not have the M bit set are information-only (i.e., they can be ignored by the receiver if they are not understood). When the "P" bit (or the protected bit) is set it indicates the need for encryption for end-to-end security. The "r" bits are unused flags reserved for future use.

19.7   Error handling

Diameter errors fall into two categories: protocol errors and application errors. Protocol errors occur at the base protocol level and are treated on a per-hop basis (e.g., errors in message routing may direct a downstream agent to take special action). Application errors are generated due to problems with specific Diameter applications.

When a request message generates a protocol error an answer message is sent back with the "E" bit set in the Diameter header, indicating a protocol error. The appropriate protocol error value is then placed in a Result-Code AVP in the answer message. On the other hand, when a Diameter entity generates an application error it sends back an answer message with the appropriate Result-Code AVP included. Since application errors do not require any Diameter agent involvement, the message is forwarded directly back to the originator of the request.

Table 19.2 describes the classes used in the Result-Code AVP data field: a 32-bit namespace for errors, maintained by IANA. The Result-Code AVP can also be accompanied by other elements that give additional information about the error message (e.g., a human-readable error message must be included for all protocol errors in an Error-Message AVP).

19.8   Diameter services

Basically, the Diameter Base Protocol provides two types of services to Diameter applications: authentication and/or authorization, and accounting. An application

Table 19.2 Diameter result codes.

Result code	Class	Description
1xxx	Information	Additional action required to complete the request
2xxx	Success	Request was successfully completed
3xxx	Protocol errors	Errors which are treated on a hop-by-hop basis
4xxx	Transient failures	Request failed but may succeed later
5xxx	Permanent failures	Request failed and should not be attempted again

can make use of authentication or authorization only or it can combine the two by requesting authentication with authorization. The accounting service can be invoked irrespective of whether authentication or authorization was requested first.

Each individual Diameter application defines its own authentication and authorization Command-Codes and AVPs. The accounting application described by the base protocol is shared among the applications – only the accounting AVPs are specific to an application.

19.8.1 Authentication and authorization

Authentication and authorization services are interlinked in Diameter. An auth request is issued by the client to invoke a service and, depending on the AVPs carried in the auth request, either authentication or authorization (or both) are performed on it.

Authentication clearly either succeeds or fails, whereas the Diameter Base Protocol provides authorization services in either stateless or statefull mode. In statefull authorization the server maintains a session state and the authorization session has a finite length. The total lifetime of a session consists of an authorization lifetime and a grace period, which together represent the maximum length of a session the server is willing to take responsibility for.[5] The authorization session can, of course, be terminated by the client or aborted by the server and, at the end of the authorization lifetime, it can also be re-authorized. These functions are provided by the Diameter Base Protocol.

The two authorization modes of operation correspond to a statefull authorization FSM and a stateless authorization FSM, with which all Diameter nodes that support authentication and authorization services have to comply.

19.8.2 Accounting

The Diameter Base Protocol provides accounting services to Diameter applications. When an accounting session is not active, there are no resources reserved for it in either the Diameter client or the Diameter server.[6] A successful Accounting Request (ACR) activates an accounting session, in which the accounting records exchanged fall into two

---

[5] This usually means financial responsibility.

[6] In this case the accounting server.

categories, based on the accounting service type:

- Measurable length services have clearly defined beginnings and ends. An accounting record is created when the service begins and another is sent when the service ends. Optionally, interim accounting records can be produced at certain intervals within the measurable length session.
- One-time events are services without a measurable service length. In a one-time event accounting record the beginning of the service and the end of the service actually coincide; therefore, a one-time event only produces a single accounting record.

The accounting strategy of a session is dictated by the accounting server or the server authorizing a user session: this is called a "server-directed model for accounting". The accounting server directs the client to use either measurable length service accounting or one-time event accounting. It also optionally specifies the time interval to use when generating interim accounting records.

The Diameter Accounting Protocol has built-in fault resilience to overcome small message loss and temporary network faults, as well as real-time delivery of accounting information. Accounting records are correlated with the Session-Id AVP, which is a globally unique identifier and present in all AAA messages. Alternatively, if a service consists of several different sessions each with a unique Session-Id AVP, then an Accounting-Multi-Session-Id AVP is used for correlation of accounting records.

19.9 Specific Diameter applications used in 3GPP

This section lists the two Diameter applications used in 3GPP Internet Protocol Multimedia Subsystem (IMS). The first is the Diameter Session Initiation Protocol (SIP) application [Draft-ietf-aaa-diameter-sip-app], which is used in the Cx, Dx, Sh and Dh interfaces (described in Section 2.3). The second is the Diameter credit control application [Draft-ietf-aaa-diameter-cc], which is utilized in the Ro interface for online charging functionality (see Section 3.12.5.2).

19.10 Diameter SIP application

The Diameter SIP application [Draft-ietf-aaa-diameter-sip-app] defines a Diameter application that can be used by a SIP server to authenticate users and to authorize usage of different SIP resources. The Diameter SIP application is close to the 3GPP IMS Cx interface in functionality, but it is designed to be generic enough for other SIP deployment scenarios to be able to benefit from it. Table 19.3 illustrates a mapping between Cx interface parameters and Diameter SIP application AVPs.

The general architecture used as the basis for the Diameter SIP application is illustrated in Figure 19.3. Proxy A and proxy B have different roles: proxy A is configured as an edge proxy to a domain and proxy B as the home proxy of a domain. Proxy A closely resembles the Interrogating Call Session Control Function (I-CSCF) function and proxy B the Serving-CSCF (S-CSCF) function in IMS. For redundancy and fault

Table 19.3 Mapping Cx parameters to the Diameter SIP application.

Cx parameter	Diameter SIP application AVP Name[a]
Visited network identifier	SIP-Visited-Network-Identifier
Public user ID	SIP-AOR
Private user ID	User-Name
S-CSCF name	SIP-Server-URI
S-CSCF capabilities	SIP-Server-Capabilities SIP-Mandatory-Capability SIP-Optional-Capability
Result	Result-Code SIP-Reason-Code
User profile	SIP-User-Data SIP-User-Data-Request-Type SIP-User-Data-Already-Available
Server assignment type	SIP-Server-Assignment-Type
Authentication data	SIP-Auth-Data-Item
Item number	SIP-Item-Number
Authentication scheme	SIP-Authentication-Scheme
Authentication information	SIP-Method SIP-Authenticate SIP-Authentication-Context
Authorization information	SIP-Authorization SIP-Authentication-Info
Confidentiality key	SIP-Confidentiality-Key
Integrity key	SIP-Integrity-Key
Number authentication items	SIP-Number-Auth-Items
Reason for de-registration	SIP-Deregistration-Reason SIP-Reason-Info
Charging information	SIP-Accounting-Information SIP-Accounting-Server-URI SIP-Credit-Control-Server-URI
Route information	Destination-Host
Type of authorization	SIP-User-Authorization-Type

[a] Note that those AVP names without the SIP-prefix are Diameter Base Protocol AVPs.

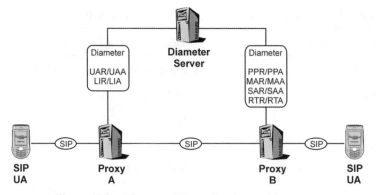

Figure 19.3 Diameter SIP application architecture.

tolerance, there may be additional Diameter Subscriber Locator (SL) nodes that either implement the Diameter relay or Diameter redirect agent functionality.

The Diameter SIP application defines a set of Command-Codes that are an extension to the Diameter Base Protocol (see Table 19.4). These Command-Codes perform the following functions:

- UAR/UAA – determine whether a user is authorized to receive a certain service and, if so, indicate the local server capable of providing that service.
- SAR/SAA – assign a specific SIP server to a particular user and deliver the user profile to it in a synchronous way.
- LIR/LIA – determine the next-hop SIP entity at an edge proxy.
- MAR/MAA – authenticate and authorize a user for a specific SIP service (e.g., SIP registration). Authentication can either be performed in the Diameter server or delegated to the SIP server.
- RTR/RTA – de-registration initiated by the Diameter server.
- PPR/PPA – asynchronously delivers the user profile to the SIP server from the Diameter server.

Table 19.4 Diameter SIP application command codes.

Command name	Abbreviation
User-Authorization-Request	UAR
User-Authorization-Answer	UAA
Server-Assignment-Request	SAR
Server-Assignment-Answer	SAA
Location-Info-Request	LIR
Location-Info-Answer	LIA
Multimedia-Auth-Request	MAR
Multimedia-Auth-Answer	MAA
Registration-Termination-Request	RTR
Registration-Termination-Answer	RTA
Push-Profile-Request	PPR
Push-Profile-Answer	PPA

Table 19.5 Diameter credit control application command codes.

Command Name	Abbreviation
Credit-Control-Request	CCR
Credit-Control-Answer	CCA

19.11 Diameter credit control application

The Diameter credit control application [Draft-ietf-aaa-diameter-cc] is used for real-time credit and cost control for a variety of different services. As opposed to non-real-time accounting, which is post-paid, this is the pre-paid mode of operation that has proved a great success in mobile phone markets: instead of, say, a monthly invoice, users buy credit up front which is then debited in real time as services are used. The credit control application specifies a pair of new Diameter Command-Codes for credit control requests (see Table 19.5). The application has two main parts:

- Cost and credit control – services need to be rated in real time, which means that the cost of a given service needs to be assessed in real time. The user's account also needs to be debited or credited in real time, based on the accrued service cost.
- Credit authorization – the user's account may also need to be inspected to see that her balance is sufficient to cover the service. Credit needs to be reserved and unused reservations need to be refunded. If the user's balance is insufficient to cover the service charges, the service element needs to be able to deny the service altogether.

As in Diameter Base Protocol accounting, credit control sessions have two distinct models: credit authorization with credit reservation and credit authorization with direct debiting. The process of authorization with reservation is similar to the process of accounting servicesof measurable length. First, an initial interrogation is performed, where the credit control server rates the service request and reserves a sufficient amount of credit from the user's account. This may optionally be followed by any number of intermediate interrogations, in which the server deducts used credit from the user's account, and may involve the server performing service rating and credit reservation to accommodate a service that is continuing. A final interrogation reports back to the credit control server that the service has been terminated, whereupon the user's account is debited for the used amount.

The second model is similar to accounting one-time events. On receiving a credit authorization request, the server deducts a suitable amount of credit from the user's account. This model can also be seen as a credit control session, except that the start and end of the session happen to coincide. The stateless mode also supports operations, such as service price queries, user account balance checks and credit refunds.

19.12 Summary

The Diameter Protocol is used to provide AAA services for a range of access technologies. It is loosely based on an existing AAA protocol called RADIUS, which has been used widely for dial-up and terminal server access. The Diameter Protocol uses a binary header format and is capable of transporting a range of data units called AVPs. The Diameter Base Protocol specifies the delivery mechanisms, capability negotiation, error handling and extensibility of the protocol, whereas individual Diameter applications specify service-specific functions and AVPs. In addition to the base protocol which includes accounting, 3GPP IMS currently makes use of two other Diameter applications: namely, the Diameter SIP application and the Diameter credit control application.

20

MEGACO

This chapter is an overview of the Media Gateway Control Protocol (MEGACO) Version 1 [RFC3525], also known as H.248 [ITU.H248.1]. This protocol is used in the Third Generation Partnership Project's (3GPP) Internet Protocol Multimedia Subsystem (IMS) Mn and Mp reference points, as explained in Sections 2.3.15 and 2.3.18, respectively.

20.1 Introduction

MEGACO is used between a media gateway and the media gateway controller to handle signalling and session management during a multimedia conference. The media gateway controller and the media gateway share a master/slave relationship.

20.2 Connection model

The connection model for the protocol describes the main objects within a media gateway as terminations and contexts that can be controlled by the media gateway controller. A termination sources or sinks[1] one or more streams, and each termination holds information about the actual media streams.

Different terminations are linked together by a context. The set of terminations that are not associated with other terminations are defined as being represented by a special type of context (namely, the null context). A context describes the topology of terminations associated with it: for example, it includes parameters about mixing in case the context contains more than two terminations.

20.3 Protocol operation

The protocol contains a set of commands that are used to manipulate the logical entities described in the connection model (namely, terminations and contexts). Specifically, the

---

[1] In layman terms: either originates or terminates.

The IMS: IP Multimedia Concepts and Services Second Edition
Miikka Poikselkä, Georg Mayer, Hisham Khartabil and Aki Niemi © 2006 John Wiley & Sons, Ltd

set of commands offered by MEGACO are:

- Add – adds a termination to a context. It is also used to implicitly create a context (a context is created as soon as the first termination is added to it).
- Modify – modifies the state properties of a termination and properties specific to media streams.
- Subtract – removes a termination from a context. It is also used to implicitly delete a context. Similar to creating one, subtracting the last termination of a context deletes the context.
- Move – moves a specific termination from one context to another.
- AuditValue – returns the current termination properties as well as the events,[2] signals[3] and statistics of a termination.
- AuditCapabilities – returns all possible values for the termination properties as well as the signals and events allowed by a particular media gateway.
- Notify – used by the media gateway to report to the media gateway controller when certain events occur.
- ServiceChange – used by the media gateway to inform the media gateway controller of specific changes in service (e.g., when a particular termination has been taken out of service or is returned to service).

Each command can carry a number of parameters, called "descriptors". A command may also return descriptors as output. Descriptors consist of a name and a list of items, some of which may have values. The list of descriptors in MEGACO is shown in Table 20.1.

---

[2] "Events" describe situations in a stream which the media gateway can detect and report (e.g., on-hook, off-hook).
[3] "Signals" are media generated by the media gateway (e.g., announcements and tones).

Table 20.1 Gateway Control Protocol descriptors.

Name	Description
Audit	Indicates the information requested in Audit commands
DigitMap	Describes patterns for a sequence of events that, when matched, should be reported
Error	Indicates an error code and, optionally, text describing the protocol error
EventBuffer	Describes events for the media gateway to detect while event buffering is used
Events	Describes events for the media gateway to detect and instructions about what to do when that happens
Local	Describes the media flow properties of flows received from a remote end
LocalControl	Describes properties that are of interest to the media gateway and the media gateway controller
Media	Specifies the parameters for all media streams
Modem	Identifies the modem type and other properties for a termination
Mux	Describes the multiplex type and identifies the terminations forming the input mux
ObservedEvents	Reports observed events
Packages	Identifies the properties, events, signals and statistics of a termination, which are grouped into packages
Remote	Describes the media flow properties of flows sent to a remote end
ServiceChange	Describes the details of a service change
Signals	Describes signals applied to terminations
Statistics	Describes the statistics kept about a termination
Stream	A list of descriptors for a single stream, consisting of Local, Remote and LocalControl descriptors
TerminationState	Describes the properties of a termination
Topology	Describes the topology of a context (i.e., flow directions between terminations)

21

COPS

This chapter is an overview of the Common Open Policy Service (COPS) [RFC2748] protocol and its use in supporting policy provisioning (COPS-PR) [RFC3084]. It also describes the specific extensions to COPS-PR used in the Go interface, as explained in Section 2.3.19.

21.1 Introduction

The COPS protocol is an Internet Engineering Task Force (IETF) protocol used for the general administration, configuration and enforcement of policies. It defines a simple query and response protocol for exchanging policy information between a policy server and its clients. The clients are denoted as Policy Enforcement Points (PEPs) and the server as a Policy Decision Point (PDP),[1] respectively. The protocol employs a client/server model in which a PEP sends requests, updates and deletes to the PDP, which in turn returns policy decisions back to the PEP. A special type of PDP is the Local Policy Decision Point (LPDP), which is used by PEPs to request local policy decisions when there is no available PDP with which to communicate. Figure 21.1 illustrates the model.

There are two main models for COPS policy control:

- Outsourcing – the PEP assigns (outsources) responsibility for authorizing certain events at the PEP to an external entity (PDP). This model assumes a one-to-one correlation between events at a PEP and decisions from a PDP.
- Configuration – unlike the previous model, there exists no direct mapping between events at the PEP and decisions from the PDP. The PDP may proactively configure the PEP based on any external events as well as events originating at the PEP. This may be performed by the PDP in bulk or in portions, but the overall timing is more flexible than the outsourcing model's.

The COPS protocol uses persistent Transmission Control Protocol (TCP) connections between the PEPs and the PDP for reliable transport of protocol messages. The COPS

---

[1] In this chapter PDP stands for Policy Decision Point and should not be confused with Packet Data Protocol.

The IMS: IP Multimedia Concepts and Services Second Edition
Miikka Poikselkä, Georg Mayer, Hisham Khartabil and Aki Niemi © 2006 John Wiley & Sons, Ltd

Figure 21.1   COPS model.

protocol is statefull since the request/decision state is shared between the client and the server. Also, unlike many other client/server protocols, the state of a given request/ decision pair can be associated with another request/decision pair. The server is also able to push state to the client and remove it once it is no longer valid. COPS has a built-in extensibility: policy objects are self-identifying and support vendor-specific objects. Extensions may describe the message formats and objects that carry the policy data without requiring any changes to the protocol itself.

21.2   Message structure

COPS is a binary protocol whose messages consist of the COPS header followed by typed objects. The COPS header (Figure 21.2) consists of the following fields:

- Version – indicates the COPS version number (its default is 1).
- Flags – denotes certain message flags. The only one currently defined is 0x1, which indicates whether the message was solicited by another COPS message.
- OP code – denotes the COPS operation. Table 21.1 lists the defined COPS operation codes.
- Client type – identifies the policy client and determines the interpretation of the accompanying typed objects, with the exception of a Keep-Alive message that always has client type set to zero.
- Message length – denotes the octet length of the message.

```
0                   1                   2                   3
0 1 2 3 4 5 6 7 8 9 0 1 2 3 4 5 6 7 8 9 0 1 2 3 4 5 6 7 8 9 0 1
```

Version	Flags	OP Code	Client-Type
Message Length				

Figure 21.2   COPS header.

Table 21.1 COPS operation codes.

Value	Short	Where	Name	Description
1	REQ	PEP → PDP	Request	Requests a decision from a PDP and establishes a client handle that identifies the specific state associated with this PEP
2	DEC	PDP → PEP	Decision	Returns one or more decisions as a response to a request
3	RPT	PEP → PDP	Report state	Reports back to the PDP the PEP's success or failure in carrying out the PDP's decision, or reports a change in state
4	DRQ	PEP → PDP	Delete request state	Indicates to the PDP that the state associated with the client handle is no longer relevant
5	SSQ	PDP → PEP	Synchronize state request	Indicates that the remote PDP wishes the PEP to re-send its state
6	OPN	PEP → PDP	Client-Open	Specifies the PEP's supported client types, the last PDP connection for a given client type and client-specific feature negotiation
7	CAT	PDP → PEP	Client-Accept	Responds positively to the Client-Open message and returns a timer object for the Keep-Alive timer
8	CC	PEP → PDP, PDP → PEP	Client-Close	Indicates that a particular client type is no longer supported
9	KA	PEP → PDP, PDP → PEP	Keep-Alive	Resets the Keep-Alive timers
10	SSC	PEP → PDP	Synchronize complete	Reports successful completion of a state synchronization

The format of COPS-specific objects is illustrated in Figure 21.3 and consists of:

- Length – object length in octets.
- C-Num – typically denotes the class of the information contained in the object.
- C-Type – typically identifies the subtype or version of the information contained in the object.

The remainder of the object encapsulates actual object information, which varies from object to object. COPS-specific objects are described in Table 21.2. Objects related to the use of COPS for provisioning are described in the following sections.

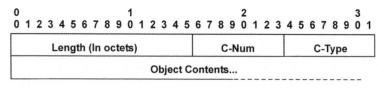

Figure 21.3   COPS-specific objects.

Table 21.2   COPS-specific object description.

C-Num	Name	Where	Description
1	Handle	Most	Unique value to identify an installed state
2	Context	REQ, DEC	Type of event that triggered a particular query
3	In interface	REQ	The address and interface identifier for the incoming interface on the PEP
4	Out interface	REQ	The address and interface identifier for the outgoing interface on the PEP
5	Reason code	DRQ	The reason for deleting the request state
6	Decision	DEC	Decision made by the PDP
7	LPDP decision	DEC	Decision made by the LPDP
8	Error	CC	Identifies a particular protocol error
9	Client-specific info	REQ, DEC, RPT, OPN	Client-specific information
10	Keep-Alive timer	CAT	Keep-Alive timer value
11	PEP identification	OPN	Identifies the PEP client to the PDP
12	Report type	RPT	The type of report on the request state, which is associated with a specific handle
13	PDP redirect address	CC	Address to which the PDP may optionally redirect the PEP
14	Last PDP address	OPN	Address of the PDP the PEP last connected
15	Accounting timer	CAT	Optionally indicates the timer value between periodic reports of accounting type
16	Message integrity	Any	Sequence number and a message authentication checksum used for integrity protection of the requests

21.3   COPS usage for policy provisioning (COPS-PR)

COPS-PR defines COPS usage for policy provisioning and is independent of the provisioned policy. The PEP describes its configurable parameters in policy requests, denoted as configuration requests. If there is a change in these parameters, an update request is sent. The requests do not necessarily map directly to any decisions, which are typically issued by the PDP only when the PDP responds to external events: for example, the PDP would issue a configuration decision only when a Service Level Agreement (SLA) is changed. Hence, both policy requests and decisions may be fairly infrequent.

The data model in COPS-PR is based on something called a Policy Information Base (PIB). Policy-provisioning data are identified by a PIB, and each area of policy provisioning may have one or more PIBs defined. In other words, a given PEP–PDP connection may see multiple PIBs in transit, but each PIB would be associated with a specific client type.

As the name suggests, it is quite similar to the Management Information Base (MIB) [RFC3418] used for the Simple Network Management Protocol (SNMP) [RFC3411]. In fact, they share a common data structure, as both use Structure for Management Information Version 2 (SMIv2) definitions. In COPS-PR this is used for Encoded Provisioning Instance Data (EPID) encoding rules.

The policy data are conceptually structured as a tree, with the PIB being the root. As such a PIB represents a virtual database for the contents of the tree. Each sub-tree represents a Provisioning Class (PRC), whereas the leaves represent instantiations of this class – namely, Provisioning Instances (PRIs). There may be multiple PRIs per PRC and there may be multiple PRCs associated with a PIB. Each PRI has an identifier, a Provisioning Instance Identifier (PRID), which is the name carried by the COPS object that identifies that particular instance of a provisioning class.

21.4   The PIB for the Go interface

As described in Section 2.3.19, the Go interface uses the COPS protocol to perform media authorization and charging correlation. It also employs COPS-PR extensions. To transmit Go-specific data, the Third Generation Partnership Project (3GPP) has defined a PIB for the Go interface. The details of the 3GPP Go PIB can be found in [3GPP TS 29.207].

21.5   Summary

The COPS protocol is an IETF protocol used for general administration, configuration and enforcement of policies. COPS-PR extensions define COPS usage for policy provisioning as well as the concept of a PIB. The 3GPP Go defines such a PIB for representing information exchanged over the Go interface.

22

IPsec

22.1 Introduction

Internet Protocol Security (IPsec) provides for various security services on the IP layer, in IPv4 as well as IPv6, thus offering protection for protocols in upper layers. IPsec is typically used to secure communications between hosts and security gateways. The set of security services that IPsec provides includes:

- access control;
- data integrity protection;
- data origin authentication;
- anti-replay protection;
- confidentiality;
- limited traffic flow confidentiality.

IPsec can operate in two modes: tunnel mode and transport mode. Transport mode is mainly used to provide security services for upper layer protocols. Tunnel mode is typically used to tunnel IP traffic between two security gateways. The difference is that in transport mode IPsec offers limited protection to IP headers, whereas in tunnel mode the full IP datagram is protected. More specifically, in the case of an Encapsulated Security Payload (ESP) there are no security services provided for the IP headers that precede the ESP header, whereas an Authentication Header (AH) does extend protection to some parts of the IP header.

The components of the IPsec security architecture [RFC2401] are:

- Security protocols – the AH [RFC2402] and the ESP [RFC2406].
- Security associations – definition of the Security Policy Database (SPD) and the Security Association Database (SAD), as well as the management and usage of security associations.
- Key management – the distribution of cryptographic keys for use with the security protocols (namely, the Internet Key Exchange, or IKE [RFC2409]).
- Algorithms used for encryption and authentication.

In the following sections we will review each of these components.

The IMS: IP Multimedia Concepts and Services Second Edition
Miikka Poikselkä, Georg Mayer, Hisham Khartabil and Aki Niemi © 2006 John Wiley & Sons, Ltd

22.2 Security associations

The concept of a security association is germane to IPsec. Security protocols make use of Security Associations (SAs) as they provide security services (the main responsibility of key management is to establish and manage SAs). An SA is a relationship between two entities that defines how they are going to use security services to secure their communications. It includes information on authentication and/or encryption algorithms, cryptographic keys and key lengths as well as the Initialization Vectors (IVs) that are shared between the entities. An SA is unidirectional; so, typically two SAs are needed for a bidirectional flow of traffic – one for inbound (read) traffic and one for outbound (write) traffic. An SA is uniquely identified by the following three items:

- Security Parameter Index (SPI);
- destination IP address;
- security protocol (either AH or ESP).

The management of SAs involves two databases: the SPD and the SAD. The SPD contains the policies by which all inbound and outbound traffic is categorized on a host or a security gateway. The SAD is a container for all active SAs and related parameters. A set of selectors – IP layer and upper layer (e.g., TCP and UDP) protocol field values – is used by the SPD to map traffic to a specific SA.

22.3 Internet Security Association and Key Management Protocol (ISAKMP)

Internet Security Association and Key Management Protocol (ISAKMP) is used for negotiating, establishing, modification and deletion of SAs and related parameters. It defines the procedures and packet formats for peer authentication creation and management of SAs and techniques for key generation. It also includes mechanisms that mitigate certain threats – e.g., Denial Of Service (DOS) and anti-replay protection.

In ISAKMP, SA and key management are separate from any key exchange protocols; so, in a sense ISAKMP is an "abstract" protocol – it provides a framework for authentication and key management and supports many actual key exchange protocols (e.g., IKE). ISAKMP defines header and payload formats, but needs an instantiation to a specific set of protocols. Such an instantiation is denoted as the ISAKMP Domain Of Interpretation (DOI): an example of this for the IPsec/IKE is the IPsec DOI [RFC2407].

ISAKMP operates in two phases. During phase 1, peers establish an ISAKMP SA – namely, they authenticate and agree on the used mechanisms to secure further communications. In phase 2 this ISAKMP SA is used to negotiate further protocol SAs (e.g., an IPsec/ESP SA). After the initial establishment of an ISAKMP SA, multiple protocol SAs can be established.

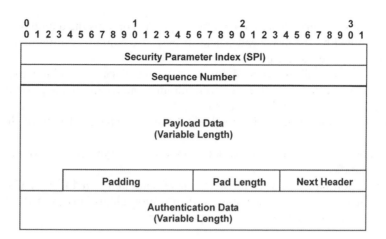

Figure 22.1 ESP packet format.

22.4 Internet Key Exchange (IKE)

Internet Key Exchange (IKE) is a key exchange protocol which, in conjunction with ISAKMP, negotiates authenticated keying material for SAs. IKE can use two modes to establish a phase 1 ISAKMP SA: main mode and aggressive mode. Both modes use the Ephemeral Diffie–Hellman key exchange algorithm[1] to generate keying material for the ISAKMP SA. The difference between these modes is that in main mode, while consuming more message round trips, the identities of the negotiating entities are protected, whereas in aggressive mode they are revealed to the outside world. After establishing the ISAKMP SA in phase 1, protocol SAs can be negotiated, while negotiation is secured using an ISAKMP SA.

22.5 Encapsulated Security Payload (ESP)

Encapsulated Security Payload (ESP) is used to provide security services in IPv4 and IPv6. It can be used alone or in unison with an AH. It can provide either confidentiality (i.e., encryption) or integrity protection (i.e., authentication), or both. As mentioned previously, ESP can operate in transport mode and in tunnel mode.

The ESP header is inserted into the IP datagram after the IP header and before any upper layer protocol headers in transport mode, or before an encapsulated IP datagram in tunnel mode. Figure 22.1 illustrates the ESP packet format.

The fields in the ESP header are as follows:

- The SPI – a unique and random 32-bit value that, together with the destination IP address and security protocol, uniquely identifies the SA for the packet.
- The sequence number – a monotonically increasing 32-bit counter used to protect against replay attacks. When an SA is established the sequence number is reset to 0.

---

[1] In the Ephemeral Diffie-Hellman algorithm the public key (out of the temporary D–H key pair) is tasked with guarding against Man-In-The-Middle (MITM) attacks.

- Payload data – a variable length field that typically contains the data payload, whose type is denoted by the next header field. It may also contain cryptographic synchronization data, such as an IV.
- Padding – used to fill the payload data to a specific block size multiple required by a particular encryption algorithm, or to randomize the length of the payload in order to protect against traffic flow analysis.
- Pad length – an 8-bit field whose value indicates in bytes the length of the padding field.
- Next header – an 8-bit field whose value indicates the type of data contained in the payload datafield.
- Authentication data – a variable length field containing an Integrity Check Value (ICV), which is computed (using an authentication algorithm) from the rest of the ESP packet, to provide data integrity protection.

To process outbound traffic, a host or security gateway first uses a set of selectors in the SPD to determine the outbound SA used. It then follows a set of steps to process the outbound packet:

1. Either the entire original outbound IP datagram is encapsulated in an ESP payload field (tunnel mode) or just the original upper layer protocol information from the outbound IP datagram is encapsulated (transport mode).
2. Appropriate padding is added to the payload data.
3. The results are encrypted using an encryption key and an algorithm.
4. The sequence number is incremented as appropriate.[2]
5. If authentication is enabled, then the ICV is calculated.
6. Possible fragmentation of the IP datagram is performed.

On receiving an IP datagram the recipient follows the following steps to process the packet:

1. Possible re-assembly of the IP datagram is performed.
2. Using the SPI, security protocol and destination IP address, an appropriate SA is looked up from the SAD.
3. If anti-replay protection is enabled, the sequence number is inspected.
4. If authentication is enabled, then the ICV is verified.
5. The packet is decrypted, padding is removed and the original IP datagram is reconstructed.

22.6 Summary

IPsec provides security services in the IP layer, in both IPv4 and IPv6, offering protection for protocols at higher layers. The IKE is used for key exchange – creating and managing SAs and related security parameters – and the ESP is used for confidentiality and integrity protection.

---

[2] In case this was the first packet to be protected using a new SA, the sequence number would start from 0.

23

Signalling Compression

Signalling Compression (SigComp) is a mechanism that application protocols use to compress messages before sending them across the network. It is presented to applications as a layer between application protocols and transport protocols. SigComp uses a Universal Decompressor Virtual Machine (UDVM) to decompress messages. Figure 8.1 demonstrates SigComp's location in the protocol stack. SigComp is defined in [RFC3320].

A message that is compressed using SigComp is referred to as a SigComp message.

23.1 SigComp architecture

SigComp's architecture can be broken down into five entities (see Figure 23.1). The entities are described as follows:

- *Compressor dispatcher* – this is the interface between the application and the SigComp system. It invokes a compressor, indicated by the application using a compartment identifier. The compressor dispatcher forwards the returned compressed message to its destination.
- *Decompressor dispatcher* – this is the interface between the SigComp system and the application. It invokes a UDVM that decompresses the message. The decompressor dispatcher then passes the decompressed message to the application. If the application wishes the decompressor to retain the state of the message, it returns what is called a "compartment identifier".
- *Compressors* – this entity compresses the application message. It uses a compartment that is identified using the compartment identifier. The compressed message is passed to the compressor dispatcher. DEFLATE is an example of a compression algorithm.
- *UDVM* – This entity decompresses a compressed message. A new instance is invoked for every new SigComp message. UDVM uses the state handler to create a state for a new message or make use of an existing state.
- *State handler* – this holds information that is stored between SigComp messages (referred to as the "state of the messages"). It can store and recover the state.

The IMS: IP Multimedia Concepts and Services Second Edition
Miikka Poikselkä, Georg Mayer, Hisham Khartabil and Aki Niemi © 2006 John Wiley & Sons, Ltd

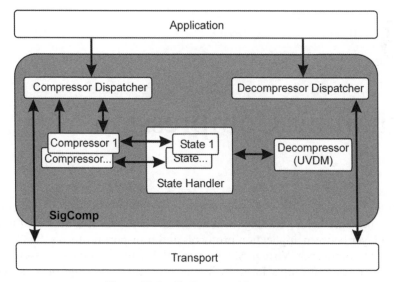

Figure 23.1 SigComp architecture.

23.2 Compartments

An application like Session Initiation Protocol (SIP) can group messages that are related. Depending on the signalling protocol, this grouping is done according to how the protocol relates messages: for instance, in SIP, messages are grouped using dialogs or, in cases where there is a configured next-hop address, messages are grouped by the same next-hop address. The application allocates a compressor per compartment and state memory in which to store state information. It also determines when a compartment should be created or destroyed. A compartment is identified uniquely using a compartment identifier.

It is the application's responsibility to determine the compartment identifier for a decompressed message. When an application receives a decompressed message, it determines or creates the compartment for the message and supplies the compartment identifier to the SigComp system.

23.3 Compressing a SIP message in IMS

23.3.1 Initialization of SIP compression

During the registration phase the User Equipment (UE) and the Proxy Call Session Control Function (P-CSCF) announce their willingness to perform compression by providing details about their compression capabilities, such as memory size and processing power, upload states and compression instructions (see Section 12.13.15). Due to the strong security requirements in the Third Generation Partnership Project's (3GPP) Internet Protocol Multimedia Subsystem (IMS), announcements and state creations are only allowed after a security association has been established. Otherwise, a malicious user could upload false states that would make compression vulnerable.

23.3.2 Compressing a SIP message

When the UE or P-CSCF wants to send a compressed SIP message, it follows the framework described in [RFC3320], which states that a SIP application in the UE should pass a message to a compressor dispatcher. The compressor dispatcher invokes a compressor, fetches the necessary compression states using a compartment identified by its ID and supplied by the application, and uses a certain compression algorithm to encode the message. Finally, the compressor dispatcher relays the compressed message to the transport layer to be delivered to the remote end (the P-CSCF).

The compressor is responsible for ensuring that the remote end can decompress the generated message. It is possible to include all the needed information in every SigComp message (i.e., every bytecode) to decompress the message. However, this would reduce the archived compression ratio significantly; so, it is better to ask the other end to create states. The information saved in these state items can then be accessed for future SigComp message decompression for messages that arrive from the same source and are related, avoiding the need to upload the data on a per-message basis.

23.3.3 Decompressing a compressed SIP message

When a decompressor dispatcher receives a message it inspects the prefix of the incoming message. As all SigComp messages contain a prefix (the five most significant bits of the first byte are set to 1) the decompression dispatcher is able to identify that the message is compressed. This prefix does not occur in UTF-8-encoded text messages. The decompressor dispatcher forwards the message to the UDVM, which requests previously created states from the state handler and uses the provided states (or provided bytecode in the message if no states exist) to decompress the message. After decompression the UDVM returns the uncompressed message to the decompression dispatcher which further passes the message to an application.

24

DHCPv6

The Dynamic Host Configuration Protocol for Internet Protocol (IP) Version 6 (DHCPv6), as defined in [RFC3315], is a client–server protocol that allows devices to be configured with management configuration information. DHCP specifically provides clients with dynamically assigned IP addresses using a DHCP server. It also provides other configuration information carried as options: options are extensions to DHCP. "Extending DHCP" means defining a new option.

DHCP messages are exchanged between clients and servers using User Datagram Protocol (UDP) for transport. Clients listen on port 546 for messages, while servers listen on port 547. A client uses a link-local address for transmitting and receiving DHCP messages. A client sends DHCP messages to servers using the link-scoped multicast address "FF02 : : 1 : 2", also known as All_DHCP_Relay_Agents_and_Servers. DHCP relay agents on the client's link allow a DHCP client to send a message to a DHCP server that is not attached to the same link.

A client needs, first, to locate a DHCP server before requesting an IP address and other configuration information. It does so by sending a DHCP solicit message to the multicast address identified earlier. A server willing to answer the request answers with a DHCP advertise message. The client then chooses one of the servers and a DHCP request message. The server responds with a relay message confirming the assigned IP addresses and other configuration information. Because assigned IP addresses expire, the client needs to send a DHCP renew message to the server in order to extend the assigned IP address lifetime. Figure 24.1 presents the format of a client–server DHCP message. The fields are described as follows:

- Message type – this is the DHCP message type, such as solicit, advertise, request, confirm, renew, bind, reply, release, decline, reconfigure, information-request, relay-forw and relay-repl.

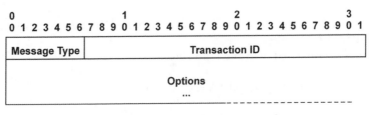

Figure 24.1   Client–server DHCP message format.

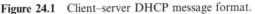

The IMS: IP Multimedia Concepts and Services Second Edition
Miikka Poikselkä, Georg Mayer, Hisham Khartabil and Aki Niemi © 2006 John Wiley & Sons, Ltd

- Transaction ID – an ID for the message exchange.
- Options – zero or more options.

24.1 DHCP options

As described earlier, options are used to hold additional information. The format of DHCP options is depicted in Figure 24.2. The fields are described as follows:

- Option code – the specific option type carried in this option.
- Option length – the length, in octets, of the option data.
- Option data – the data for the option.

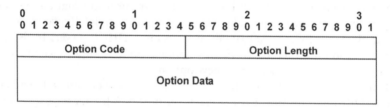

Figure 24.2 DHCP options format.

24.2 DHCP options for SIP servers

[RFC3319] specifies two DHCPv6 options that allow Session Initiation Protocol (SIP) clients to locate their outbound SIP proxy server. One option carries a list of SIP server domain names and the other carries a list of 128-bit IPv6 addresses.

The "SIP server domain name list" option can carry multiple domain names. Each domain name has a Naming Authority Pointer (NAPTR) record that differs from any other. The client tries the records in the order listed, as specified in Section 12.12.1.1, trying the next record if the previous one fails.

The option is populated with the following:

- Option code – OPTION_SIP_SERVER_D (21).
- Option length – variable.
- Option data – SIP server domain name list.

The "SIP server IPv6 address list option" can carry multiple IP addresses. Servers are listed in order of preference.

The option is populated with the following:

- Option code – OPTION_SIP_SERVER_A (22).
- Option length – variable.
- Option data – IPv6 addresses of SIP servers.

25

XCAP

In many services today the service provider needs – in order to carry out a request – to have access to information that can only be set by users. Such services include presence, messaging and conferencing.

In Extensible Markup Language (XML) Configuration Access Protocol (XCAP) a user is able to upload information to an XCAP server, which provides this uploaded information to application servers that use this information to satisfy a request demanded by the user. With XCAP, the user is also allowed to manipulate, add and delete such data. An example of the data that a user can upload is the user's resource list for presence (defined in Section 4.3).

XCAP uses the Hypertext Transfer Protocol (HTTP) to upload and read the information set by users and the Information is represented using XML.

25.1 XCAP application usage

Applications, like presence, need to define an XCAP application usage, which defines the way that a unique application can make use of XCAP. It defines the XML document for the application, along with the following:

- Application Usage ID (AUID) – this ID uniquely identifies an application usage.
- Additional constraints – these cover data constraints that are not possible to represent using an XML schema.
- Data semantics – the semantics for each element and attribute in an application usage XML document.
- Naming convention – defines how an application constructs the Uniform Resource Identifier (URI) representing the document that is to be read or written by the application so that it can carry out its tasks.
- Resource interdependence – in some XCAP application usages, many entities need to contribute to populating the XML document element and attribute values to make the XML document ready. This interdependence is specified in a section of the XCAP application usage document.

The IMS: IP Multimedia Concepts and Services Second Edition
Miikka Poikselkä, Georg Mayer, Hisham Khartabil and Aki Niemi © 2006 John Wiley & Sons, Ltd

- Authorization policy – By default, the creator of the XML document is the sole resource with access to the document. Application usages can define a different authorization policy to override the default.

Examples of application usages can be seen in Sections 4.4 and 4.5.

26

Common Policy

26.1 Introduction

Many applications enable access to information about a user. This can be in the form of presence information or location information, which reveals personal details about the user's status and whereabouts. The richness of such detailed information is both an extraordinary opportunity for enabling communications as well as a considerable threat to privacy. So, as a result of such applications, there needs to exist a robust and similarly rich system to control the privacy settings of information.

Common Policy [Draft.ietf.geopriv-common-policy] defines an authorization policy markup language that can be used to describe the detailed access rights pertaining to an application. Its origins lie in describing privacy settings that relate to geolocation information but, in the spirit of reuse, Common Policy is also the basis for describing presence authorization policies.

In fact, any application that deals with access to a resource – be it in the form of a subscription, a fetch or an invitation of sorts – can take the basic tools of Common Policy and extend its permission statements to suit the application's special needs.

26.2 Model and rule structure

The most important aspect of the Common Policy model is that of additive permissions. This means that the privacy settings for a resource always start out with no permissions at all; the settings are then added with grants of permission by the user based on some form of stimulus – e.g., a user might find that another user wants to subscribe to her presence information via the watcher information [RFC3857] mechanism.

This is in contrast to a traditional blacklist model where everyone basically has access, except the blacklisted individuals. Blacklists have a major drawback in that simply changing one's identity – called "identity minting" – is enough to circumvent access controls.

Similarly to a whitelist model, additive permissions make the system much more privacy-safe and resistant to identity minting. Simply minting an identity is no longer

The IMS: IP Multimedia Concepts and Services Second Edition
Miikka Poikselkä, Georg Mayer, Hisham Khartabil and Aki Niemi © 2006 John Wiley & Sons, Ltd

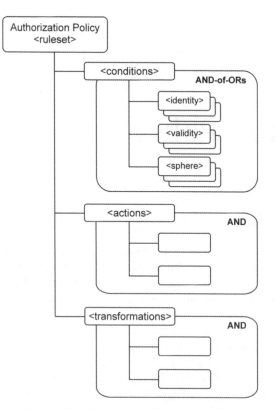

Figure 26.1 Common Policy data model.

enough to gain access; instead, a malicious user would actually need to get on another user's buddy list. Assuming identities are authentic and cannot be faked,[1] getting on someone else's buddy list is hard to do; it would normally require an exchange of business cards or an earlier communication of one sort or another.

Of course, applications benefit immensely from the ability to receive or grant some level of information to others. This is also the case with Common Policy. Everyone (with an authentic identity) can be granted low-level permission – e.g., to receive the most basic level of presence information or civic location information revealing only the country in which the user currently is – or permission to make requests. More trusted parties, on the other hand, can be given higher level access rights.

Common Policy defines an authorization policy as a set of rules – called a "ruleset" – that govern the access rights to information. Each rule grants permission based on certain matching criteria, like the identity of the user accessing the information, or the date and time of day. These matching criteria are called "conditions". Permission is further divided into an action and a transformation. The action part dictates in which way the system should act, while the transformation part dictates the exact way in which the action should be carried out. This structure is illustrated in Figure 26.1.

---

[1] This is not an assumption that is easy to make – as the abundance of email spam has taught us.

26.3   Data types and permission processing

As already mentioned, an authorization policy consists of a set of rules, each granting a permission. These rules are independent of each other and are processed in one pass – i.e., the ordering of the rules is not important. After all rules are processed, something called a "resultant" of all granted permissions is generated. This is another important distinction between Common Policy and other access right systems. For instance, the *nix file system makes use of Access Control Lists (ACLs) and three lists – namely, *user*, *group* and *others* – are visited in order; the first one to match determines the access rights.

The Common Policy processing model means that a user may get granted several permissions, with different levels of access rights. To resolve this ambiguity, an algorithm is provided that combines the different permissions allowed by the various rules. It depends on the data types of permissions – *boolean, integer* and *set*.

The boolean type can have only two values: TRUE or FALSE. When combining boolean-type permissions, the OR operation is applied across them. In other words, if one or more of the permissions are TRUE, then the resultant is TRUE; if none of the permissions is TRUE, the resultant for that permission is FALSE. The default is always FALSE; so, in the absence of any matching rule, an automatic FALSE is granted.

The integer type assigns each permission an integer value. The higher the value, the higher the rights assigned to that permission – 0 being the default granting the fewest rights. When combining integer-type permissions, the highest value of granted permissions is selected. In the absence of any rules, the default is taken, which constitutes a setting with the most privacy – i.e., releases little or no information, or blocks sharing of information.

The set type assigns each permission an attribute in a set. Each attribute names a certain permission – the empty set is the default. When combining set-type permissions, a union of all permissions is undertaken. For example, if two permissions grant "yellow" and "green", the resultant is a union of the two: {"yellow", "green"}.

After permission processing, the system is in a position to determine which permissions to apply to the requested information.

References

Third Generation Partnership Project (3GPP)

3GPP TR 23.815	Charging implications of IMS architecture.
3GPP TR 23.981	Interworking aspects and migration scenarios for IPv4-based IP Multimedia Subsystem (IMS) implementations.
3GPP TS 21.111	USIM and IC card requirements.
3GPP TS 22.101	Service principles.
3GPP TS 22.228	Service requirements for the IP multimedia core network subsystem.
3GPP TS 22.340	IP Multimedia Subsystem (IMS) messaging; Stage 1, March 2003.
3GPP TS 23.002	Network architecture.
3GPP TS 23.003	Technical Specification Group Core Network; Numbering, addressing and identification.
3GPP TS 23.060	General Packet Radio Service (GPRS); Service description; Stage 2.
3GPP TS 23.107	Quality of Service (QoS) concept and architecture.
3GPP TS 23.125	Overall high-level functionality and architecture impacts of flow-based charging.
3GPP TS 23.207	End-to-end QoS concept and architecture.
3GPP TS 23.218	IP Multimedia (IM) session handling; IM call model; Stage 2.
3GPP TS 23.221	Architectural requirements.
3GPP TS 23.228	IP Multimedia (IM) Subsystem; Stage 2.
3GPP TS 23.279	Combining CS and IMS services; Stage 2.
3GPP TS 23.803	Evolution of policy control and charging.
3GPP TS 24.008	Mobile Radio Interface Layer 3 specification; Core Network protocols; Stage 3.
3GPP TS 24.141	Presence service using the IP Multimedia (IM) Core Network (CN) subsystem; Stage 3.
3GPP TS 24.147	Conferencing using the IP Multimedia (IM) Core Network (CN) subsystem; Stage 3.
3GPP TS 24.228	Signalling flows for the IP multimedia call control based on SIP and SDP; Stage 3.

The IMS: IP Multimedia Concepts and Services Second Edition
Miikka Poikselkä, Georg Mayer, Hisham Khartabil and Aki Niemi © 2006 John Wiley & Sons, Ltd

3GPP TS 24.229   IP Multimedia Call Control based on SIP and SDP; Stage 3.

3GPP TS 24.247   Messaging service using the IP Multimedia (IM) Core Network (CN) subsystem; Stage 3.

3GPP TS 26.234   Transparent end-to-end packet-switched streaming service (PSS); Protocols and codecs.

3GPP TS 26.235   Packet-switched multimedia applications; Default codecs.

3GPP TS 26.236   Packet-switched conversational multimedia applications; Transport protocols.

3GPP TS 29.198   Open Service Access (OSA); Application Programming Interface (API), Multiple parts.

3GPP TS 29.207   Policy control over Go interface, June 2003.

3GPP TS 29.208   End-to-end Quality of Service (QoS) signaling flows.

3GPP TS 29.209   Policy control over Gq interface.

3GPP TS 29.228   IP Multimedia (IM) Subsystem Cx and Dx Interfaces; Signaling flows and message contents.

3GPP TS 29.229   Cx and Dx interfaces based on the Diameter protocol; Protocol details, June 2003.

3GPP TS 29.329   Sh interface based on the Diameter protocol, June 2003.

3GPP TS 31.102   Characteristics of the USIM application.

3GPP TS 31.103   Characteristics of the IP Multimedia Services Identity Module (ISIM) application.

3GPP TS 32.240   Charging architecture and principles.

3GPP TS 32.295   Charging management; Charging Data Record (CDR) transfer.

3GPP TS 32.296   Online Charging System (OCS): Applications and interfaces.

3GPP TS 32.299   Diameter charging applications.

3GPP TS 33.102   3G security; Security architecture, June 2003.

3GPP TS 33.120   Security objectives and principles, April 2001.

3GPP TS 33.203   3G security; Access security for IP-based services, June 2003.

3GPP TS 33.210   3G security; Network Domain Security (NDS); IP network layer security, June 2003.

3GPP TS 33.220   3G security; Generic Authentication Architecture (GAA); Generic Bootstrapping Architecture, 3GPP, December 2003.

3GPP TS 33.222   Generic Authentication Architecture (GAA); Access to network application functions using Secure Hypertext Transfer Protocol (HTTPS), February 2004.

3GPP TS 33.310   Network Domain Security; Authentication Framework (NDS/AF), December 2003.

3GPP TS 33.978   Security aspects of early IP Multimedia Subsystem (IMS).

Internet Engineering Task Force (IETF)

RFC0791   Postel, J., Internet Protocol, STD 5, September 1981.

RFC0793   Postel, J., Transmission Control Protocol, STD 7, September 1981.

RFC1305	Mills, D., Network Time Protocol (Version 3) Specification, Implementation, March 1992.
RFC1321	Rivest, R., The MD5 Message-Digest Algorithm, April 1992.
RFC1633	Braden, R., Clark, D. and S. Shenker, Integrated Services in the Internet Architecture: An Overview, June 1994.
RFC1851	Karn, P., Metzger, P. and W. Simpson, The ESP Triple DES Transform, September 1995.
RFC2104	Krawczyk, H., Bellare, M. and R. Canetti, HMAC: Keyed-Hashing for Message Authentication, February 1997.
RFC2246	Dierks, T., Allen, C., Treese, W., Karlton, P., Freier, A. and P. Kocher, The TLS Protocol Version 1.0, January 1999.
RFC2327	Handley, M. and V. Jacobson, SDP: Session Description Protocol, April 1998.
RFC2396	Berners-Lee, T., Fielding, R., Irvine U.C. and L. Masinter, Uniform Resource Identifiers (URI): Generic Syntax, August 1998.
RFC2401	Kent, S. and R. Atkinson, Security Architecture for the Internet Protocol, November 1998.
RFC2402	Kent, S. and R. Atkinson, IP Authentication Header, November 1998.
RFC2403	Madson, C. and R. Glenn, The Use of HMAC-MD5-96 within ESP and AH, November 1998.
RFC2404	Madson, C. and R. Glenn, The Use of HMAC-SHA-1-96 within ESP and AH, November 1998.
RFC2406	Kent, S. and R. Atkinson, IP Encapsulating Security Payload (ESP), November 1998.
RFC2407	Piper, D., The Internet IP Security Domain of Interpretation for ISAKMP, November 1998.
RFC2408	Maughan, D., Schneider, M. and M. Schertler, Internet Security Association and Key Management Protocol (ISAKMP), November 1998.
RFC2409	Harkins, D. and D. Carrel, The Internet Key Exchange (IKE), November 1998.
RFC2412	Orman, H., The OAKLEY Key Determination Protocol, November 1998.
RFC2451	Pereira, R. and R. Adams, The ESP CBC-Mode Cipher Algorithms, November 1998.
RFC2475	Blake, S., Black, D., Carlson, M., Davies, E., Wang, Z. and W. Weiss, An Architecture for Differentiated Service, December 1998.
RFC2477	Aboba, B. and G. Zorn, Criteria for Evaluating Roaming Protocols, January 1999.
RFC2486	Aboba, B. and M. Beadles, The Network Access Identifier, January 1999.
RFC2543	Handley, M., Schulzrinne, H., Schooler, E. and J. Rosenberg, SIP: Session Initiation Protocol, March 1999.
RFC2578	McCloghrie, K., Perkins, D., Schoenwaelder, J., Case, J., McCloghrie, K., Rose, M. and S. Waldbusser, Structure of Management Information Version 2 (SMIv2), STD 58, April 1999.

RFC2616   Fielding, R., Gettys, J., Mogul, J., Nielsen, H., Masinter, L., Leach, P. and
               T. Berners-Lee, Hypertext Transfer Protocol – HTTP/1.1, June 1999.

RFC2617   Franks, J., Hallam-Baker, P., Hostetler, J., Lawrence, S., Leach, P.,
               Luotonen, A. and L. Stewart, HTTP Authentication: Basic and Digest
               Access Authentication, June 1999.

RFC2633   Ramsdell B., S/MIME Version 3 Message Specification, June 1999.

RFC2748   Durham, D., Boyle, J., Cohen, R., Herzog, S., Rajan, R. and A. Sastry, The
               COPS (Common Open Policy Service) Protocol, January 2000.

RFC2753   Yavatkar, R., Pendarakis, D. and R. Guerin, A Framework for Policy-
               based Admission Control, January 2000.

RFC2782   Gulbrandsen, A., Vixie, P. and L. Esibov, A DNS RR for specifying the
               Location of Services (DNS SRV), February 2000.

RFC2806   Vaha-Sipila, A., URLs for Telephone Calls, April 2000.

RFC2810   Kalt, C., Internet Relay Chat: Architecture, April 2000.

RFC2833   Schulzrinne, H. and S. Petrack, RTP Payload for DTMF Digits, Telephony
               Tones and Telephony Signals, May 2000.

RFC2865   Rigney, C., Willens, S., Rubens, A. and W. Simpson, Remote Authentica-
               tion Dial In User Service (RADIUS), June 2000.

RFC2915   Mealling, M. and R. Daniel, The Naming Authority Pointer (NAPTR)
               DNS Resource Record, September 2000.

RFC2960   Stewart, R., Xie, Q., Morneault, K., Sharp, C., Schwarzbauer, H., Taylor,
               T., Rytina, I., Kalla, M., Zhang, L. and V. Paxson, Stream Control Trans-
               mission Protocol, October 2000.

RFC2977   Glass, S., Hiller, T., Jacobs, S. and C. Perkins, Mobile IP Authentication,
               Authorization, and Accounting Requirements, October 2000.

RFC3084   Chan, K., Seligson, J., Durham, D., Gai, S., McCloghrie, K., Herzog, S.,
               Reichmeyer, F., Yavatkar, R. and A. Smith, COPS Usage for Policy Pro-
               visioning (COPS-PR), March 2001.

RFC3169   Beadles, M. and D. Mitton, Criteria for Evaluating Network Access Server
               Protocols, September 2001.

RFC3261   Rosenberg, J., Schulzrinne, H., Camarillo, G., Johnston, A., Peterson, J.,
               Sparks, R., Handley, M. and E. Schooler, SIP: Session Initiation Protocol,
               June 2002.

RFC3262   Rosenberg, J. and H. Schulzrinne, Reliability of Provisional Responses in
               the Session Initiation Protocol (SIP), June 2002.

RFC3263   Rosenberg, J. and H. Schulzrinne, SIP: Locating SIP Servers, June 2002.

RFC3264   Rosenberg, J. and H. Schulzrinne, An Offer/Answer Model with SDP, June
               2002.

RFC3265   A. B. Roach, Session Initiation Protocol (SIP)-specific Event Notification,
               June 2002.

RFC3310   Niemi, A., Arkko, J. and V. Torvinen, Hypertext Transfer Protocol (HTTP)
               Digest Authentication Using Authentication and Key Agreement (AKA),
               September 2002.

RFC3311	Rosenberg, J., The Session Initiation Protocol (SIP) UPDATE Method, October 2002.
RFC3312	Camarillo, G., Marshall, W. and Rosenberg, J., Integration of Resource Management and Session Initiation Protocol, October 2002.
RFC3313	Marshal, W., Private Session Initiation Protocol (SIP) Extensions for Media Authorization, January 2003.
RFC3315	Droms, R., Bounds, J., Volz, B., Lemon, T., Perkins, C. and M. Carney, Dynamic Host Configuration Protocol for IPv6 (DHCPv6), July 2003.
RFC3319	Schulzrinne, H. and B. Volt, Dynamic Host Configuration Protocol (DHCPv6) Options for Session Initiation Protocol (SIP) Servers, July 2003.
RFC3320	Price, R., Bormann, C., Christoffersson, J., Hannu, H., Liu, Z. and J. Rosenberg, Signaling Compression (SigComp), January 2003.
RFC3323	Peterson, J., A Privacy Mechanism for the Session Initiation Protocol (SIP), November 2002.
RFC3325	Jenning, C., Peterson, J. and M. Watson, Private Extensions to the Session Initiation Protocol (SIP) for Asserted Identity within Trusted Networks, November 2002.
RFC3327	Willis, D. and B. Hoeneisen, Session Initiation Protocol (SIP) Extension Header Field for Registering Non-Adjacent Contacts, December 2002.
RFC3329	Arkko, J., Torvinen, V., Camarillo, G., Niemi, A. and T. Haukka, Security Mechanism Agreement for the Session Initiation Protocol (SIP), January 2003.
RFC3388	Camarillo, G., Eriksson, G., Holler, J. and H. Schulzrinne, Grouping of Media Lines in the Session Description Protocol (SDP), December 2002.
RFC3411	Harrington, D., Presuhn, R. and B. Wijnen, An Architecture for Describing Simple Network Management Protocol (SNMP) Management Frameworks, STD 62, December 2002.
RFC3418	Presuhn, R., Management Information Base (MIB) for the Simple Network Management Protocol (SNMP), STD 62, December 2002.
RFC3428	Campbell, B. and J. Rosenberg, Session Initiation Protocol Extension for Instant Messaging, September 2002.
RFC3455	Garcia-Martin, M., Henrikson, E. and D. Mills, Private Header (P-Header) Extensions to the Session Initiation Protocol (SIP) for the 3rd-Generation Partnership Project (3GPP), January 2003.
RFC3485	Garcia-Martin, M., Bormann, C., Ott, J., Price, R. and A.B. Roach, The Session Initiation Protocol (SIP) and Session Description Protocol (SDP) Static Dictionary for Signaling Compression (SigComp), February 2003.
RFC3486	Camarillo, G., Compressing the Session Initiation Protocol (SIP), February 2003.
RFC3515	Sparks, R., The Session Initiation Protocol (SIP) REFER Method, April 2003.
RFC3520	Hamer, L-N., Gage, B., Kosinski, B. and H. Shieh, Session Authorization Policy Element, April 2003.
RFC3524	Camarillo, G. and A. Monrad, Mapping of Media Streams to Resource Reservation Flows, April 2003.

RFC3525   Groves, C., Pantaleo, M., Anderson, T. and T. Taylor, Gateway Control
                   Protocol Version 1, June 2003.

RFC3539   Aboba, B. and J. Wood, Authentication, Authorization and Accounting
                   (AAA) Transport Profile, June 2003.

RFC3550   Schulzrinne, H., Casner, S., Frederick, R. and V. Jacobson, RTP: A Trans-
                   port Protocol for Real-Time Applications, July 2003.

RFC3551   Schulzrinne, H. and S. Casner, RTP Profile for Audio and Video Confer-
                   ences with Minimal Control, July 2003.

RFC3588   Calhoun, P., Loughney, J., Guttman, E., Zorn, G. and J. Arkko, Diameter
                   Base Protocol, September 2003.

RFC3589   Loughney, J., Diameter Command Codes for Third Generation Partnership
                   Project (3GPP) Release 5, September 2003.

RFC3608   Willis, D. and B. Hoeneisen, Session Initiation Protocol (SIP) Extension
                   Header Field for Service Route Discovery during Registration, December
                   2002.

RFC3680   Rosenberg, J., A Session Initiation Protocol (SIP) Event Package for Regis-
                   trations, March 2004.

RFC3761   Faltstrom, P. and M. Mealling, The E.164 to URI DDDS Application
                   (ENUM), April 2004.

RFC3764   Peterson, J., enumservice registration for SIP Addresses-of-Record, April
                   2004.

RFC3862   Atkins, D. and G. Klyne, Common Presence and Instant Messaging:
                   Message Format, August 2004.

RFC3863   Sugano, H. and S. Fujimoto, Presence Information Data Format (PIDF),
                   August 2004.

RFC3860   Peterson, J., Common Profile for Instant Messaging (CPIM), August 2004.

RFC3966   Schulzrinne, H. and A. Vaha-Sipila, The tel URI for Telephone Numbers,
                   December 2004.

RFC3856   Rosenberg, J., A Presence Event Package for the Session Initiation Protocol
                   (SIP), August 2004.

RFC3857   Rosenberg, J., A Watcher Information Event Template-package for the
                   Session Initiation Protocol (SIP), August 2004.

RFC3903   Niemi, A., Session Initiation Protocol (SIP) Extension for Event State
                   Publication, October 2004.

Draft-ietf-aaa-   Hakala, H., Diameter Credit-control Application, Work In Progress,
diameter-cc   October 2003.

Draft-ietf-aaa-   Calhoun, P., Farrell, S. and W. Bulley, Diameter CMS Security Applica-
diameter-cms-sec   tion, Work In Progress, March 2002.

Draft-ietf-aaa-   Calhoun, P., Johansson, T. and C. Perkins, Diameter Mobile IPv4 Applica-
diameter-mobileip tion, Work In Progress, October 2003.

Draft-ietf-aaa-   Calhoun, P., Zorn, G., Spence, D. and D. Mitton, Diameter Network
diameter-nasreq   Access Server Application, Work In Progress, October 2003.

Draft-ietf-aaa-   Garcia-Martin, M., Diameter Session Initiation Protocol (SIP) Application,
diameter-sip-app   Work In Progress, October 2003.

Draft-ietf-aaa-eap   Eronen, P., Hiller, T. and G. Zorn, Diameter Extensible Authentication
                      Protocol (EAP) Application, Work In Progress, October 2003.

Draft.ietf-geopriv-   Schulzrinne, H., A Document Format for Expressing Privacy Preferences,
common-policy   draft-ietf-geopriv-common-policy-05 (work in progress), July 2005.

Draft-ietf-   Handley, M., Jacobson, V. and C. Perkins, SDP: Session Description
mmusic-sdp-new   Protocol, Work In Progress, July 2005.

Draft.ietf-ngtrans-   Templin, F., Gleeson, T., Talwar, M., and D. Thaler, Intra-Site Automatic
isatap   Tunnel Addressing Protocol (ISATAP), draft-ietf-ngtrans-isatap-24 (work
                      in progress), January 2005.

Draft-ietf-simple-   Rosenberg, J., An Extensible Markup Language (XML) Representation for
common-policy-   Expressing Policy Capabilities, Work In Progress, July 2005.
caps

Draft-ietf-simple-   Roach, A., Rosenberg, J. and B. Campbell, A Session Initiation Protocol
event-list   (SIP) Event Notification Extension for Resource Lists, Work In Progress,
                      December 2004.

Draft-ietf-simple-   Rosenberg, J., Presence Authorization Rules, Work In Progress, July 2005.
presence-rules

Draft-ietf-simple-   Rosenberg, J., An Extensible Markup Language (XML) Representation for
pres-policy-caps   Expressing Presence Policy Capabilities, Work In Progress, July 2005.

Draft-ietf-simple-   Rosenberg, J., Presence Authorization Rules, Work in Progress, October
rules   2004.

Draft-ietf-simple-   Rosenberg, J., The Extensible Markup Language (XML) Configuration
xcap   Access Protocol (XCAP), Work In Progress, October 2005.

Draft-ietf-sipping-   Rosenberg, J. and H. Schulzrinne, A Session Initiation Protocol (SIP) Event
conference-   Package for Conference State, Work In Progress, July 2005.
package

Draft.ietf-simple-   Campbell, B., The Message Session Relay Protocol, draft-ietf-simple-
message-sessions   message-sessions-10 (work in progress), February 2005   .

Open Mobile Alliance (OMA)

OMA PoC AD   Push to talk over Cellular (PoC) – Architecture.

OMA PoC RD   Push to Talk over Cellular Requirements.

OMA POC   OMA PoC Control Plane.
Control Plane

OMA-TS-POC-   PoC XDM Specification.
XDM

International Telecommunications Union (ITU)

ITU.H248.1   Gateway Control Protocol: Version 1, ITU-T Recommendation H.248.1, March 2002.

ITU-T G.711   Pulse code modulation (PCM) of voice frequencies.

ITU-T G.726   40, 32, 24, 16 kbit/s adaptive differential pulse code modulation (ADPCM).

Abbreviations

1G	First generation
2G	Second generation
3G	Third generation
3GPP	Third Generation Partnership Project
3GPP2	Third Generation Partnership Project 2
AAA	Authentication, Authorization and Accounting
AAAA RR	IPv6 Address Resource Record
AAL	ATM Adaptation Layer
AAR&AAA	AA-Request/AA-Answer
ACA	Accounting Answer
ACL	Access Control List
ACR	Accounting Request
ADSL	Asynchronous Digital Subscriber Line
AH	Authentication Header
AKA	Authentication and Key Agreement
ALG	Application Level Gateway
AMR	Adaptive Multi-Rate
AMR-WB	Adaptive Multi-Rate WideBand
AOR	Address Of Record
API	Application Programming Interface
APN	Access Point Name
ARIB	Association of Radio Industries and Businesses (Japan)
ARR	IPv4 Address Resource Record
AS	Application Server
ASR&ASA	Abort-Session-Request/Abort-Session-Answer
ATM	Asynchronous Transfer Mode
AUC	AUthentication Centre
AUID	Application Usage ID
AUTN	Authentication token
AUTS	Synchronization token
AV	Authentication Vector
AVP	Attribute Value Pair; Audio Video Profile
B2BUA	Back to Back UA
BCF	Bearer Charging Function
BER	Bit Error Ratio
BGCF	Breakout Gateway Control Function
BICC	Bearer Independent Call Control

BNF	Backus-Naur Form grammar
BS	Bearer Service; Billing System
BSF	Bootstrapping Server Function
BTS	Base Transceiver Station
CA	Certificate Authority
CAMEL	Customized Applications for Mobile network Enhanced Logic
CAP	CAMEL Application Part
CCA	Credit-Control-Answer
CCF	Charging Collection Function
CCR	Credit-Control-Request
CCSA	China Communications Standards Association
CDF	Charging Data Function
CDMA	Code Division Multiple Access
CDR	Charging Data Record; Call Data Record
CGF	Charging Gateway Function
CGI	Cell Global ID
CK	Ciphering (Cipher) Key
CN	Core Network
COPS	Common Open Policy Service
COPS-PR	Common Open Policy Service Usage for Policy PRovisioning
CPCP	Conference Policy Control Protocol
CPIM	Common Presence and Instant Messaging
CPS	Conference Policy Server
CRF	Charging Rule Function
CRLF	Carriage Return–Line Feed pair
CS	Circuit Switched
CS CN	Circuit Switched Core Network
CSCF	Call Session Control Function
CSE	CAMEL Service Environment
CSRC	Contributing source
CTF	Charging Trigger Function
DDDS	Dynamic Delegation Discovery System
DES	Data Encryption Standard
DHCP	Dynamic Host Configuration Protocol
DL	Downlink
DNS	Domain Name System
DOI	Domain Of Interpretation
DOS	Denial Of Service
DSL	Digital Subscriber Line
DTMF	Dual-Tone Multi-Frequency
EAP	Extensible Authentication Protocol
ECF	Event Charging Function
EDGE	Enhanced Data Rates for Global Evolution
ENUM	Telephone E.164 Number Mapping
EPID	Encoded Provisioning Instance Data
ESP	Encapsulating (Encapsulated) Security Payload
ETSI	European Telecommunications Standards Institute
FMC	Fixed to Mobile Convergence
FQDN	Fully Qualified Domain Name
FRS	Family Radio Service

FSM	Finite State Machine
G-CDR	GGSN-CDR
GAA	Generic Authentication Architecture
GBA	Generic Bootstrapping Architecture
GBR	Guaranteed Bit Rate
GCID	GPRS Charging IDentifier
GGSN	Gateway GPRS Support Node
GPRS	General Packet Radio Service
GSM	Global System for Mobile Communications
GTP	GPRS Tunnelling Protocol
HLR	Home Location Register
HLR/AUC	Home Location Register and AUthentication Center
HSS	Home Subscriber Server
HTTP	Hypertext Transfer (or Transport) Protocol
I-CSCF	Interrogating-CSCF
IAB	Internet Architecture Board
IANA	Internet Assigned Numbers Authority
ICID	IMS Charging IDentifier
ICV	Integrity Check Value
IESG	Internet Engineering Steering Group
IETF	Internet Engineering Task Force
IK	Integrity Key
IKE	Internet Key Exchange
IM-SSF	IP Multimedia Service Switching Function
IMS	IP Multimedia Subsystem
IMS-GWF	IMS-Gateway Function
IMS-MGW	IP Multimedia Subsystem-Media Gateway Function
IMSI	International Mobile Subscriber Identifier
IOI	Inter-Operator Identifier
IP	Internet Protocol
IP-CAN	IP-Connectivity Access Network
IPsec	Internet Protocol security
IPv4	Internet Protocol Version 4
IPv6	Internet Protocol Version 6
IRC	Internet Relay Chat
ISAKMP	Internet Security Association and Key Management Protocol
ISATAP	Intra-Site Automatic Tunnel Addressing Protocol
ISB	Incoming Session Barring
ISC	IMS Service Control
ISDN	Integrated Services Digital Network
ISIM	IP Multimedia Services Identity Module
ISP	Internet Service Provider
ISUP	ISDN User Part
IV	Initialization Vector
K	Shared secret
L1	Layer 1
LCS	Location services
LIA	Location-Info-Answer
LIR	Location-Info-Request
LMR	Land Mobile Radio

LPDP	Local Policy Decision Point
M3UA	SS7 MTP3-User Adaptation layer
MAA	Multimedia-Auth-Answer
MAC	Message Authentication Checksum
MAP	Mobile Application Part
MAR	Multimedia-Auth-Request
Mbone	Multicast backbone
MBR	Maximum Bit Rate
MCC	Mobile Country Code
MDS	Multimedia Delivery Service
MEGACO	Media Gateway Control Protocol
MGCF	Media Gateway Control Function
MGW	Media Gateway Function
MIB	Management Information Base
MID	Media stream IDentification
MIME	Multipurpose Internet Mail Extension
MITM	Man In The Middle
MMD	MultiMedia Domain
MMS	Multimedia Messaging Service
MMSC	Multimedia Messaging Service Centre
MNC	Mobile Network Code
MOBILE IP	Mobile Internet Protocol
MPV	Music Photo Video
MRFC	Multimedia Resource Function Controller
MRFP	Media Resource Function Processor
MSC	Mobile Switching Centre
MSIN	Mobile Subscriber Identification Number
MSISDN	Mobile Subscriber Integrated Services Digital Network; Mobile Subscriber International ISDN; Mobile Station International ISDN
MSRP	Message Session Relay Protocol
MTP	Message Transfer Part
MTPN	Message Transfer Part level n
MTU	Maximum Transfer Unit
NAF	Network Application Function
NAI	Network Access Identifier
NAPTR	Naming Authority PoinTeR
NAS	Network Access Server
NASREQ	Network Access Server REQuirements
NAT	Network Address Translator
NAT-PT	Network Address Translator–Protocol Translator
NDS	Network Domain Security
NDS/AF	NDS Authentication Framework
NTP	Network Time Protocol
OCS	Online Charging System
OMA	Open Mobile Alliance
OSA	Open Services Architecture
OTA	Over The Air
P-CSCF	Proxy-CSCF
P2P	Peer to peer

PA	Presence Agent
PCMU	Pulse Code Modulation μ-law
PDF	Policy Decision Function
PDP	Packet Data Protocol; Policy Decision Point
PEF	Policy Enforcement Function
PEP	Policy Enforcement Point
PIB	Policy Information Base
PKI	Public Key Infrastructure
PLMN	Public Land Mobile Network
PMR	Professional Mobile Radio
PNA	Push-Notification-Answer
PNR	Push-Notification-Request
POC	Push to talk over the Cellular service
POTS	Plain Old Telephone Service
PPA	Push-Profile-Answer
PPR	Push-Profile-Request
PRACK	Provisional Response ACKnowledgement
PRC	PRovisioning Class
PRI	PRovisioning Instance
PRID	PRovisioning instance IDentifier
PS	Packet Switched; Presence Server
PSI	Public Service Identity
PSK	Pre-shared Secret Key
PSTN	Public Switched Telephone Network
PUA	Presence User Agent; Profile-Update-Answer
PUR	Profile-Update-Request
QOS	Quality of Service
RADIUS	Remote Authentication Dial In User Service
RAN	Radio Access Network
RAND	Random challenge
RAR&RAA	Re-Auth-Request/Re-Auth-Answer
RES	Response
RFC	Requests For Comments
RLS	Resource List Server
RNC	Radio Network Controller
ROAMOPS	Roaming operations
RR	Receiver Report; Resource Record
RSVP	Resource ReserVation setup Protocol
RTA	Registration-Termination-Answer
RTCP	RTP Control Protocol; Real-time Transport Control Protocol
RTP	Real-time Transport Protocol
RTP/AVP	RTP Audio and Video Profile
RTR	Registration-Termination-Request
S-CDR	SGSN-CDR
S-CSCF	Serving-CSCF
S/MIME	Secure MIME
SA	Security Association
SAA	Server-Assignment-Answer
SAD	Security Association Database
SAR	Server-Assignment-Request

SBC	Session Border Controller
SBLP	Service-Based Local Policy
SCF	Session Charging Function
SCS	Service Capability Server
SCTP	Stream Control Transmission Protocol
SDP	Session Description Protocol
SDU	Service Data Unit
SEG	Security Gateway
SGSN	Serving GPRS Support Node
SGW	Signalling Gateway
SHA	Secure Hash Algorithm
SigComp	Signalling Compression
SIM	Subscriber Identity Module
SIP	Session Initiation Protocol
Sip-Sec-Agree	SIP Security Mechanism Agreement
SIPS	Secure SIP
SL	Subscriber Locator
SLA	Service-Level Agreement
SLF	Subscription Locator Function
SMG	Special Mobile Group
SMI	Structure for Management Information
SMS	Short Messaging Service
SMTP	Simple Mail Transfer Protocol
SNA	Subscribe-Notifications-Answer
SNMP	Simple Network Management Protocol
SNR	Subscribe-Notifications-Request
SPD	Security Policy Database
SPI	Security Parameter Index
SPT	Service Point Trigger
SQN	Sequence number
SR	Sender Report
SRF	Single Reservation Flow
SRV RR	Service record
SS7	Signaling System No. 7
SSF	Service Switching Function
SSRC	Synchronization source
STR&STA	Session-Termination-Request/Session-Termination-Answer
TBCP	Talk Burst Control Protocol
TCP	Transmission Control Protocol
TCP/IP	TCP/IP stack
TD-CDMA	Time Division/Code Division Multiple Access
THIG	Topology Hiding Inter-network Gateway
TIA	Telecommunications Industry Association (North America)
TLS	Transport Layer Security
TPF	Traffic Plane Function
TTA	Telecommunications Technology Association (South Korea)
TTC	Telecommunications Technology Committee (Japan)
TTL	Time To Live
TU	Transaction User
UA	User Agent

UAA	User-Authorization-Answer
UAC	User Agent Client
UAR	User-Authorization-Request
UAS	User Agent Server
UDA	User-Data-Answer
UDP	User Datagram Protocol
UDR	User-Data-Request
UDVM	Universal Decompressor Virtual Machine
UE	User Equipment
UICC	Universal Integrated Circuit Card
UL	Uplink
UMTS	Universal Mobile Telecommunications System
URI	Uniform Resource Identifier
URL	Universal Resource Locator
URN	Uniform Resource Name
USIM	Universal Subscriber Identity Module
UTRAN	UMTS Terrestrial Radio Access Network
VHE	Virtual Home Environment
VOIP	Voice over IP
WAP	Wireless Application Protocol
WB	WideBand
WCDMA	Wideband Code Division Multiple Access
WLAN	Wireless Local Area Network
WWW	World Wide Web obvious
XCAP	XML Configuration Access Protocol
XCON	Centralized Conferencing
XDM	XML Document Management
XDMS	XML Document Management Server
XML	Extensible Markup Language
XRES	Expected response

Index